CLIMATE EXTREMES AND SOCIETY

The past few decades have brought extreme weather and climate events to the forefront of societal concerns. Ordinary citizens, industry, and governments are concerned about the apparent increase in the frequency of weather and climate events causing extreme, and in some instances, catastrophic, impacts.

Climate Extremes and Society focuses on the recent and potential future consequences of weather and climate extremes for different socioeconomic sectors. The book also examines actions that may enable society to better respond and adapt to climate variability, regardless of its source. It provides examples of the impact of climate and weather extremes on society – how these extremes have varied in the past, and how they might change in the future – and of the types of effort that will help society adapt to potential future changes in climate and weather extremes.

This review volume is divided into two sections: one examining the evidence for recent and projected changes in extremes of weather and climate events, and the other assessing the impacts of these events on society and on the insurance industry. Chapters examine a variety of climatic extremes using both the analysis of observational data and climate model simulations. Other chapters highlight recent innovative efforts to develop institutional mechanisms and incentives for integrating knowledge on extremes and their economic impacts.

The book will appeal to all scientists, engineers, and policymakers who have an interest in the effects of climate extremes on society.

DR HENRY F. DIAZ is a Research Meteorologist in the Earth System Research Laboratory at the National Oceanic and Atmospheric Administration (NOAA). He has worked on a variety of climate issues at NOAA over the past 15 years, particularly the impact of climatic variation on water resources of the western United States. He is recognized as an expert on the

El Niño – Southern Oscillation (ENSO) phenomenon and coedited *El Niño: Historical and Paleoclimatic Aspects of the Southern Oscillation*, also published by Cambridge University Press (1992).

DR RICHARD MURNANE is the Program Manager for the Risk Prediction Initiative (RPI) and a Senior Research Scientist at the Bermuda Institute of Ocean Sciences (BIOS), where he leads RPI's efforts to transform science into knowledge for assessing risk from natural hazards. Dr Murnane's own research focuses on tropical cyclones, climate variability, and the global carbon cycle. Before joining the RPI and BIOS in 1997, Dr Murnane was on the research staff of Princeton University in the Program in Atmospheric and Oceanic Sciences.

CLIMATE EXTREMES
AND SOCIETY

Edited by

HENRY F. DIAZ
National Oceanic and Atmospheric Administration Boulder, USA

RICHARD J. MURNANE
Bermuda Institute of Ocean Sciences, USA

CAMBRIDGE
UNIVERSITY PRESS

CAMBRIDGE UNIVERSITY PRESS
Cambridge, New York, Melbourne, Madrid, Cape Town,
Singapore, São Paulo, Delhi, Tokyo, Mexico City

Cambridge University Press
The Edinburgh Building, Cambridge CB2 8RU, UK

Published in the United States of America by Cambridge University Press, New York

www.cambridge.org
Information on this title: www.cambridge.org/9780521298483

First published 2008
First paperback edition 2011

A catalogue record for this publication is available from the British Library

ISBN 978-0-521-87028-3 Hardback
ISBN 978-0-521-29848-3 Paperback

Additional resources for this publication at www.cambridge.org/9780521298483

Contents

The colored plates will be found between pages 144 and 145*.

*The plates found between pages 144 and 145 are available for
download in color from www.cambridge.org/9780521870283

Contributors

M. Beniston
Climate Research, University of Geneva, 7 chemin de Drize, CH-1227
Carouge, GE, Switzerland

A. Boissonnade
Risk Management Solutions (RMS), 7015 Gateway Boulevard, Newark,
CA 94560, USA

H. Brooks
NOAA National Severe Storms Laboratory, 1313 Halley Circle, Norman,
OK 73069, USA

M. J. C. Crabbe
Luton Institute of Research in the Applied Natural Sciences, Faculty of
Creative Arts, Technologies and Science, University of Bedfordshire, Park
Square, Luton LU1 3JU, UK

S. L. Cutter
Department of Geography, University of South Carolina, Columbia,
SC 29208, USA

H. F. Diaz
NOAA Earth System Research Laboratory, 325 Broadway, Boulder,
CO 80305, USA

A. Dlugolecki
17 Craigie Place, Perth PH2 0BB, Scotland, UK

N. Dotzek
DLR-IPA, Department of Atmospheric Dynamics, Oberpfaffenhofen, 82234 Wessling, Germany

H. Douville
Centre National de Recherches Météorologiques, Météo-France, 42 Avenue G Coriolis, 31057 Toulouse cedex 1, France

D. R. Easterling
National Climatic Data Center, NOAA, 151 Patton Ave, Asheville, NC 28801, USA

J. B. Elsner
Department of Geography, The Florida State University, Tallahassee, FL 32306, USA

C. T. Emrich
Department of Geography, University of South Carolina, Columbia, SC 29208, USA

M. Gall
Department of Geography, University of South Carolina, Columbia, SC 29208, USA

A. Gershunov
Scripps Institution of Oceanography, UCSD, La Jolla, CA 92093-0224, USA

P. Grossi
Risk Management Solutions (RMS), 7015 Gateway Boulevard, Newark, CA 94560, USA

T. H. Jagger
Department of Geography, The Florida State University, Tallahassee, FL 32306, USA

M. E. Johnson
Department of Statistics and Actuarial Science, University of Central Florida, Orlando, FL 32816-2370, USA

A. Knap
Bermuda Institute of Ocean Sciences, Ferry Reach, St. George's, GE 01
Bermuda

T. R. Knutson
Geophysical Fluid Dynamics Laboratory, NOAA, P.O. Box 308, Princeton,
NJ 08542, USA

G. A. Meehl
National Center for Atmospheric Research, P.O. Box 3000, Boulder,
CO 80307, USA

S. Miller
Development Studies Institute, London School of Economics, Houghton
Street, London WC2 A 2AE, UK

R. Muir-Wood
Risk Management Solutions (RMS), 7015 Gateway Boulevard, Newark,
CA 94560, USA

R. J. Murnane
RPI/BIOS, P.O. Box 405, Garrett Park, MD 20896, USA

M. A. Saunders
Department of Space and Climate Physics, University College London,
Holmbury St. Mary, Dorking, Surrey RH5 6NT, UK

D. B. Stephenson
Department of Meteorology, University of Reading, Early Gate, Reading
RG6 6BB, UK

H. von Storch
Institute for Coastal Research, GKSS-Research Centre, 21502 Geesthacht,
Germany

C. Tebaldi
National Center for Atmospheric Research, P.O. Box 3000, Boulder,
CO 80307, USA

R. E. Tuleya
Center for Coastal Physical Oceanography, Old Dominion University,
768 W. 52nd St, Norfolk, VA 23529, USA

E. L. L. Walker
Department of Meteorology, University of Reading, Early Gate, Reading
RG6 6BB, UK

C. C. Watson
Kinetic Analysis Corporation, 330 Columbus Drive, Savannah, GA 31405,
USA

R. Weisse
Institute for Coastal Research, GKSS-Research Centre, 21502 Geesthacht,
Germany

Foreword

The Intergovernmental Panel on Climate Change (IPCC, 2007) definition of *climate change* refers to any change in climate over time, whether it is due to natural variability or is a result of human activity. Climate and non-climatic factors interact to produce both opportunities and disasters. It is the goal of good adaptation practices to take advantage of such opportunities and to reduce associated risks. There are substantial vulnerabilities to hurricanes along the Atlantic seaboard of the United States. The major concentrations of vulnerable economic activity and capital (with capital stock greater than US$100 billion) are located in areas affected by hurricanes, including the coastal cities of Miami, New Orleans, Houston, and Tampa, three of which have been hit by major storms in the past 15 years. The year 2005 was an economic outlier because the cost per unit of hurricane power was high, but not because the power was extraordinarily high.

Adaptation to climate change must account for a variety of timescales if it is to be effectively embedded into development plans. In any event, a more robust formulation for integrating disaster risks into sustainable development is important in an uncertain and changing environment. Key informational needs for mainstreaming climate change and adaptation into development plans include knowledge of the spatial and temporal characteristics of climatic extremes and the integration of loss estimates with projections of such extremes. Given the recent experience with increasing losses, early assessments of potential high-impact locations and the magnitude of potential extreme events are as important as early warning of an impending physical event. In addition, an effective basis for integrating monitoring, research, and management must exist. The National Science Board has observed that "the imperative to act has never been clearer, nor have the science technology, and intellectual capacity needed to address the challenge been more capable of rising to the occasion."

This volume offers assessments and new knowledge related to the problem of integrating our knowledge of weather and climate extremes, and its effects into decisions related to development and adaptation. To date, most adaptation practices have been observed in the insurance sector and have focused on property damage. This risk management component is well represented in this volume. Over time, insurance mechanisms have fostered risk prevention through: (i) implementing and strengthening building standards, (ii) planning risk prevention measures and developing best practices, and (iii) raising awareness of policyholders and public authorities. As this volume shows, the demand for insurance products is expected to increase, while climate change impacts could reduce insurability and threaten insurance schemes in places of high risk. In the longer term, it is hoped that climate change may also induce insurers to adopt forward-looking pricing methods in order to maintain insurability.

A fundamental problem within many economic impact studies lies in the unlikely assumption that there are no other influences on the macro-economy during the period analyzed for each disaster. There is already a great deal of reliable knowledge about reactive and anticipatory approaches to natural hazards and attendant disasters: much of the information was summed up by I. Burton, R. Kates, and G. White a few years ago in a paper entitled "Knowing better and losing even more." Increasingly small changes are producing disproportionately larger impacts, especially when they are aggregated over time. A major goal of this book is to provide an understanding of where and how critical conditions for significant losses arise. As such, this volume offers key methodological insights into integrating loss estimates over time with past and future projections of climate extremes. More important, the chapters represent the efforts of guides on convergent paths. The small group of people contributing to these pages is illustrative of the intellectual capacity identified as needed by the National Science Board. Much is to be learned from them.

While extremes and sequences of extremes, exposure of population, and economic assets matter, it is really our choices about what risks are acceptable – and to whom – that change a hazard into a disaster. A major stumbling block has been our limited understanding of the costs associated with such choices for past as well as future risks. The chapters together are a clarion call for national and international clearinghouses to maintain economic and environmental infrastructure databases and loss inventories. A "grand" challenge is to understand how different knowledge systems can be integrated into risk management information in support of resilience strategies. This volume is a significant contribution to meeting the challenge.

Roger S. Pulwarty
National Oceanic and Atmospheric Administration
Boulder, Colorado

Preface

Extreme events are critical determinants in the evolution and character of many natural and human-influenced systems. From such a perspective, extreme climatic events, in particular, present society with significant challenges in the context of a rapidly warming world. The societal impacts of recent extreme climatic events around the world motivated us to bring together in one book a scientific exploration of the nature of climatic extremes – past, present, and future – and examples of efforts aimed at making these events more comprehensible and manageable.

Extreme climatic events can affect both natural systems (e.g., coastal and riparian ecosystems) and human systems (e.g., the city of New Orleans). Despite having one of the most effective emergency response systems in the world, the United States has experienced months, and will likely continue to experience years, of difficulties in coping with the aftermath of Hurricane Katrina. Furthermore, while Hurricane Katrina may not be classified as an "extreme" hurricane in terms of its wind intensity at landfall, or a rare event in terms of the wind speed return period, the consequences of its landfall along the northern Gulf Coast would likely qualify as an extreme and, one hopes, rare event.

The capacity of society to respond optimally to climatic events such as active hurricane periods or long droughts depends on its ability to understand, anticipate, prepare for, and respond to extremes. Two years of intense hurricane landfalls in the southeastern United States in 2004 and 2005 illustrate that socioeconomic links enhance opportunities for extreme events to produce cascading consequences, such as: damaged oil production facilities leading to soaring fuel prices; large insured losses leading to rapidly rising insurance premiums and even the withdrawal of insurers from the marketplace; long-term displacement of residents, fomenting civil and political unrest; and other unforeseen or poorly foreseen outcomes. As was witnessed in the aftermath of Hurricane Mitch in Central America in 1998 and the devastating December 1999 floods in Venezuela, the destruction of much of New Orleans by Hurricane Katrina highlights the fact that disenfranchised groups tend to be disproportionately vulnerable to the impacts of extreme events.

These events, and the potential impacts of future events, motivate a major aim of this book: a survey of extreme climatic events that attempts to integrate a variety of disciplines and approaches. Extreme climate and weather events often have severe impacts as a result of interactions between different types of systems. The net impact of an extreme climate or weather event is a reflection of society's vulnerability, and the ultimate impact of extreme events, such as Katrina, on people reflects underlying vulnerabilities – whether or not they were evident prior to the event. We therefore include chapters that describe the application of analytical tools (e.g., statistical and numerical climate models, and complex risk models) used by the scientific community and insurance industry to investigate the physical aspects of extreme events and their economic and social impacts. An important aspect of this effort is quantifying the economic impacts of extremes and how they might change in the future as a result of climate variability. We therefore offer several chapters that attempt to quantify the losses produced by extreme events and consider how such losses might increase in the future.

Opportunities to reduce vulnerabilities are often created by extreme events. The thousands of excess deaths in Europe that resulted from the extreme heat waves in the summer of 2003 pointed out glaring deficiencies in the mitigation response plans of several western European countries. The shock from this event spurred the governments in the affected region to develop plans to mitigate the effects of future heat waves. Hot summer months in this region have recurred since 2003, without anywhere near the impacts of that event.

Similarly, opportunities for collaboration in the development of decision support tools for a variety of actors – from emergency managers to catastrophe reinsurance companies – have arisen as a result of the apparent increase in extreme climatic events and rising costs of damage resulting from these events. Because decision making is inherently forward-looking, scientific predictions supported by research from public, private, and public–private partnerships have the potential to benefit the decision process. This potential is especially relevant in the case of extreme climatic events, because of their rarity and the potential severity of their impacts. It is important, however, that prediction uncertainties be clearly articulated (and understood) so that users can be aware of their implications and consequences for actions in response to this information, or the lack thereof. An understanding of the exposure and distribution of vulnerability in a community is a critical component in the effort to develop integrative analysis, prediction, and assessment systems for use by emergency management decision makers. Such knowledge may help determine the types of information flows needed to reduce vulnerability, and the manner in which such information is best communicated.

Finally, we note the importance of generating high-quality long-term data-sets for extreme value analyses. The analysis of extreme, infrequent events requires the highest quality climate data. The current debate about the adequacy of our hurricane event datasets prior to the advent of high-resolution satellite data is a case in point. We are now on the cusp of an era when scientists expect a change in the intensity of tropical cyclones as a result of changes in sea surface and upper atmosphere temperatures and atmospheric humidity. Unfortunately, the limited quality of our data archives almost precludes our ability to assess unambiguous interdecadal changes in intensity. The use of synthetic hurricane datasets can provide some degree of insight about the recurrence probability of certain types of extreme events and how they might respond to changes in climate. However, synthetic datasets will never replace quality observational data.

In summary, this book examines a variety of climatic extremes using both the analysis of observational data and climate model simulations. It also illustrates some comprehensive approaches to understanding and responding to extreme events through the application of catastrophe risk models, and it highlights recent innovative efforts to develop institutional mechanisms and incentives for integrating knowledge on extremes and their economic impacts. We believe that the recent occurrence of severe climatic episodes – including intense droughts, floods, heat waves, and cyclones – has drawn much greater attention to the impact that climatic change may be having on the occurrence of extreme events. We hope this book contributes to the examination of some of these issues, and helps provide some perspective on these extremes.

The significance of weather and climate extremes to society: an introduction

HENRY F. DIAZ AND RICHARD J. MURNANE

Events over the past few decades have brought extreme weather and climate events to the fore of societal concerns. Ordinary citizens, individuals in the private sector, and people at the highest levels of government worry about the apparent increase in the frequency of weather and climate events causing extreme, and in some instances catastrophic, impacts. We differentiate between weather events – relatively short-term phenomena associated with, for instance, tropical cyclones (hurricanes and typhoons, for example), severe floods, and the like – and climate events – longer-lived and/or serial phenomena such as drought, season-long heat waves, record wildfire seasons, multiple occurrences of severe storms in a single season or year, etc. The differentiation is related to the distinction between weather, which can be forecast on short timescales of less than 1–2 weeks, and climate, which can be forecast on monthly, seasonal, and annual timescales. The adage "Climate is what you expect and weather is what you get" probably originates from the fact that climate is the statistical average of the weather over a specified time period. Regardless of whether an extreme event is weather- or climate-related, it could have significant and numerous implications for society.

This book summarizes our knowledge of different aspects of weather and climate extremes and then focuses on their recent and potential future consequences for different socioeconomic sectors. We also examine some actions that may enable us to better respond to and adapt to climate variability regardless of its source – for example, the development of public–private research and applications partnerships, and the development of state-supported public hurricane risk models for decision support. The book is divided into two parts: Part I, titled "Defining and modeling the nature of weather and climate extremes," where we examine evidence for recent and projected changes in extremes of weather and climate events, and Part II, titled "Impacts of weather and climate extremes," where we assess the impacts of

Climate Extremes and Society, ed. H. F. Diaz and R. J. Murnane. Published by Cambridge University Press. © Cambridge University Press 2008.

these events on the insurance industry. The chapters in Part I progress through the description of extremes and an assessment of recent changes in climate through an examination of how extremes might change in the future. Those in Part II evaluate the changing socioeconomic impacts of extremes and provide examples of how public and private enterprises are attempting to understand and respond to ongoing changes in extreme events.

The likely connection between climate change and extreme event frequency on multiple timescales has been recognized for some time (see Wigley, 1985). Wigley's paper illustrated for arbitrary climate variables the high sensitivity of low-probability occurrences to shifts in the mean. It seems likely that our experience and response to changes in weather and climate extremes will be a function of physical and temporal factors, for example: the intensity of the extreme; the temporal scale of the extreme (short-lived or persistent); the frequency of the extreme (rare or common); and the sensitivity and resiliency of our societies to a range of typical, and potentially new, extremes. Extremes in weather and climate are an inherent part of nature. Nature, and in many cases society, have a built-in resiliency to extreme events. Often, natural systems require extreme events in order for a species to reproduce or survive. Problems may arise when the frequency, intensity, distribution, or other characteristic of an extreme changes either beyond a threshold, or too rapidly. Therefore, it seems appropriate that this book opens with a chapter by Stephenson, which examines the subject from a taxonomic viewpoint, considering both statistical descriptions of extreme events and the fundamental climate patterns that give rise to them.

Increases in heat waves and intense precipitation are two of the most probable consequences of anthropogenic climate change. The increasing risk of severe heat waves (Schär *et al.*, 2004; Stott *et al.*, 2004) and flooding (Milly *et al.*, 2002, 2005) as global climate change progresses is a major concern for insurers, public health managers, and policy makers. Chapter 2, by Easterling, provides an overview of recent changes in temperature and precipitation derived from observational data. Temperature and precipitation are often the focus of studies on changes in extremes, in large part because of the relatively high quality of the data record. Chapter 3, by Brooks and Dotzek, illustrates the limitations of weather and climate data and provides an example of how people deal creatively with data inhomogeneities.

The fourth chapter, by von Storch and Weisse, examines how wind and wave extremes have changed over the past decades; Gershunov and Douville's chapter (Chapter 5) provides a unique assessment of extremes that accounts for the spatial scale of climate extreme events. Gershunov and Douville also examine how the probability distribution of seasonal extreme temperature values is changing and how it is likely to change based on projections from global climate models. Chapter 6, by Tebaldi and Meehl, provides the reader with our best estimates of how temperature and precipitation extremes are likely to change under high atmospheric concentrations of greenhouse gases.

The final chapter in Part I by Knutson and Tuleya (Chapter 7), examines how the intensities of tropical cyclones are likely to change under specific future emission scenarios of greenhouse gases. The theory relating tropical cyclone intensity to climate is well established (see Emanuel, 1987 and Holland, 1997). The likely impact of climate warming due to increased atmospheric CO_2 and other so-called greenhouse gases will be to increase, on average, tropical cyclone intensity (i.e., stronger maximum winds and lower central pressures). Recent studies support the contention that greenhouse forcing is already having an effect on tropical cyclone intensity in the North Atlantic (Emanuel, 2005; Elsner, 2006) as well as in the other ocean basins (Webster *et al.*, 2005; Hoyos *et al.*, 2006). There are a number of issues related to data quality (Chan, 2006; Landsea *et al.*, 2006) that raise questions about the robustness, or even existence, of recently observed changes. However, recent work (Kossin *et al.*, 2007) suggests that recent changes in hurricane intensity observed in the Atlantic are real.

The salient messages conveyed by the chapters in Part I are that climate is not static, that the frequency and intensity of extremes have changed over the past decades, and that we can expect to see similar changes in the future due in part to anthropogenic climate change. The newsworthiness of recent extreme events naturally makes people anxious about the future and leads to questions of how recent changes in weather and climate extremes are related to increases in greenhouse gases emitted through fossil fuel burning and other societal activities, and whether these events are harbingers of the future. Of course, without a decrease in our vulnerability, the losses of life and property to extreme events will continue to increase as long as the number of people and amount of property exposed to extremes continues to increase. An increase in the intensity and frequency of an extreme will only exacerbate the situation, sometimes in very nonlinear ways. Many of these issues are examined in Part II, which considers some of the impacts that recent extreme climate events have had on society, and some implications for the future.

The material in Part II attempts to assess the impact of changes in weather and climate extremes on society in general and the insurance industry in particular. These chapters provide case studies of how changes in climate extremes can influence different parts of our society. However, one should not forget that the impacts of extreme weather and climate events are by no means limited to the examples presented here. More recent, larger impact events are often in the news. The 2004 and 2005 Atlantic hurricane seasons, for example, were very active, with multiple landfalls affecting the United States. Because the United States was struck repeatedly, the cumulative impact was very costly to insurers. Different parts of the world are prone to different types of weather and climate hazards. In the western United States, for example, a widespread and intense 5-year drought was punctuated in 2003 by extreme wildfires in southern California that caused losses worth billions of dollars. Three years later, in 2006, the wildfire season set a new record for acreage burned. In fact, in the past decade or two, several new records have been set for acreage burned in the United States, while in Europe the summer of 2003 saw record-breaking heat that contributed to the premature deaths of thousands of people.

The first two chapters in Part II provide examples of how weather and climate extremes can affect ecosystems and society. Beniston's chapter (Chapter 8) examines regional-scale changes in temperature and precipitation in the European Alps, an area with sharp climatic gradients driven by changes in elevation. The author then evaluates climate change projections for the region in the context of observed climate changes over the past century. Chapter 9, by Crabbe *et al.*, examines how temperature and wave extremes influence coral growth. Crabbe *et al.* show that for many coral communities, an increase in temperature of only a few degrees may result in significant reductions in growth rates and widespread mortality. In addition, an increase in tropical cyclone intensity will also have a direct effect on coral reef systems, as stronger wave action can also result in reef damage.

Loss data are one of the few benchmarks that can be used to assess the impact of weather and climate extremes over time. However, issues related to the collection and quality of loss data generally make temperature and precipitation data appear to be ideal. This is an unfortunate situation, as monetary losses provide one of the few measures of the impacts of extremes that easily conform to a typical decision-making process.

Chapter 14, by Cutter *et al.*, discusses some of the issues related to the collection of loss data and presents an overview of a public archive of losses from extremes. This information is difficult to collect, in part because there is no formal mechanism for collecting or identifying loss data.

The insurance industry commonly collects the most complete information on losses from extreme climate and weather events. Unfortunately, this information is often proprietary. Nevertheless, a number of companies produce publicly available reports that aggregate and analyze losses on regional and global scales. The total amount of insured losses arising from the top 40 extreme weather and climate extremes worldwide from 1970 to 2004 was approximately US$142 billion (in 2004), with US hurricanes causing the lion's share of these losses (Murnane and Diaz, 2006). In December 2005, the Munich Re Foundation reported that 2005 was the costliest year on record for economic losses due to natural disasters, with about US$200 billion in economic losses from weather-related disasters. These losses surpassed the previous record of about US$145 billion set the previous year (see also Chapter 13 by Dlugolecki, this volume).

The data on overall natural hazards and disaster losses for the past 35 years suggest a rather steady increase in the level of monetary losses adjusted for inflation (Cutter and Emrich, 2005). Inflation-adjusted economic losses from catastrophic events – those that resulted in economic losses that exceeded some arbitrary criterion – rose 8-fold in the last four decades of the twentieth century, and insured losses rose by 17-fold (Mills, 2005). But these studies do not account for changes in other factors, such as population and wealth, that also play a role in loss (Diaz and Pulwarty, 1997; Choi and Fisher, 2003; Simpson, 2003). Chapter 12, by Miller *et al.*, analyzes loss data on national and global scales and accounts for changes in population and wealth. Miller *et al.* find no statistically significant trend in these normalized losses.

Although there is no definitive trend in normalized losses, available records indicate a significant increase in the size and frequency of insured losses. Dlugolecki (Chapter 13) discusses the implications of these increases from a global perspective. In particular, the author argues that recent rapid increases in insurance losses may in part reflect the rather rapid pace of global warming in the past few decades. An insurer's rational response to the potential for large losses would be to estimate the probability of the loss and then plan appropriately. We include three chapters that offer examples of potential planning tools that could be useful as risk management tools. Chapter 10 by Jagger *et al.* provides a novel approach for directly forecasting annual insured losses for US landfalling hurricanes as a function of seasonal and interannual climate variability. Chapter 11, by Watson and Johnson, discusses an approach for integrating climate model simulations into catastrophe risk models commonly used by insurers for estimating losses. Chapter 15 by Muir-Wood and Grossi describes the impact that Hurricane Katrina had on the catastrophe insurance industry in general and from the perspective of one company, Risk Management Solutions (RMS).

We end our discussion of the impacts of weather and climate extremes with two examples of institutional actions to manage our response to extremes. In Chapter 15 Muir-Wood and Grossi discuss the aftermath of Hurricane Katrina from the point of view of the catastrophe modeling community. The final chapter, by Murnane and Knap (Chapter 16), discusses a science–business partnership – funded by companies active in the catastrophe risk insurance industry – that supports research on weather and climate extremes of interest to the insurance industry.

This book provides examples of the impacts of climate and weather extremes on society; how these extremes have varied in the past, and how they might change in the future; and the types of efforts that will help society adapt to future changes in climate and weather extremes. This is obviously a huge subject that is evolving rapidly. We hope that we provide the reader with a snapshot of our understanding of extremes and how our society is attempting to respond to them.

References

Chan, J. C. L. (2006). Comment on "Changes in tropical cyclone number, duration, and intensity in a warming environment." *Science*, **311**, 1713, doi:1710.1126/science.1121522.

Choi, O., and Fisher, A. (2003). The impacts of socioeconomic development and climate change on severe weather catastrophe losses: mid-Atlantic region (MAR) and the U.S. *Climatic Change*, **58**, 149–70.

Cutter, S. L., and Emrich, C. (2005). Are natural hazards and disaster losses in the U.S. increasing? *Eos, Transactions, American Geophysical Union*, **86**, 381.

Diaz, H. F., and Pulwarty, R. S. (eds) (1997). *Hurricanes, Climate and Socioeconomic Impacts*. Berlin: Springer.

Elsner, J. B. (2006). Evidence in support of the climate change–Atlantic hurricane hypothesis. *Geophysical Research Letters*, **33**, L16705, doi:10.1029/2006GL026869.

Emanuel, K. (1987). The dependence of hurricane intensity on climate. *Nature*, **326**, 483–5.

Emanuel, K. (2005). Increasing destructiveness of tropical cyclones over the past 30 years. *Nature*, **436**, 686–8.

Holland, G. J. (1997). The maximum potential intensity of tropical cyclones. *Journal of the Atmospheric Sciences*, **54**, 2519–41.

Hoyos, C. D., Agudelo, P. A., Webster, P. J., and Curry, J. A. (2006). Deconvolution of the factors contributing to the increase in global hurricane intensity. *Science*, **312**, 94–7.

Kossin, J. P., Knapp, K. R., Vimont, D. J., Murnane, R. J., and Harper, B. A. (2007). A globally consistent reanalysis of hurricane variability and trends. *Geophysical Research Letters*, **34**, L04815, doi:10.1029/2006GL028836.

Landsea, C. W., Harper, B. A., Hoarau, K., and Knaff, J. A. (2006). Climate change: can we detect trends in extreme tropical cyclones. *Science*, **313**, 452–4, doi:410.1126/science.1128448.

Mills, E. (2005). Insurance in a climate of change. *Science*, **309**, 1040–4.

Milly, P. C. D., Dunne, K. A., and Vecchia, A. V. (2005). Global pattern of trends in streamflow and water availability in a changing climate. *Nature*, **438**, 347–50.

Milly, P. C. D., Wetherald, R. T., Dunne, K. A., and Delworth, T. L. (2002). Increasing risk of great floods in a changing climate. *Nature*, **415**, 514–17.

Murnane, R. J., and Diaz, H. F. (2006). Assessing, modeling, and monitoring the impacts of extreme climate events. *Eos, Transactions, American Geophysical Union*, **87**, 25.

Schär, C., Vidale, P. L., Lüthi, D., *et al.* (2004). The role of increasing temperature variability in European summer heatwaves. *Nature*, **427**, 332–6.

Simpson, R. (ed.) (2003). *Hurricane!* Washington, DC: American Geophysical Union.

Stott, P. A., Stone, D. A., and Allen, M. R. (2004). Human contribution to the European heatwave of 2003. *Nature*, **432**, 610–14.

Webster, P. J., Holland, G. J., Curry, J. A., and Chang, H. R. (2005). Changes in tropical cyclone intensity in a warming environment. *Science*, **309**, 1844–6.

Wigley, T. M. L. (1985). Impact of extreme events. *Nature*, **316**, 106–7.

I

Defining and modeling the nature of weather and climate extremes

1

Definition, diagnosis, and origin of extreme weather and climate events

DAVID B. STEPHENSON

Condensed summary

Extreme weather and climate events are a major source of risk for all human societies. There is a pressing need for more research on such events. Various societal changes, such as increased populations in coastal and urban areas and increasingly complex infrastructure, have made us potentially more vulnerable to such events than we were in the past. In addition, the properties of extreme weather and climate events are likely to change in the twenty-first century owing to anthropogenic climate change.

The definition, classification, and diagnosis of extreme events are far from simple. There is no universal unique definition of what is an extreme event. This chapter discusses these issues and presents a simple framework for understanding extreme events that will help enable future work in this important area of climate science and global reinsurance.

1.1 Introduction

Human society is particularly vulnerable to severe weather and climate events that cause damage to property and infrastructure, injury, and even loss of life. Although generally rare at any particular location, such events cause a disproportionate amount of loss.

In this chapter, I attempt to describe a framework for classifying, diagnosing, and understanding extreme weather and climate events. The multi-dimensional aspect of complex extreme events will be discussed, and statistical methods will be presented for understanding simple extreme events. Some preliminary ideas about the origin of extreme events will also be presented.

Climate Extremes and Society, ed. H. F. Diaz and R. J. Murnane. Published by Cambridge University Press. © Cambridge University Press 2008.

1.2 Definition of extreme events

Extreme events are generally easy to recognize but difficult to define. This is due to several reasons. First, there is no unique definition for what is meant by the word "extreme": several definitions are in common use. Second, the concept of "extremeness" is relative and so strongly depends on context. Third, the words "severe," "rare," "extreme," and "high-impact" are often used interchangeably.

1.2.1 Severe, rare, extreme, or high-impact?

In an attempt to alleviate some of the confusion, here are some definitions of these terms.

- **Severe events** are events that create large losses in measures such as number of lives, financial capital, or environmental quality (e.g., loss of species). The severity can be measured by the expected long-term loss, which is known as the risk. Risk depends on the product of the probability of the event (the hazard), the exposure to the hazards (e.g., how many people are exposed), and the vulnerability (i.e., how much damage ensues when someone is hit by the event). In other words, severity is a function of not only the meteorological hazard but also the human state of affairs. For example, the severity of US landfall hurricanes has increased considerably in recent years, mainly owing to increased numbers of people settling in the US Gulf states (increased exposure).
- **Rare events** are events that have a low probability of occurrence. Because of the rarity of these events, human societies (and other ecosystems) are often not well adapted to them and so suffer large amounts of damage when they do occur. Hence, despite their rarity, the large vulnerability associated with such events can often lead to large mean losses (and hence they are a type of severe event).
- **Extreme events** are events that have extreme values of certain important meteorological variables. Damage is often caused by extreme values of certain meteorological variables, such as large amounts of precipitation (e.g., floods), high wind speeds (e.g., cyclones), high temperatures (e.g., heat waves), etc. Extreme is generally defined as either taking maximum values or exceedance above pre-existing high thresholds. Such events are generally rare; for example, extreme wind speeds exceeding the 100-year return value, which have a probability of only 0.01 of occurring in any particular year.
- **High-impact events** are severe events that can be either short-lived weather systems (e.g., severe storms) or longer-duration events such as blocking episodes that can lead to prolonged heat waves and droughts. The World Meteorological Organization (WMO) program THORPEX uses the phrase "high-impact weather" rather than "severe weather" to help people avoid confusing the term severe with only short-lived events such as individual storms (D. Burridge, personal communication).

1.2.2 Multidimensional nature of extreme events

In addition to this potential source of confusion, extreme events have a variety of different attributes and so cannot be completely described by a single number. The multidimensional nature of extreme events is often overlooked in rankings of the events based on only one of the attributes (e.g., the category numbers for hurricanes based solely on maximum surface wind speed).

Extreme events have attributes such as:

- rate (probability per unit time) of occurrence
- magnitude (intensity)
- temporal duration and timing
- spatial scale (footprint)
- multivariate dependencies

For example, a major hurricane is rare, has large-magnitude surface wind speeds, has a generally large spatial scale (with the exception of certain small-spatial-footprint events such as Hurricane Camille in 1969), and develops over synoptic timescales ranging from hours to several days. In addition, the severity of such events can also depend on the combination of extreme behavior in more than one variable; for example, much hurricane damage is due to extreme precipitation as well as extreme wind speeds. For example, a severe ice storm can involve conditions for all three variables: temperature, wind, and precipitation amount. The Intergovernmental Panel on Climate Change (IPCC, 2001) defines "complex extreme" events as "severe weather associated with particular climatic phenomena, often requiring a critical combination of variables." The magnitude of the losses depends on several meteorological variables and hence dependencies between them.

The temporal duration of extreme events plays an important role in the exposure and hence total losses. For example, the long duration of the flooding in New Orleans caused by Hurricane Katrina led to large insurance losses due to business interruption in addition to property damage losses. Temporal duration also provides a useful way of classifying extreme events. The duration is implicit when one describes an event as a "climate" extreme event rather than a "weather" extreme event. The medical illness concepts of chronic and acute can be usefully applied to weather and climate events:

- **Acute extremes**: events that have a rapid onset and follow a short but severe course. Examples are short-lived weather systems such as tropical and extratropical cyclones, polar lows, and convective storms with extreme values of meteorological variables such as wind speed and precipitation that can lead to devastating wind, flood, and ice damage. In addition to these obvious examples of high-impact extreme events, there are less obvious acute extreme events, such as fog that causes major transport disruption (e.g., at airports).

• **Chronic extremes**: events that last for a long period of time (e.g., longer than 3 months) *or* are marked by frequent recurrence. Examples are heat waves and droughts that can lead to such impacts as critical water shortages, crop failure, heat-related illness and mortality, and agricultural failure. Because of their extended duration and generally lower intensity, chronic extreme events can often be harder to define than acute extreme events, but they have the advantage that there is more time to issue warnings and take protective actions. Note that not all high-impact weather events are acute; for example, blocking weather events that last several days are chronic events.

1.2.3 A simple taxonomy

One approach towards creating a definition is to try to list and classify all events that one considers to be extreme. For example, the following events are often cited as examples of extreme weather/climate events.

• **Tropical cyclones and hurricanes** (e.g., Typhoon Tracy, Hurricane Hugo, etc.). These storms are the major source of global insured catastrophe loss after earthquakes.
• **Extratropical cyclones** (e.g., the "Perfect Storm" that hit the northeast coast of the United States, October 28–30, 1991). These storms are generally referred to as "windstorms" by the reinsurance industry.
• **Convective phenomena** such as tornadoes, waterspouts, and severe thunderstorms. These phenomena can lead to extreme local wind speeds and precipitation amounts on horizontal scales of up to about 10 km. Deep convection often leads to precipitation in the form of hail, which can be very damaging to crops, cars, and property.
• **Mesoscale phenomena** such as polar lows, mesoscale convective systems, and sting jets. These features can lead to extreme wind speeds and precipitation amounts on horizontal scales from 100 to 1,000 km.
• **Floods** of rivers, lakes, coasts, etc., due to severe weather conditions; for example, river floods caused by intense precipitation over a short period (e.g., flash floods) and persistent/recurrent precipitation over many days (e.g., wintertime floods in northern Europe), river floods caused by rapid snowmelt due to a sudden warm spell, or coastal floods caused by high sea levels due to wind-related storm surges.
• **Drought**. Meteorological drought is defined usually on the basis of the degree of dryness (in comparison to some "normal" or average amount) and the duration of the dry period. Simple definitions relate actual precipitation departures to average amounts on monthly, seasonal, or annual timescales. However, meteorological

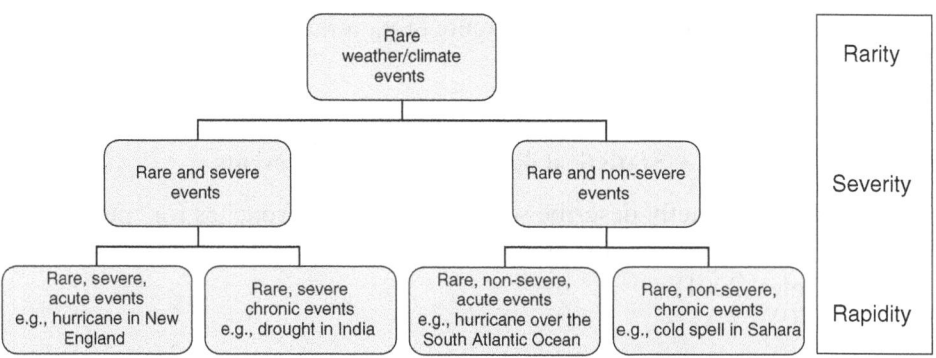

Figure 1.1. Simple taxonomy of extreme weather and climate events.

drought also depends on other quantities such as evaporation that depend on variables such as temperature.

- **Heat waves**. Periods of exceptionally warm temperatures can have profound impacts on human health and agriculture. Duration is a key component determining the impact.
- **Cold waves/spells** (e.g., extremely cold days or a succession of frost days with minimum temperatures below 0 °C).
- **Fog**. Extremely low visibility has major impacts on various sectors such as aviation and road transport.

In order to make more sense of this set of diverse events, it is useful to try to classify them into smaller subgroups. A simple binary taxonomy of weather and climate events can be based on the attributes of rarity, severity, and duration as shown in Figure 1.1.

First, for any particular location an event can be considered to be either rare or not rare depending on how often such an event happens. For example, events can be considered rare if they happen less frequently than once every 250 years (a return period often used by the reinsurance industry to assess acceptable levels of risk).

Second, depending upon its impacts, an event can be considered either severe or not severe. Severity depends not only upon the characteristics of an event but also upon the exposure and vulnerability of the system it impacts. For example, a heat wave in the Gobi Desert is not a severe event in terms of human impact, because there is very little human exposure (i.e., very few people live there).

Finally, events can be classified by their longevity into either acute or chronic events. In contrast to the medical situation, where severe syndromes can generally not be sustained over a long period of time and therefore tend to be acute rather than chronic (e.g., severe acute respiratory syndrome [SARS]), severe weather events can be either acute (e.g., a major hurricane) or chronic (e.g., a major drought).

1.3 Statistical diagnosis of extreme events

This section will briefly describe some statistical approaches for interpreting extreme events. A more comprehensive discussion is given in the excellent book by Coles (2001).

1.3.1 Point process modeling of simple extreme events

In order to make the analysis more amenable to mathematical modeling, it is useful to neglect (important!) attributes such as temporal duration, spatial scale, and multivariate dependencies. The IPCC (2001) defined "simple extreme" events to be "individual local weather variables exceeding critical levels on a continuous scale."

This highly simplified view of a complex extreme event is widely used in weather and climate research. However, one can always consider an event as a simple extreme in overall loss (i.e., severity) no matter how complex the underlying meteorological situation may be.

Simple extreme events, defined as having exceedances above a high threshold, are amenable to various types of statistical analysis. Because exceedances occur at irregular times and the excesses tend to be strongly skewed, such series are not amenable to the usual methods of time series analysis. However, exceedances can be considered to be a realization of a stochastic process known as a *marked point process*: a process with random magnitude *marks* (the excesses above the threshold) that occurs at random *points* in time (see Diggle, 1983; Cox and Isham, 2000). Rare exceedances above a sufficiently high threshold can be described by a nonhomogenous Poisson process (Coles, 2001).

Point process methods have been widely used in various areas of science; for example, in providing a framework for earthquake risk assessment and prediction in seismology (Daley and Vere-Jones, 2002). Point process methods can be used to explore and summarize such records and are invaluable for making inferences about the underlying process that gave rise to the record. Broadly speaking, this analysis is performed by considering statistical properties of the points, such as the number of events expected to occur per unit time interval (the *rate/intensity* of the process), statistical properties of the marks (the probability distribution of the excesses), and joint properties such as how the marks depend on the position and spacing of the points, the magnitude of preceding events, etc.

1.3.2 Example: central England temperature observations

Typical time series resemble the example of 3,080 monthly means of observed central England temperature (CET) shown in Figure 1.2.

The 308 warm extreme events with temperatures in excess of the 90th quantile of 15.6 °C (the dots above the solid line in the upper panel) form a realization of a marked point process. Rather than being recorded continuously, this example and many other meteorological variables are generally recorded and stored on computers at regular discrete time intervals. Strictly speaking, this is a special type of marked point process in which the points occur on a discrete set of regularly spaced times (e.g., daily values) rather than at any possible time. The occurrence of an exceedance in such cases can be modeled by using a discrete-time Markov chain (see section 5.2 of Lindsey, 2004).

1.3.3 Choice of threshold

The high threshold used to define the extreme events can be chosen in many different ways. The simplest approach is to choose a constant absolute threshold related to impacts; for example, the threshold of 25 °C widely used to

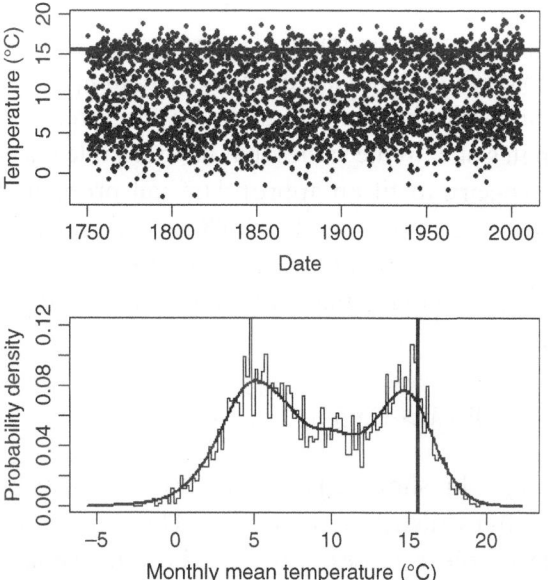

Figure 1.2. Monthly mean values of the central England temperature series from 1750 to 2006 (upper panel). Lower panel: the histogram of the monthly values (*x*-axis, in °C) together with a smooth kernel estimate of the probability density function. Ten percent of the values exceed the 90th quantile of 15.6 °C (indicated by the solid line in both panels).

define extreme heat wave indices (e.g., Alexander *et al.*, 2006; New *et al.*, 2006). Such extreme events can lead to severe health situations, which are likely to become more prevalent due to global warming (McGregor *et al.*, 2005). A more relative approach is to choose a constant threshold based on the empirical distribution of the variable at each location; for example, the 90th quantile shown in the central England temperature example. This approach is useful in that it ensures that a given fraction (e.g., 10 %) of events will by definition be "extreme." In other words, it defines "extremeness" in terms of "rarity." In addition to these two approaches, one can also consider time-varying thresholds. For example, "record-breaking" events can be defined by choosing the threshold to be the maximum value of all previously observed values. One can also choose trending thresholds to help take account of non-stationarities such as the changing baseline caused by global warming. Such definitions of extreme events can help us avoid the paradoxical situation whereby "extreme events will become the norm" (as was stated by Deputy Prime Minister John Prescott after the autumn 2000 UK floods).

1.3.4 Magnitude of the extreme events (distribution of the marks)

The magnitude of the extreme events can most easily be summarized by calculating summary statistics of the sample of excesses; for example, the mean excess above the threshold (Coles, 2001). However, such an approach does not allow one to make inferences about as-yet-unobserved extreme values or provide probability estimates of extreme values that have reliable uncertainty estimates. It is therefore necessary to fit an appropriate tail probability distribution to the observed excesses (e.g., the sample of 308 excesses for the CET example).

Under rather general assumptions, a limit theorem shows that for most continuous random variables, the probability of exceedance above a large value $x > u$ is given by

$$\Pr(X>x|X>u) = \left[1 + \xi\left(\frac{x-u}{\sigma}\right)\right]^{-\frac{1}{\xi}}$$

for a sufficiently high threshold u. This two-parameter distribution is known as the generalized Pareto distribution (GPD). The dimensional parameter, σ, defines the scale of the excesses, whereas the dimensionless parameter, ξ, defines the overall shape of the tail. These two parameters can be easily estimated by using maximum likelihood estimation.

For the CET example, the excesses above the 90th quantile $u = 15.6\,°C$ give a scale parameter estimate of $1.38\,°C$ (with a standard error of $0.09\,°C$) and a shape parameter estimate of -0.30 (with a standard error of 0.04). The

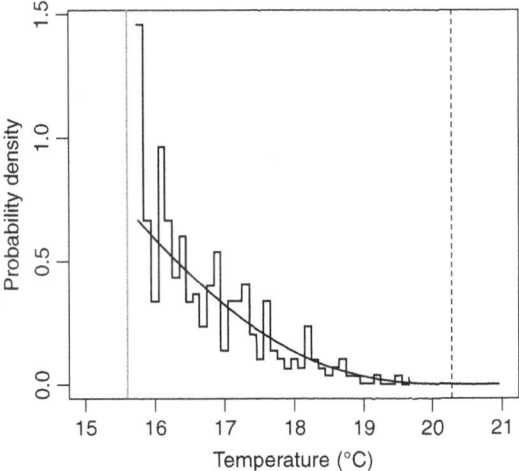

Figure 1.3. Histogram of the 308 monthly mean values of the central England temperature series from 1750 through 2006 that exceed the 90th quantile of 15.6 °C (bold line) together with the GPD fit (continuous curve). The GPD fit has a finite upper limit at 20 °C, above which there is zero probability of a temperature value occurring.

negative shape parameter implies that the tail distribution has an upper limit at $u - \sigma/\xi = 20.3$ °C.

Figure 1.3 shows that the estimated GPD provides a good fit to the histogram of exceedance values. This is even clearer in the return level plot shown in Figure 1.4. The GPD fit can be used to estimate either the probability of exceedance above a chosen temperature (or its reciprocal, the return period), or the temperature quantile (the return level) corresponding to a specific exceedance probability. The GPD fit provides a smooth interpolation between the observed extreme values and a way of extrapolating beyond the maximum value observed in the finite sample.

1.3.5 Timing of the extreme events (distribution of the points)

There are two main approaches for estimating the rate of a point process: the *counting specification*, based on counting the number of points in fixed time intervals, and the *interval specification*, based on estimating the mean time interval between successive points (see Cox and Isham, 2000, p. 11).

The simplest counting specification approach involves dividing the time axis into a set of non-overlapping, equally spaced bins and then counting the number of points that fall into each bin. This approach gives rather noisy results due to the sharp bin edges. More efficient and smoother rate estimates

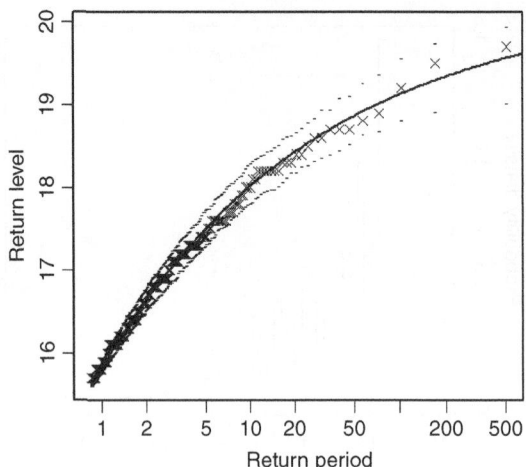

Figure 1.4. Return level plot for the central England temperature series from 1750 through 2006. The return period is the reciprocal of the probability of a monthly temperature value exceeding the return level. The crosses denote the 308 empirical quantiles and probabilities, and the solid curve is the GPD fit. Note the concavity of the curve, which is characteristic of a tail distribution having a negative shape parameter.

can be obtained by using smooth local weighting based on a smooth kernel function rather than a sharp-edged bin (Diggle, 1985). Such an approach was recently used to investigate extreme flooding events in eastern Norway as observed in paleoclimatic lake sediments (Bøe *et al.*, personal communication). Various tests can be used to test for trends in the rate of a Poisson process, but these have differing abilities to detect trends (Bain *et al.*, 1985; Cohen and Sackrowitz, 1993).

In addition to characterizing changes in the rate and magnitude of a point process, one can also investigate the temporal clustering. For example, Mailier *et al.* (2006) used simple point process overdispersion ideas to evaluate the clustering in transits of extratropical cyclones. They found that there is significant clustering of cyclones over Western Europe that can be attributed to rates varying in time due to the dependence on large-scale flow patterns. In general, time dependence in rates for an extreme event can lead to clustering. This is a very interesting area for future research on extreme weather and climate events.

1.3.6 Some ideas for future work

Statistical analysis of extreme events generally focuses on a given set of extreme events and tends to neglect how such events came into existence.

This is a strength in that it makes extreme value techniques more universally applicable to different areas of science no matter how the extremes formed. However, this approach also has a weakness in that it ignores information about the process that led to the extreme events that could help improve inference. How moderately large events evolve into extreme events can provide clues into the very nature of the extreme events. The dynamical knowledge about underlying formation processes should be exploited in the statistical analysis.

1.4 The origin of extreme events

Understanding the processes that lead to the creation of extreme events and how they might change in the future is a key goal of climate science. To help tackle the problem of the origin of extremes, I propose two guiding principles.

- **The evolutionary principle**. Extreme events do not arise spontaneously: instead, they evolve continuously from less extreme events and they stop evolving to become even more extreme events.
- **The stationary principle**. Extremes such as local maxima and minima are quasi-stationary states in which the rate of change of their amplitude is zero. This characteristic implies that there is an interesting balance between forcing and dissipation tendencies for such extreme events.

There are various processes that can give rise to extreme events:

- **Rapid growth due to instabilities** caused by positive feedbacks; for example, the rapid growth of storms due to convective and baroclinic instability.
- **Displacement** of a weather system into a new spatial location (e.g., a hurricane in Boston) or into a different time period (e.g., a late frost in spring).
- **Simultaneous coincidence** of several non-extreme conditions (e.g., freak waves caused by several waves occurring together).
- **Localization** of activity into intermittent regions (e.g., precipitation in intertropical convergence zones).
- **Persistence or frequent recurrence** of weather leading to chronic extremes as caused by slower variations in the climate system (e.g., surface boundary conditions).
- **Natural stochastic/chaotic variation** that will lead to more extreme values being recorded as the time length of the record increases.

Understanding these processes is the key to understanding how extreme events have behaved in the past and how they might behave in the future. In addition to being of interest because of their large impacts, extreme events are worth studying because they can reveal insights into key processes. For example, investigation of rapidly deepening Atlantic storms ("bombs") has helped improve scientific knowledge of fundamental baroclinic instability

D. Stephenson

mechanisms (explosive cyclogenesis). For numerical weather and climate models to correctly simulate extreme events, they will need to adequately represent such processes.

1.5 Conclusion

This chapter has addressed the perplexing issues of how to define and diagnose extreme events. It has been shown that extreme events are generally complex entities described by several different attributes: rate of occurrence, magnitude (intensity), temporal duration and timing, spatial structure, and multivariate dependencies.

Despite this complexity, extreme weather and climate events are often described by using only a single variable (e.g., maximum wind speed at land-fall). Exceedances of such a variable above a high threshold define what is known as *simple* extreme events. This simple description of complex events can be considered to be a realization of a stochastic marked point process. Point process techniques can be usefully employed to characterize properties of simple extreme events such as the rate, the magnitude, and temporal clustering.

Despite the societal relevance, estimates and predictions of extreme events are prone to large sampling uncertainty due to the inherent rarity of such events. Careful inference is needed to make definitive statements about extreme events such as the regional changes one is likely to see due to global warming (e.g., Beniston *et al.*, 2006). Inference can be improved by various approaches such as extrapolating from less extreme events (e.g., using tail distributions such as the generalized Pareto distribution), by pooling extreme events over a spatial region to reduce rarity (e.g., tropical cyclones over all the tropics), and by relating changes in extremes to changes in mean and variance (Beniston and Stephenson, 2004; Ferro *et al.*, 2006). Such approaches require careful statistical modeling that can benefit from insight gained from knowledge of dynamical processes that determine extreme events.

Acknowledgments

I thank Rick Murnane and Henry Diaz for inviting me to present these ideas in the opening seminar at the "Assessing, Modeling, and Monitoring the Impacts of Extreme Climate Events" workshop in Bermuda, October 13–14, 2005. Many of the ideas presented here have grown out of exciting discussions that I have had with colleagues over the past few years; in particular, with Dr Chris Ferro and Professor Martin Beniston. Finally, but not least, I thank the reviewers of this chapter for their useful comments.

References

Alexander, L. V., Zhang, X., Peterson, T. C., *et al.* (2006). Global observed changes in daily climate extremes of temperature and precipitation. *Journal of Geophysical Research (Atmospheres)*, **111**, D05109, doi:10.1029/2005JD006290.

Bain, L. J., Engelhardt, M., and Wright, F. T. (1985). Tests for an increasing trend in the intensity of a Poisson process: a power study. *Journal of the American Statistical Association*, **80**(390), 419–22.

Beniston, M., and Stephenson, D. B. (2004). Extreme climatic events and their evolution under changing climatic conditions. *Global and Planetary Change*, **44**, 1–9.

Beniston, M, Stephenson, D. B., Christensen, O. B., *et al.* (2006). Future extreme events in European climate: an exploration of regional climate model projections. *Climatic Change*, PRUDENCE special issue.

Cohen, A., and Sackrowitz, H. B. (1993). Evaluating tests for increasing intensity of a Poisson process. *Technometrics*, **35**(4), 446–8, doi:10.2307/1270277.

Coles, S. (2001). *An Introduction to Statistical Modeling of Extreme Values*. London: Springer-Verlag.

Cox, D. R., and Isham, V. (2000). *Point Processes*. New York: Chapman & Hall/CRC.

Daley, D. J., and Vere-Jones, D. (2002). *An Introduction to the Theory of Point Processes*, 2nd edition. Berlin: Springer-Verlag.

Diggle, P. J. (1983). *Statistical Analysis of Point Processes*. London: Chapman & Hall.

Diggle, P. J. (1985). A kernel method for smoothing point process data. *Applied Statistics*, **34**, 138–7.

Ferro, C. A. T., Hannachi, A., and Stephenson, D. B. (2006). Simple non-parametric techniques for exploring changing probability distributions of weather. *Journal of Climate*, **18**, 4344–54.

Intergovernmental Panel on Climate Change (IPCC) (2001). *Climate Change 2001: Synthesis Report*. Cambridge: Cambridge University Press.

Lindsey, J. K. (2004). *Statistical Analysis of Stochastic Processes in Time*. Cambridge: Cambridge University Press.

Mailier, P. J., Stephenson, D. B., Ferro, C. A. T., and Hodges, K. I. (2006). Serial clustering of extratropical cyclones. *Monthly Weather Review*, **134**(8), 2224–40.

McGregor, G. R., Ferro, C. A. T., and Stephenson, D. B. (2005). Projected changes in extreme weather and climate events in Europe. In *Extreme Weather and Climate Events and Public Health Responses*, ed. W. Kirch, B. Menne, and R. Bertollini. Dresden: Springer, pp. 13–23.

New, M., Hewitson, B., Stephenson, D. B., *et al.* (2006). Evidence of trends in daily climate extremes over southern and west Africa. *Journal of Geophysical Research (Atmospheres)*, **111**, D14102, doi:10.1029/2005JD006289.

2

Observed changes in the global distribution of daily temperature and precipitation extremes

DAVID R. EASTERLING

Condensed summary

Observed changes in climate extremes have been documented for both temperature and precipitation in many parts of the globe. These changes include decreases in frost days and a lengthening of the frost-free season, increases in the number of days with temperatures above percentile-based thresholds, and increases in heavy precipitation events. These changes are generally consistent with observed warming in mean annual temperatures, and with observed changes in annual precipitation.

2.1 Introduction

This chapter reviews the scientific literature on variability and change in observed climate extremes over the globe. Observed changes in extremes should be considered in light of observed changes in mean quantities, including observed changes in annual average temperature, and changes in maximum and minimum temperatures and the diurnal temperature range (DTR) (Easterling *et al.*, 1997; Vose *et al.*, 2005). The globally averaged annual temperature (Fig. 2.1) shows a linear increase of approximately 0.6 degrees per century since the late 1800s, but the rate of increase since the mid 1970s has itself increased to approximately 2 degrees per century. Seasonally, the strongest increases have occurred in the boreal winter (December–January–February, DJF) and spring (March–April–May, MAM). Figure 2.2 shows that, regionally, the southeastern United States and northern Atlantic continue to show a trend to cooling, but the southeastern US trend appears to be getting smaller with each additional year of data; the largest increases in temperature have occurred in the high latitudes of the Northern Hemisphere.

Changes in daily maximum and minimum temperatures averaged annually across the globe for 1950–2004 show minimum temperature increasing at a

Climate Extremes and Society, ed. H. F. Diaz and R. J. Murnane. Published by Cambridge University Press. © Cambridge University Press 2008.

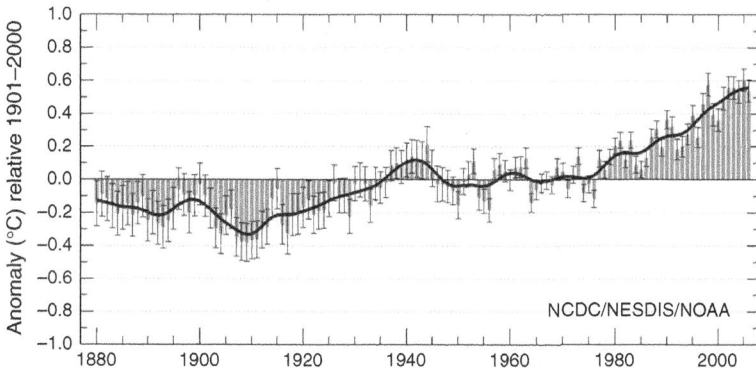

Figure 2.1. Globally averaged annual temperatures (over land and ocean) for the period 1880–2005. The 95% confidence limits are shown for each year; the heavy black line is smoothed by using a 13-point binomial filter. (Data from Smith and Reynolds, 2005.)

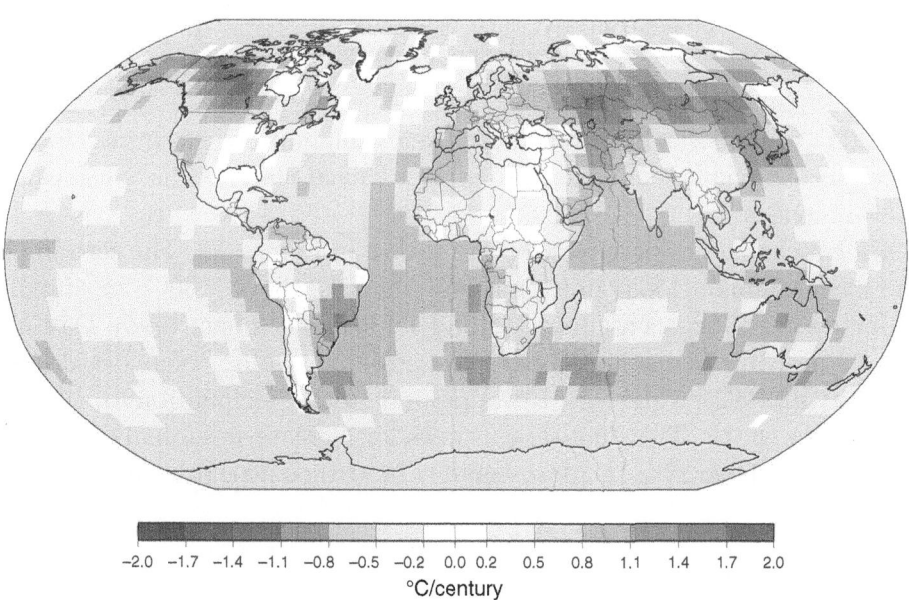

Figure 2.2. Linear trends in average annual temperature for the period 1901–2005. Areas in gray are excluded due to a lack of reliable data. (Data is from Smith and Reynolds, 2005.) For color version, see plate section.

faster rate than maximum temperature, resulting in a decrease in the diurnal temperature range (maximum minus minimum). However, since the late 1970s both maximum and minimum temperatures have been increasing at approximately the same rate, resulting in little change in the DTR (Vose *et al.*, 2005). Similarly, for precipitation there have been increases averaged

D. Easterling

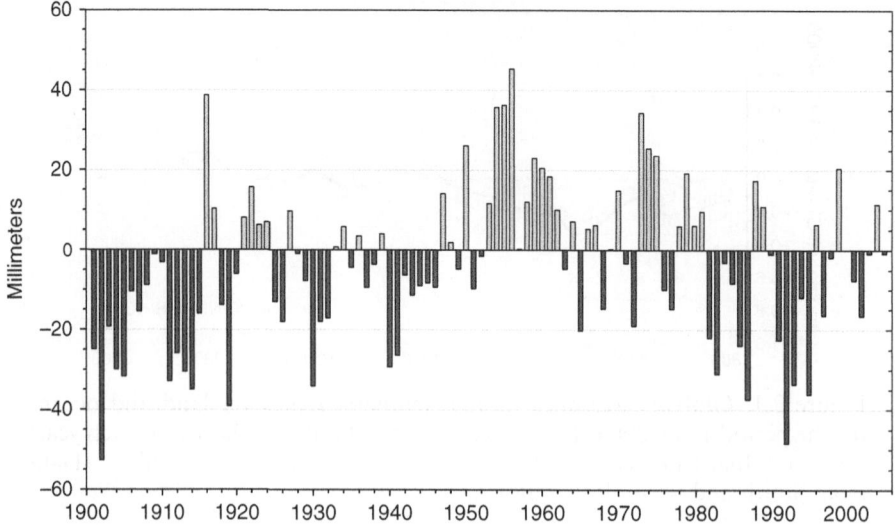

Figure 2.3. Globally averaged, land-based annual precipitation anomalies (1900–2005) from the 1961–90 base period mean.

over the entire globe; however, there are regional differences in trends, and many smaller regions, such as the southwest United States or the Sahel region, show decreases.

Global changes in precipitation are shown in Figure 2.3. The zero line is the mean annual global precipitation, which is about 1,000 mm, but the estimates range between about 785 and 1,130 mm (Hulme, 1995). Data for this figure come from the Global Historical Climatology Network dataset (Vose *et al.*, 1992), which consists of monthly total precipitation for stations around the globe. The time series shows an increase in globally averaged precipitation from 1901 to approximately 1950, then little trend in the period since 1950.

2.2 Climate extremes and data issues

Lack of long-term climate data suitable for analysis of extremes has been the single biggest obstacle to quantifying whether extreme events changed over the twentieth century, either worldwide or on a more regional basis (Easterling *et al.*, 1999). For many parts of the world, we lack high temporal and spatial resolution observations of temperature, precipitation, humidity, winds, and atmospheric pressure, and the shortages create problems for examining and quantifying observed changes in extremes. This problem has been partially addressed by a series of regional workshops that have produced analyses of climate extremes for many countries for which information was previously unavailable (e.g., Easterling *et al.*, 2003; Peterson *et al.*, 2002).

2.3 Changes in temperature extremes

On a regional basis, a number of studies of extremes have been completed as a part of a series of regional workshops coordinated by the joint World Meteorological Organization (WMO) Climate Variability and Predictability/ Climate Change Indices (CLIVAR/CCI) Expert Team on Climate Change Detection, Monitoring and Indices (ETCCDMI). Each of these workshops used a set software that calculated extremes defined in Frich *et al.* (2002). These workshops included the Caribbean (Peterson *et al.*, 2002), southern South America (Vincent *et al.*, 2005), Central America and northern South America (Aguilar *et al.*, 2005), central and northern Africa (Easterling *et al.*, 2003), southern and western Africa (New *et al.*, 2007), the Middle East (Zhang *et al.*, 2005), Australasia and southeast Asia (Griffiths *et al.*, 2005), and central and southern Asia (Klein Tank *et al.*, 2006). In addition, other researchers have performed regional studies for North America (Vincent and Mekis, 2007), the Arctic (Groisman *et al.*, 2003), Western Europe and East Asia (Kiktev *et al.*, 2003), and China (Zhai and Pan, 2003). Results from these studies are consistent with the observed increases in global temperatures, showing evidence of changes in temperature and precipitation extremes defined by the 10th and 90th percentiles.

More specifically, in the United States, two studies focused on the north-eastern United States support the notion that changes in the number of days exceeding certain thresholds have occurred. Cooter and LeDuc (1995) showed that the start of the frost-free season in the northeastern United States occurred 11 days earlier in the 1990s than in the 1950s. In an analysis of 22 stations in the northeastern United States for 1948–93, DeGaetano (1996) found significant trends to fewer extreme cold days, but trends to fewer warm maximum temperatures as well. More recently, Easterling (2002) found decreases in the number of days where the minimum temperature fell below freezing (0 °C) in the United States for the period 1948–99, with the largest decreases occurring in the western United States.

Results for the frost-free season showed increases in length for all areas of the continental United States, being driven mainly by changes to earlier dates of the last freeze in the spring season. Kunkel *et al.* (2004), using newly available data for the pre-1948 period, found that the average length of the frost-free season during 1895–2000 for the United States has increased by almost 2 weeks (Fig. 2.4). The change is characterized by four distinct regimes, with decreasing

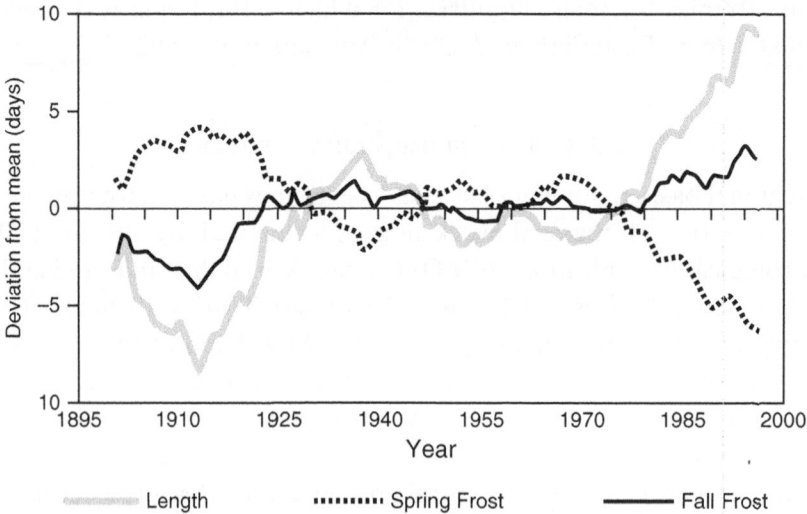

Figure 2.4. Time series of frost-free season length; Julian date of the first fall frost and last spring frost for the conterminous United States. (From Kunkel *et al.*, 2004.)

frost-free season length from 1895 to 1910, an increase in length of about 1 week from 1910 to 1930, little change during 1930 to 1980, and strong increases since 1980. Both Easterling (2002) and Kunkel *et al.* (2003) found that the frost-free season length has increased more in the western United States than in the eastern United States, which is consistent with results of Cayan *et al.* (2001), who found the spring pulse of snowmelt water in the western United States now comes as much as 7–10 days earlier than in the 1950s. Nemani *et al.* (2001) showed that frost days in the Napa/Sonoma region of California diminished from 28 days in 1950 to about 8 days by 1997, together with a substantial increase (66 days) in the length of the continuously frost-free season – from 254 to 320 days per year – which has benefited the premium wine industry.

An analysis of multiday extreme heat and cold episodes when the temperature exceeded the 10-year return period did not show any overall trend for the period 1931–98 (Kunkel *et al.*, 1999). The most notable feature of the temporal distribution of these very extreme heat waves is the high frequency in the 1930s compared with the rest of the record. DeGaetano and Allen (2002) examined daily exceedances of 90th, 95th, and 99th percentile thresholds (defined monthly) in the United States. The number of days exceeding these thresholds has increased in recent years, although they were also dominated earlier in the twentieth century by the extreme heat and drought of the 1930s. Changes in cold extremes (days below the 10th, 5th, and 1st percentile threshold temperatures) have shown decreases, particularly since 1960. Bonsal *et al.* (2001) used daily data adjusted for inhomogeneities to examine changes in temperature extremes in Canada. They found fewer cold extremes in winter, spring, and summer in southern Canada and more high temperature extremes in winter and spring, but little change in warm extremes in summer. Robeson (2004) examined changes in daily maximum and minimum temperatures by percentile for the United States and Canada and found that the largest increases in temperature have been occurring in the colder days of each month.

Globally, recent work by Alexander *et al.* (2006) showed that changes in the number of days (for night-time temperatures) exceeding selected temperature percentile thresholds corresponds well to areas with strong warming in minimum temperatures. In particular, the number of cold nights (less than the 10th percentile threshold temperature) has decreased for most of the areas examined, including the middle and higher latitudes of both the Northern and Southern Hemispheres, and the number of warm nights (nights warmer than the 90th percentile threshold) have increased in the same areas. There is less consistency with changes in the number of warm or cold days (daytime highs above the 90th percentile or below the 10th percentile); however, the global trends in both the night-time and daytime temperatures have changed in a similar fashion, with fewer cold days (nights) and more warm days (nights).

2.4 Extreme precipitation

Increased temperatures lead to an increase in the water-holding capacity of the atmosphere, such that with each 1 °C increase we have an increase of about 7% in the water-holding capacity (Trenberth *et al.*, 2003). Furthermore, observations suggest that as this capacity increases with increased temperatures, relative humidity remains more or less constant, resulting in an increase in water vapor in the atmosphere owing to enhanced drying of the surface. Figure 2.5 shows

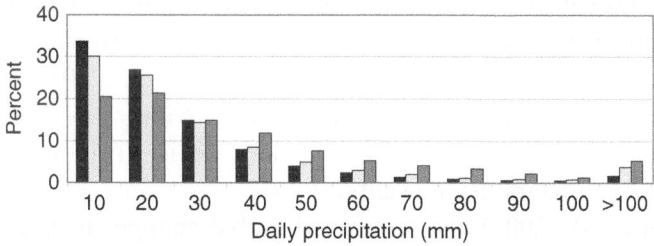

Figure 2.5. Percentage of annual precipitation from various daily rainfall rates for a cold climate (black), temperate climate (light gray), and warm climate (dark gray). Cold climates receive more precipitation from lighter amounts, and warm climates from heavier amounts.

typical rainfall rates for three climates: a warm climate, a temperate climate, and a cold climate. For the warm climate, more of the rainfall distribution falls in heavier amounts, compared with the temperate and cold climates, suggesting that, as the climate warms, more rainfall would be expected to occur in heavier events.

Trends in one-day and multiday extreme precipitation events in the United States and other countries show a tendency to more days with extreme 24 h precipitation totals (Karl and Knight, 1998). The number of days annually exceeding 50.8 mm (2 inches) of precipitation has been increasing in the United States (Karl *et al.*, 1996). Also, the frequency of 1- to 7-day precipitation totals exceeding station-specific thresholds for one in 1 year and one in 5 year recurrences, as well as for the upper 5 percentiles, have been increasing (Karl and Knight, 1998; Kunkel *et al.*, 1999). Increases are largest for the Southwest, Midwest, and Great Lakes regions of the United States, and increases in extreme events are responsible for a disproportionate share of the observed increases in total annual precipitation (Groisman *et al.*, 2005).

The tendency in most countries that have experienced an increase in monthly or seasonal precipitation has been for this increase to be directly related to an increase in the amount of precipitation falling during the heavy and extreme precipitation events. On the other hand, Akinremi *et al.* (1999) found that, although the Canadian prairie has experienced increased precipitation over the past 40 years, this increase appears to be mainly due to an increase in the number of lighter (<5 mm) daily rainfall totals.

Recent work by Groisman *et al.* (2005) provides more evidence that heavy precipitation events did, indeed, increase over the United States during the twentieth century. This increase was accompanied by increases in streamflow, and in the eastern United States there is evidence for increases in heavy streamflow as well. In a similar analysis, Kunkel *et al.* (2003) found increases in heavy precipitation, defined as those days exceeding the 1-year recurrence threshold; however, they also found evidence of a period of increased heavy rainfall events in the late 1890s, particularly in the western and central United States (Figure 2.6). However, in Canada, Zhang *et al.* (2001) found no evidence for increases in heavy precipitation for the country as a whole. Only in eastern Canada, during the spring, is there an identifiable trend toward increasing heavy precipitation events.

Alexander *et al.* (2006) also examined global changes in heavy daily precipitation events by using gridded daily data. The globally averaged trends in the percent contribution of heavy precipitation days (amounts above the 95th percentile) to the annual total have shown small, but statistically significant, increases since 1951.

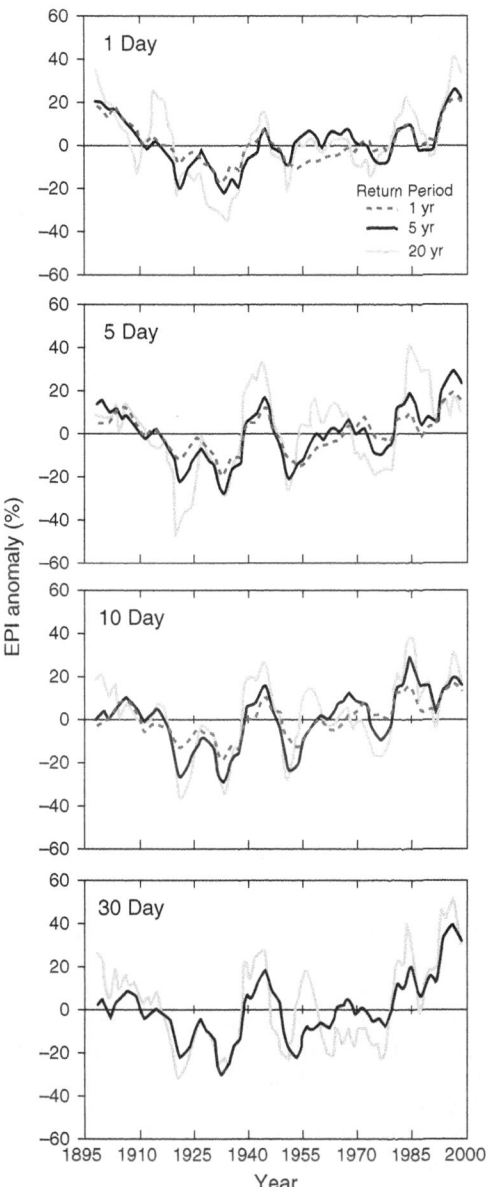

Figure 2.6. Time series of heavy precipitation events as defined by the Extreme Precipitation Index (EPI) from Kunkel *et al.* (2003) for the contiguous United States; anomalies from the long-term mean for 1-day, 5-day, 10-day, and 30-day totals.

2.5 Summary

The preceding review has raised a number of important points regarding potential changes in extreme events. For many regions analyzed, there have been significant changes in temperature and precipitation extremes. However, for many others it is unclear whether changes in temperature and precipitation extremes have occurred. As far as attribution of trends is concerned, both in mean values and in extreme events, one of the most critical issues relates to the hypothesis that with greenhouse gas-enhanced climate change the hydrologic cycle should intensify (see Figure 2.5). In recent years, evidence that this is occurring has become more prevalent on a regional basis. The globally averaged annual time series of precipitation shows evidence of increased precipitation over the twentieth century; however, during the latter half of the century, which is the period most closely linked to anthropogenic influences on climate, there was little or no observed trend in global precipitation (Figure 2.3).

In the past, one of the biggest problems in analyzing extreme climate events for many parts of the globe was a lack of access to high-quality, long-term climate data with the time resolution appropriate for analyzing extreme events. Fortunately, this situation has been improving, with a recent study by Alexander *et al.* (2006) being able to perform analyses of climatic extremes for approximately 70% of the global landmass.

References

Aguilar, E., *et al.* (2005). Changes in precipitation and temperature extremes in Central America and northern South America, 1961–2003. *Journal of Geophysical Research*, **110**, D23107, doi:10.1029/2005JD006119.

Akinremi, O. O., McGinn, S. M., and Cutforth, H. W. (1999). Precipitation trends on the Canadian prairies. *Journal of Climate*, **12**, 2996–3003.

Alexander, L., *et al.* (2006). Global observed changes in daily climate extremes of temperature and precipitation. *Journal of Geophysical Research*, **111**, D05109, doi:10.1029/2005JD006290.

Bonsal, B. R., Zhang, X., Vincent, L., and Hogg, W. (2001). Characteristics of daily and extreme temperatures over Canada. *Journal of Climate*, **14**, 1959–76.

Cayan, D. R., *et al.* (2001). Changes in the onset of spring in the western United States. *Bulletin of the American Meteorological Society*, **82**, 399–415.

Cooter, E., and LeDuc, S. (1995). Recent frost date trends in the northeastern United States. *International Journal of Climatology*, **15**, 65–75.

DeGaetano, A. T. (1996). Recent trends in maximum and minimum temperature threshold exceedances in the northeastern United States. *Journal of Climate*, **9**, 1646–57.

DeGaetano, A. T., and Allen, R. J. (2002). Trends in twentieth-century temperature extremes across the United States. *Journal of Climate*, **15**, 3188–205.

Easterling, D. R. (2002). Recent changes in frost days and the frost-free season in the United States. *Bulletin of the American Meteorological Society*, **83**, 1327–32.

Easterling, D. R., *et al.* (1997). Maximum and minimum temperature trends for the globe. *Science*, **277**, 364–7.

Easterling, D. R., *et al.* (1999). Long-term observations for monitoring climate extremes in the Americas. *Climatic Change*, **42**, 285–308.

Easterling, D. R., *et al.* (2003). CCl/CLIVAR workshop to develop priority climate indices. *Bulletin of the American Meteorological Society*, **84**, 1403–7.

Frich, P., *et al.* (2002). Observed coherent changes in climatic extremes during the second half of the twentieth century. *Climate Research*, **19**, 193–212.

Griffiths, G. M., *et al.* (2005). Change in mean temperature as a predictor of extreme temperature change in the Asia-Pacific region. *International Journal of Climatology*, **25**, 1301–30.

Groisman, P. Ya, *et al.* (2003). Contemporary climate changes in high latitudes of the Northern Hemisphere: daily time resolution. In *Proceedings of the International Symposium on Climate Change*, Beijing, China, March 31–April 3, 2003, World Meteorological Organization Publication 1172, pp. 51–5.

Groisman, P. Ya, *et al.* (2005). Trends in precipitation intensity in the climate record. *Journal of Climate*, **18**, 1326–50.

Hulme, M. (1995). Estimating global changes in precipitation. *Weather*, **50**(2), 34–42.

Karl, T. R., and Knight, R. W. (1998). Secular trends of precipitation amount, frequency, and intensity in the United States. *Bulletin of the American Meteorological Society*, **79**, 1107–19.

Karl, T. R., *et al.* (1996). Indices of climate change for the United States. *Bulletin of the American Meteorological Society*, **77**, 279–91.

Kiktev, D., *et al.* (2003). Comparison of modeled and observed trends in indices of daily climate extremes. *Journal of Climate*, **16**, 3560–71.

Klein Tank, A. M. G., *et al.* (2006). Changes in daily temperature and precipitation extremes in Central and South Asia. *Journal of Geophysical Research*, **111**, D16105, doi:10.1029/2005JD006316.

Kunkel, K. E., Easterling, D. R., Hubbard, K., and Redmond, K. (2004). Temporal variations in frost-free season in the United States: 1895–2000. *Geophysical Research Letters*, **31**, L03201, doi:10.1029/2003GL018624.

Kunkel, K. E., Easterling, D. R., Redmond, K., and Hubbard, K. (2003). Temporal variations of extreme precipitation events in the United States: 1895–2000. *Geophysical Research Letters*, **30**(17), 1900, doi:10.1029/2003/GL018052.

Kunkel, K. E., Pielke, R. A., Jr., and Changnon, S. A. (1999). Temporal fluctuations in weather and climate extremes that cause economic and human health impacts: a review. *Bulletin of the American Meteorological Society*, **80**, 1077–98.

Nemani, R. R., *et al.* (2001). Asymmetric warming over coastal California and its impact on the premium wine industry. *Climate Research*, **19**, 24–5.

New, M. B., *et al.* (2007). Evidence of trends in daily climate extremes over southern and west Africa. *Journal of Geophysical Research* (in press).

Peterson, T. C. *et al.* (2002). Recent changes in climate extremes in the Caribbean region. *Journal of Geophysical Research*, **107**, 4601, doi:10.1029/2002JD002251.

Robeson, S. (2004). Trends in time-varying percentiles of daily minimum and maximum temperature over North America. *Geophysical Research Letters*, **31**, L04203, doi:10.1029/2003GL019019.

Smith, T. M., and Reynolds, R. W. (2005). A global merged land and sea surface temperature reconstruction based on historical observations (1880–1997). *Journal of Climate*, **18**, 2021–36.

Trenberth, K. E., *et al.* (2003). The changing character of precipitation. *Bulletin of the American Meteorological Society*, **84**, 1205–17.

Vincent, L. A., and Mekis, E. (2007). Changes in daily and extreme temperature and precipitation indices for Canada over the twentieth century. *Atmosphere–Ocean*. (in press)

Vincent, L. A., *et al.* (2005). Observed trends in indices of daily temperature extremes in South America 1960–2000. *Journal of Climate*, **18**, 5011–23.

Vose, R. S. *et al.* (1992). The Global Historical Climatology Network: long-term monthly temperature, precipitation, sea level pressure, and station pressure data. ORNL/CDIAC-53, NDP-041. Oak Ridge, TN: Carbon Dioxide Information Analysis Center, Oak Ridge National Laboratory.

Vose, R. S., Easterling, D. R., and Gleason, B. (2005). Maximum and minimum temperature trends for the globe: an update through 2004. *Geophysical Research Letters*, **32**, L23822, doi:10.1029/2004GL024379.

Zhai, P. M., and Pan, X. H. (2003). Trends in temperature extremes during 1951–1999 in China. *Geophysical Research Letters*, **30**, 1913, doi:10.1029/203GL018004.

Zhang, X., *et al.* (2001). Spatial and temporal characteristics of heavy precipitation events over Canada. *Journal of Climate*, **14**, 1923–36.

Zhang, X., *et al.* (2005). Trends in Middle East climate extremes indices 1951–2003. *Journal of Geophysical Research*, **110**, D22104, doi:10.1029/2005JD006181.

3

The spatial distribution of severe convective storms and an analysis of their secular changes

HAROLD E. BROOKS AND NIKOLAI DOTZEK

Condensed summary

Severe convective storms are responsible for billions of US dollars in damage each year around the world. They form an important part of the climate system by redistributing heat, moisture, and trace gases, as well as by producing large quantities of precipitation.

Reporting of severe convection varies from country to country, however, so determining their distribution from the reports alone is difficult, at best. Evidence does exist that the intensity of some events, particularly tornadoes, follows similar distributions in different locations, making it possible to build statistical models of occurrence. Remote sensor observations provide some insight, but the relationships between the observable parameters and the actual events of interest limit the quality of the estimates. Another approach is to use observations of the larger-scale environments.

As has been stated, the relationship between the observation and the event limits the estimate, but global coverage is possible. Time series of the favorable environments can also be developed from such data. In order to improve the estimates, the most pressing need is for better observational data on events. Very few countries have formal systems for collecting severe thunderstorm reports. A new effort by a consortium of researchers in Europe to develop a continent-wide database offers the possibility of a significant improvement in data for that part of the world.

3.1 Introduction

Convective storms play a vital role in weather and climate. They act to redistribute heat, moisture, and trace gases in the vertical and in the horizontal. In the tropics and in the warm season in the mid-latitudes, they provide a significant part of the precipitation. Therefore, they are beneficial to society,

Climate Extremes and Society, ed. H. F. Diaz and R. J. Murnane. Published by Cambridge University Press. © Cambridge University Press 2008.

particularly in agriculture. When convection is particularly strong, however, the resulting weather can have adverse effects on life and property and is typically referred to as "severe." Although definitions of what is called severe vary from place to place, in general, hail, high winds, tornadoes, and extremely heavy precipitation leading to flash flooding are frequently considered. Doswell (2001) provides an overview of the problem of severe convection.

Thunderstorms occur all over the world.[1] Estimates of global occurrence and losses from them are not available, in general, although there are regional estimates. A recent expert survey of tornado occurrence per year in Europe, produced on a country-by-country basis, yielded an estimate of 300 tornadoes on land and an additional 400 waterspouts (tornadoes over water) (Dotzek, 2003). Munich Re estimates yearly overall losses of about €5 billion to €8 billion due to severe convective storms in Europe. Approximately 1,200 tornadoes are reported in the United States each year, with roughly 50 deaths and US$400 million in damage annually, US$10 billion in damage caused by hail, based on US National Weather Service data. It is important to note that there is wide interannual variability in those figures. The tornado that struck Oklahoma City and the surrounding area on May 3, 1999, caused US$1 billion in damage by itself. Brooks and Doswell (2001a) estimated that, adjusting for temporal changes in economic variables, the United States can expect a billion dollars in damage from a single tornado about once per decade.

Estimates of occurrences of hail, convective wind, and convectively induced flooding rainfall are much more difficult to obtain. In the United States, there are currently over 10,000 reports per year each of hail and wind. This number has increased by an order of magnitude in the past few decades, as a result of efforts to improve data collection. Attempting to know the true distribution is obviously difficult in a situation with such large temporal changes.

In this chapter, we will review the problems associated with severe convection reporting, and we will also cover a range of approaches to making estimates. Although public reports provide a straightforward way to make such estimates, the problems with these reports (discussed in Section 3.2) make those estimates dubious at best, with the possible exceptions of those for a few regions. Suitable approaches to solving the problem, including the use of proxy observations and relationships between the environments and events of interest,

---

[1] To be precise, thunderstorms imply the existence of lightning. Maddox *et al.* (1997) presented evidence for inadvertent modification of thunderstorms that limits lightning. Presumably, it is possible for natural processes to accomplish the same thing. In this chapter, "thunderstorm" will be used to imply deep, moist convection – the process of the rapid vertical ascent of potentially buoyant air and subsequent precipitation, cooling, and moistening of the planetary boundary layer and stabilizing of the atmosphere, but nonthundering convection is not excluded (Doswell *et al.*, 1996).

will be described. Finally, we will close with a discussion of possible alterations in severe thunderstorms associated with climate change and suggestions for improving data collection to address the situation.

3.2 Underlying problems and approaches

Part of the problem in defining severe convection is the susceptibility of different parts of society to effects of the same weather event. For example, a large amount of small hail (less than 2 cm in diameter) may have little impact on urban areas, but could be devastating to agricultural interests at certain times of the year and for certain crops. Similarly, strong winds may have minimal impact in an area of grassland used for cattle ranching, whereas the same winds may be damaging in a forested or urban setting. As a result, the definitions developed in different locations are necessarily arbitrary.

A further complication is that most severe thunderstorm events require the presence of an observer at the time and place of the event. Unlike temperature or precipitation, for example, for which routine observations can be collected on a regular basis by humans or by automated systems, the intermittent, isolated nature of severe thunderstorms in time and space, and the difficulty of remotely observing the events of interest, means that we depend on a fortuitous combination of the occurrence of an event and the presence of an observer to detect the event (and classify it correctly). Assuming that the event is observed is insufficient to ensure that it is recorded. In many countries, there is no systematic official process to collect reports of events that are observed.[2] In the absence of officially supported systems, it is difficult to create historical records of past events.

Even in the United States, which has had an official collection effort since the 1950s, there are still obvious reporting problems. Verbout *et al.* (2006) showed that the number of tornado reports increased by approximately 14 per year from 1954 to 2004 (Fig. 3.1). Brooks and Doswell (2001b) showed that this increase came from the increasing number of reported weak tornadoes over time, likely a result of factors such as improved public awareness and report collection procedures, increasing urbanization, or better radar identification of severe (and potentially tornadic) storms by the WSR-88D network. Further complicating the analysis is the apparent step function decrease in the number of strong and violent tornadoes that took place in the mid 1970s,

---

[2] We note that records exist in many countries, but that, in general, these are developed by interested individuals independent of any professional meteorological or climatological responsibilities they may have.

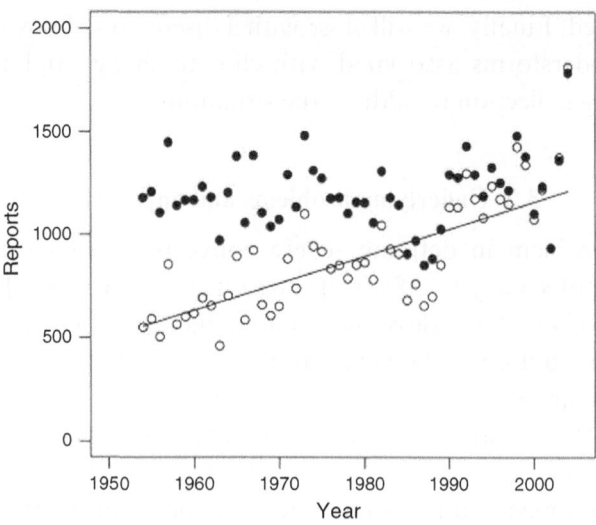

Figure 3.1. Tornado reports in official database in the United States from 1954 to 2004. Raw reports (open diamonds), solid line (linear regression), and reports adjusted to 2002 reporting system (solid diamonds) are shown.

presumably associated with adoption of the Fujita scale[3] (Fujita, 1971), linking peak wind speed v and scale class F, according to equation (3.1),

$$v(F) = 6.30 \text{ m s}^{-1}(F + 2)^{3/2}, \tag{3.1}$$

as a measure of tornado damage and the subsequent retrospective rating of tornadoes in the National Weather Service's database extending back to 1950. Thus, it seems that the reasonably reliable record of tornado occurrence is perhaps 30 years or less, depending on the variable considered.

The non-tornadic severe weather reporting problem is even more serious. Spatial and temporal changes in reports in the United States make it difficult to interpret the database. Trapp *et al.* (2006) examined individual events during the course of a field project and found that reports of convective wind gusts in the database frequently gave a misleading impression of the scope of the event. In some cases, the reports greatly underestimated the extent of damage, compared with aerial surveys; in other cases, the reports made relatively minor events seem more widespread, particularly in urban areas. Doswell *et al.* (2005) showed that differences in the practices of local National

---

[3] Tornadoes are rated by the maximum damage anywhere along the path. The scale goes from F0 (the weakest) to F5 (the strongest). Following Kelly *et al.* (1978), F0 and F1 tornadoes are referred to as "weak," F2 and F3 as "strong," and F4 and F5 as "violent." F2 and stronger tornadoes are collectively described as "significant."

Weather Service Forecast Offices led to discontinuities in the spatial patterns of hail and wind reports. Temporally, non-tornadic reports have increased exponentially over the past 50 years: from perhaps 2,000 per year in the 1950s to almost 30,000 per year now. Obviously, even a relatively large physical change of a few percent per year could be masked by such large changes in the reports.

In order to deal with some of the problems with the raw reports, Brooks *et al.* (2003a) and Doswell *et al.* (2005) put the reports onto a regularly spaced grid, considered the variable of interest to be whether or not at least one event occurred in a 24 h period in a particular grid box, and then applied spatial and temporal smoothers to produce a continuous distribution of daily climatological probability of events. This approach produces fields that appear to be physically reasonable, but because of the smoothing, detail is obviously lost. It seems plausible that the detail is on a scale in which little confidence can be placed, given the problems in the data, but if physically real differences occur on small spatial scales, those details will be lost. In addition, the methodology requires the existence of a relatively large data-base in order to produce the raw input fields. To date, such a database exists only in the United States. The European Severe Weather Database (ESWD; eswd.eu); (Fig. 3.2), which is currently under development, may produce a sufficiently high-quality database to carry out smoothing, but that has not occurred as of yet.

The inherent problems with the databases from different countries do not prevent data analysis, if it is done carefully and cautiously. Brooks and Doswell (2001b) noted that the distribution of tornadoes by intensity appeared similar in many countries, implying that there might be an underlying physical process (or set of processes) that determines the intensity of a tornado. Since that time, Dotzek *et al.* (2003, 2005) and Feuerstein *et al.* (2005) have worked to put that speculation on a firmer statistical footing. Dotzek *et al.* (2003) and Feuerstein *et al.* (2005) showed that Weibull distributions could be fit to the data for a variety of countries worldwide; they applied a two-parameter least-squares fit to observed worldwide tornado intensity distributions on both the F-scale and wind speed, v. With x denoting either of these, the Weibull distribution is given in three-parameter form for probability density $p(x)$:

$$p(x) = \frac{c}{b}\left(\frac{x-a}{b}\right)^{c-1}\exp\left(-\left(\frac{x-a}{b}\right)^{c}\right). \qquad (3.2)$$

Here, a is a fixed parameter and denotes the lower boundary of the variable x. The scaling factor b and the shape parameter c are the two model parameters to be estimated. Note that for $c=1$, Equation (3.2) includes the exponential

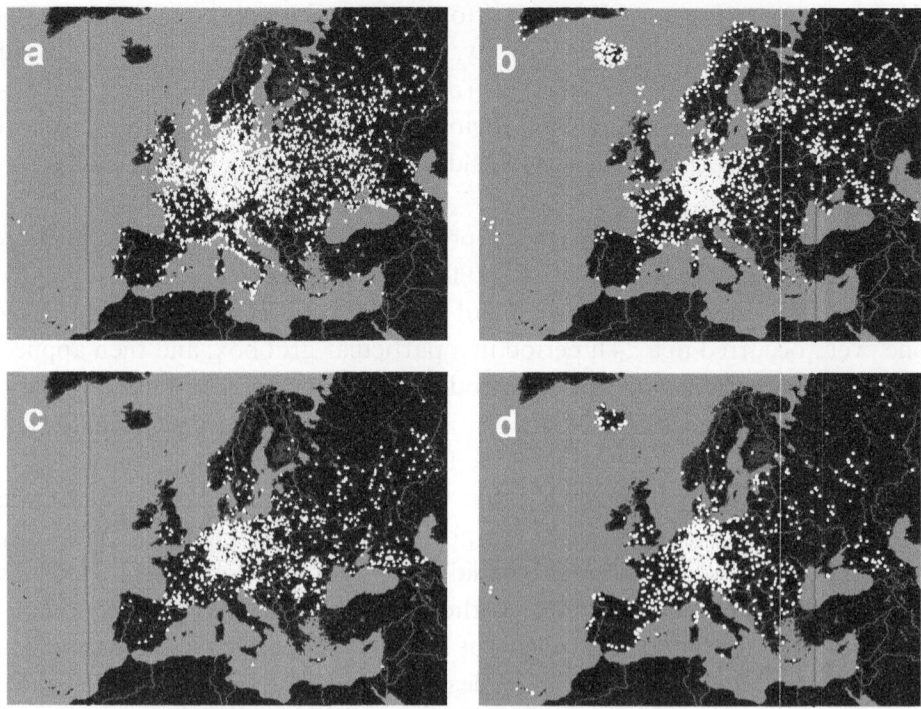

Figure 3.2. ESWD data since 1950 (as of 18 April 2007): tornadoes (a, $n = 3062$), straight-line winds (b, $n = 2408$), large hail (c, $n = 1928$), and heavy rain (d, $n = 1327$). Owing to inclusion of the TorDACH reports, data density is presently highest in Germany.

distribution discussed by Brooks and Doswell (2001b), whereas $c = 2$ yields the Rayleigh distribution:

$$p(x) = \frac{2(x - a)}{b^2} \exp\left(-\frac{(x - a)^2}{b^2}\right). \tag{3.3}$$

Using the wind speed v as the independent variable, i.e., substituting $v \equiv x-a$, and $v_0 \equiv b$, Dotzek et al. (2005) were able to corroborate and generalize results by Kurgansky (2000), who had analytically predicted a Rayleigh distribution for tornado intensity distributions over wind speed. Comparing the evolution of parameter c from general Weibull distribution fits worldwide, Dotzek et al. (2005) showed that for increasing quality and sample size of tornado intensity datasets, a trend towards $c = 2$ can indeed be concluded.

Further support for the relevance of the Rayleigh distribution for evaluating tornado intensities comes from earlier statistical modeling work in the field of wind energy assessment. Researchers have shown that any wind speeds should be Rayleigh-distributed if v can be regarded as the absolute value of

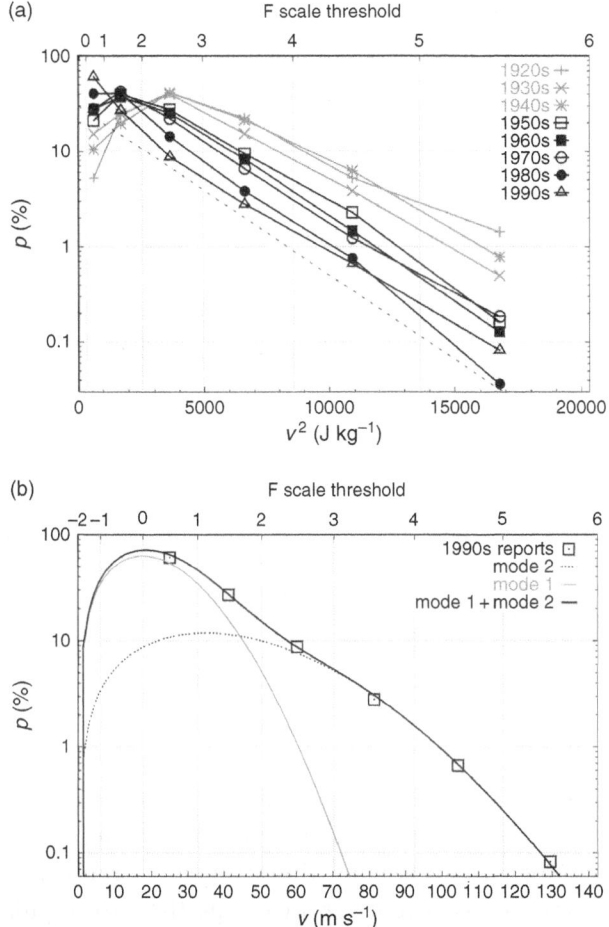

Figure 3.3. Decadal tornado intensity distributions from the United States, for 1920–99, over v^2 (a), and the US 1990s data over v (b), showing signs of emerging bimodality into separate low- and high-intensity Rayleigh modes (solid, gray and dotted, respectively). The solid black line gives the sum of both modes, and the open boxes are the observations. F-scale threshold wind speeds are also given on the upper abscissa of both panels.

a two-dimensional vector, the coordinates of which are independent and which follow a Gaussian distribution with zero mean and variance $v_0^2/2$ (see Conradsen *et al.*, 1984). These conditions are best met in isotropic conditions, when no wind direction is preferred. Clearly, winds in tornado vortices fulfill these conditions even better than ordinary winds, due to their high degree of axial symmetry.

Figure 3.3 shows the results obtained by Dotzek *et al.* (2005) for the observed decadal tornado intensity distributions in the United States from 1920 to 1999. To detect the presence of a Rayleigh distribution, it is helpful to exploit the fact that a Rayleigh distribution in v is equivalent to an exponential distribution in v^2 (e.g., Dotzek *et al.*, 2005). Thus, Figure 3.3a depicts the tornado intensity distributions over v^2. For all decades, clear exponential tails from F2 intensity on can be seen; from about the 1960s on, the Rayleigh distribution is already reproduced from F1 intensity on. Validity of a Rayleigh distribution is a consistent feature over the 80 years of data, even with the increasing number of reported weak tornadoes over time and the reporting issues discussed by Verbout *et al.* (2006).

The 1990s data are plotted separately over v in Figure 3.3b. Here, a new feature appears to emerge owing to enhanced detection efficiency for weak tornadoes: bimodality. The data are now best reproduced by a superposition of two independent Rayleigh distribution modes, characterized by different variances v_0^2. A similar result follows from the 1990s Oklahoma data alone. The separation of the modes occurs at about the F2 intensity threshold (50.4 m s^{-1}). The modes can be attributed with good confidence to non-mesocyclonic and mesocyclonic tornado activity, respectively (Dotzek *et al.*, 2005).

The tornado reports for the United States from 2000 to 2005 show an even higher percentage of weak tornadoes than those for the 1990s, and they lack any F5 events. The latter fact may not only be due to meteorological reasons, but may also be influenced by current tornado rating practice in this country. Hence, even for large tornado databases, which are probably approaching completeness in annual numbers, reporting issues remain as a limiting factor in risk assessment based on intensity distribution modeling.

One possible approach to solving the problem of reporting limitations is to develop proxies for the events from things that are regularly observed. Toracinta and Zipser (2000) looked at satellite-observed mesoscale convective systems in the tropics and subtropics by using microwave imagery and lightning detection. Levizzani and Setvák (1996) found satellite-based signatures in multispectral data associated with supercell thunderstorms in the United States. These approaches show promise, but currently either are limited in spatial extent or are labor-intensive. In addition, it is clear that they identify strong updrafts, which are logically related to severe thunderstorms, but it is possible, if not likely, that they do not cover the complete range of severe thunderstorms. In particular, the relationship between the observed variables and tornadoes, if there is one, is not obvious.

Bissolli *et al.* (2007) combined a synoptic weather pattern analysis for central Europe in comparison with the German TorDACH tornado reports[4] for the

---

[4] http://tordach.org/de/. From 2006 on, severe weather reports from Germany only add to the ESWD, and the TorDACH reports until 2005 are converted to ESWD format and added to it.

period 1950–2003. The synoptic pattern scheme used by Deutscher Wetterdienst (DWD, the German national weather service) takes into account the direction of air mass advection, the cyclonicity, and the humidity (precipitable water) of the troposphere. The majority of the tornadoes can be attributed to only three specific weather types, all with a southwesterly advection and high humidity. This feature is found for weak as well as for significant tornadoes. Anticyclonicity/cyclonicity at 950 hPa reduces/enhances the tornado frequency by roughly a factor of 2, while at 500 hPa cyclonicity shows no significant influence. Therefore, a synoptic weather type prediction may both provide early warning guidance in operational forecasting, and become a tool in regional climate modeling in which changes in the frequency or location of synoptic patterns are to be studied for future climate scenarios.

Another source of proxy data is environmental observations. Brown and Murphy (1996) explored the use of well-observed atmospheric variables in order to deduce the presence of a poorly observed forecast variable, aircraft icing. The icing forecast problem bears some similarity to the severe thunderstorm forecast problem in that the observed event of interest depends on the presence of an observer and, in many cases, the issuance of a forecast leads to observers avoiding the region. By the use of proxy variables, the problem becomes one of establishing the relationship between the observed variable and the event of interest and then comparing the forecast to the well-observed variable. The quality of the relationship between the variable and the event puts limits on the applicability of the technique. A workshop report from the Intergovernmental Panel on Climate Change (IPCC, 2002) recommended that analysis of changes in favorable environments be used to look at climate change questions.

There is a long history of attempts to determine relationships between large-scale environmental conditions and severe thunderstorms and tornadoes (e.g., Beebe, 1955, 1958; Brooks *et al.*, 1994; Rasmussen and Blanchard, 1998). The primary motivation for these studies has been to improve forecasting. Environments that favor severe convection have been identified. An important result is that the discrimination between severe events and non-events is best when only the strongest severe thunderstorms are considered (e.g., F2 or stronger tornadoes, 5 cm or larger diameter hail, wind gusts of at least 120 km h^{-1}). It appears relatively easy for the atmosphere to produce thunderstorms that are marginally severe weather, but only a limited set of conditions can lead to the most severe events. Unfortunately for this analysis, the conditions do not always produce severe thunderstorms. This result stems from a variety of causes, perhaps most importantly the problem of convective initiation. The over-forecasting that results from just considering the large-scale environment can be thought of as a by-product of forecasting a rare event with high costs associated with missed detections (Brooks, 2004). The identification of environments that are favorable for the development of severe thunderstorms is likely, however, to provide a best first guess for the distribution of severe thunderstorms. In this context, convective initiation and other problems can be thought of as noise or quasi-random processes.

3.3 Environmental conditions associated with severe thunderstorms

A number of recent studies have highlighted convective available potential energy (CAPE) and some measure of deep tropospheric wind shear of the horizontal environmental winds (deep shear) as important discriminators between environments associated with significant severe thunderstorms and those that are not (e.g., Rasmussen and Blanchard 1998; Craven and Brooks, 2004). This distinction can be seen in the estimated probability of significant severe thunderstorm occurrence given a combination of CAPE and deep shear, based on data from the operational US radiosonde network when soundings happened to be launched in the vicinity of severe thunderstorms (so-called proximity soundings); the work is described in Brooks and Craven (2002) and Craven and Brooks (2004) (Fig. 3.4). As an environment moves towards greater CAPE and deep shear, the probability of its being associated with a significant severe storm increases rapidly. To first order, the gradient in probability is perpendicular to the product of the CAPE and deep shear.

The principal parameters that discriminate between significant tornadic and significant non-tornadic environments are the height of the lifted condensation

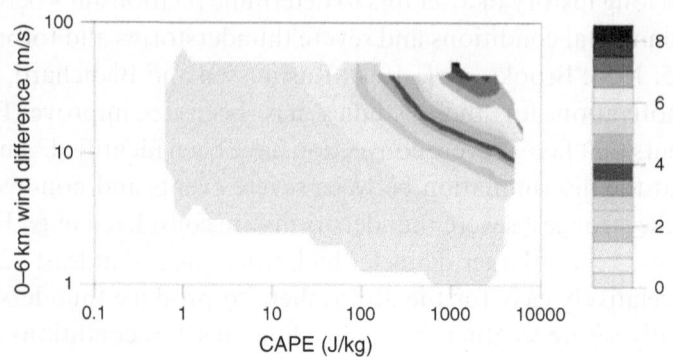

Figure 3.4. Probability in percent of environments producing severe thunderstorm with a tornado with at least F2 damage, 5 cm diameter hail, or 120 km h^{-1} wind gusts in the USA. (Based on data described by Brooks and Craven, 2002.) For color version, see plate section.

level (LCL) and measures of lower tropospheric wind shear (Rasmussen and Blanchard, 1998; Craven and Brooks, 2004). Tornadic environments are favored when the LCL height is low and the wind shear in the lowest kilometer above the ground is high.

These two steps in discrimination (severe vs. non-severe, tornadic vs. non-tornadic) can form the basis of identifying environments that are favorable for various classes of weather events. Given the sparse coverage of upper-air observations, however, carrying the discrimination to other locations is challenging. To address this problem, Brooks *et al.* (2003b) attempted to use data from the National Center for Atmospheric Research/National Centers for Environmental Prediction (NCAR/NCEP) global reanalysis dataset. The reanalysis was treated as a source of pseudo-proximity soundings, and the analysis of Brooks and Craven (2002) was repeated (Lee, 2002).

Discrimination between the severe and non-severe environments in the reanalysis dataset was found to be almost identical to the observational dataset. Discrimination was not as good, but still used the same variables in the same qualitative sense. Problems with sharp vertical gradients and the boundary layer in the reanalysis are likely sources of the differences.

Brooks *et al.* (2003b) counted the number of days per year with conditions that the reanalysis identified as favorable for significant severe thunderstorms and tornadoes from a 7-year period over the land area of the globe. Here, we present updated versions of those figures, counting the number of individual 6 h time slices per year from 1970 through 1999 (Fig. 3.5). Severe thunderstorms are concentrated downstream of high terrain and poleward of moisture sources in the form of warm water or rain forest. This is because of the high mid-tropospheric lapse rates that advect off the high terrain over the boundary layer moisture, creating a high-CAPE environment when the flow aloft is from the direction of the high terrain and the flow at low levels is poleward. The change of the wind direction with height associated with this configuration implies the presence of significant vertical wind shear, the second important parameter for creating severe thunderstorms. We have less confidence in the delineation of small regions, but the prominent areas in South America and near the Himalayas match those of Toracinta and Zipser (2000). There is a hint of the central United States area in Toracinta and Zipser, but the northern end of the analysis cuts off at 35° N, near the southern extent of the maximum seen in Figure 3.5. In passing, we note that possible problems exist in the reanalysis depiction. In particular, there is a two-point-wide line in the reanalysis just east of the Andes that appears to result from the response to the depiction of the Andes in spectral space. The regions in northeastern Mexico and near the Arabian Peninsula are also likely overestimated, given the lightning

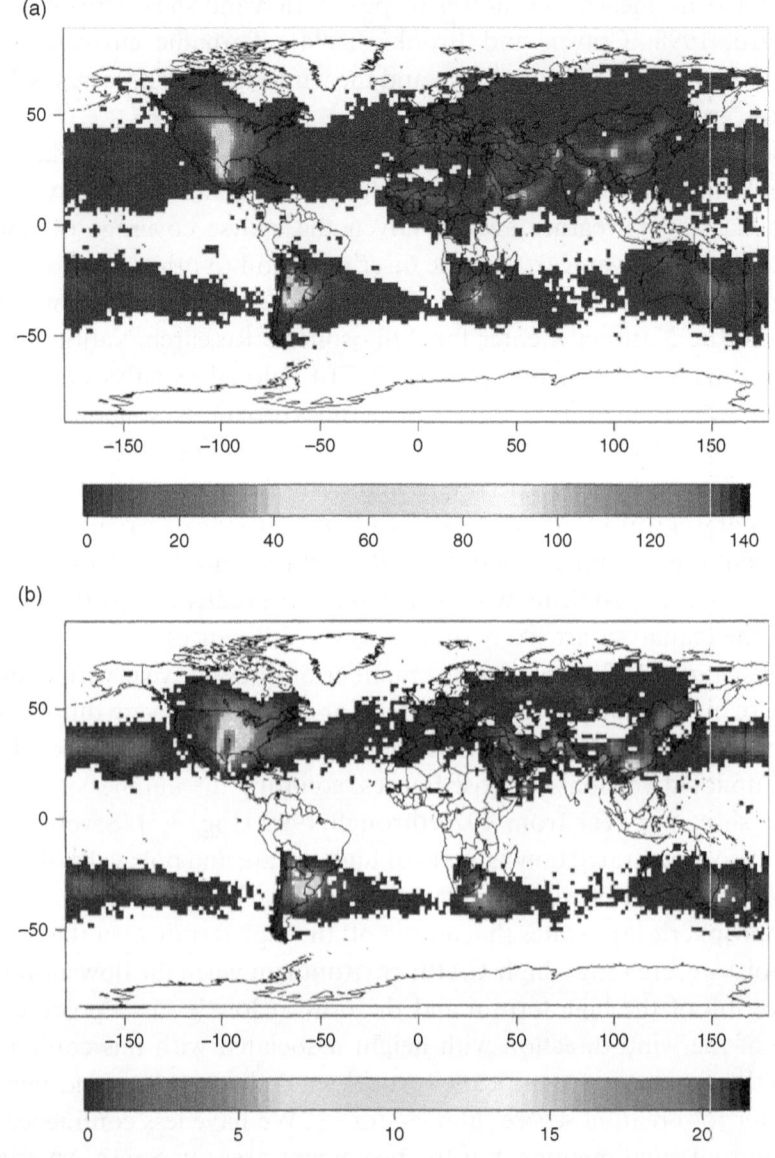

Figure 3.5. Six-hourly periods per year with environments supportive of significant severe thunderstorms (a) and significant tornadoes (b) based on NCAR/NCEP reanalysis data for 1970–1999. (Updated from Brooks *et al.* [2003b].) For color version, see plate section.

distribution derived from the Tropical Rainfall Measuring Mission lightning imaging sensor observations. Brooks *et al.* (2003b) noted problems with the representation of the capping inversion, frequently seen in observations, that suppresses the formation of convection in the atmosphere.

The reanalysis estimate of the distribution of significant tornado-producing environments is dominated by the central United States, with lesser maxima over South America and southeast China. As was mentioned earlier, the discrimination of tornadic conditions is poorer than the discrimination of severe conditions. It may be overly tuned to the conditions associated with outbreaks in the United States. Conditions that are observed relatively rarely in the United States may be sufficient for producing tornadoes, but absent from the discrimination. Given that the reanalysis would be expected to work best in depicting large-scale synoptic conditions, locally concentrated regions may be missed as well.

3.4 Historical changes in environments

Studies of the historical distribution of severe thunderstorms and tornadoes and possible changes are necessarily limited by data quality. However, efforts to date have not detected statistically significant changes. Concannon *et al.* (2000) found large variability between different periods in the record of strong and violent tornadoes in the United States. In their analysis of tornadoes in Germany, Bissolli *et al.* (2007) reported no shift of the seasonal cycle detectable for the period 1980–2003 compared with 1950–2003, and also no shift of the intensity distribution.

Severe thunderstorms and tornadoes are products of the juxtaposition of relatively rare conditions on subsynoptic scales. Attempting to estimate likely changes in their occurrence associated with climate change scenarios is difficult at best (IPCC, 2001, 2007). Considering the mean annual conditions in the mid-latitudes, we might expect CAPE to increase as a result of increasing surface temperature and boundary layer moisture in a warming world. Gaffen and Ross (1999) presented observational evidence for an increase in boundary layer moisture in the southern United States in summer. To first order, deep tropospheric wind shear might be expected to decrease from thermal wind considerations if the equator-to-pole temperature gradient decreases. The question of the balance between those changes would determine if more environments were found in the higher-probability space shown in Figure 3.4.

The mean picture is of limited value, however. Given the relationships of large-scale orographic features to the locations of the maxima in favorable environments, the locations of future events may be constrained by those large-scale features. More importantly, we are interested in the combination of conditions on a day-to-day basis. It is possible that days with the greatest increase in CAPE could be when the deep shear is small, so that the change in

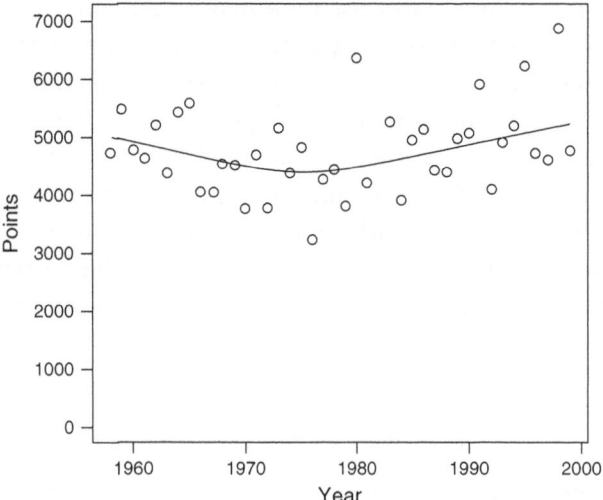

Figure 3.6. Six-hourly periods per year with environments supportive of significant severe thunderstorms in United States east of the Rocky Mountains. The line is the local regression fit to the series.

severe threat might be small. Brooks *et al.* (2007) have shown that CAPE and deep shear tend to be out of phase in their annual cycle.

Again, the reanalysis data can provide insight into historical behavior (Brooks, 2006). The annual total of points that are identified as favorable for severe thunderstorms in the United States, east of the Rocky Mountains, shows a minimum in the early 1970s (Fig. 3.6). From the time of the minimum to 1999, the annual total increased by about 0.8% per year. There is a super- ficial similarity to the US annual temperature record, but it is not statistically significant. Unfortunately, the reanalysis does not extend back in the same format for the period before 1958. Thus we cannot be sure of the long-term relationship between surface temperature and favorable storm environments in the United States.

We can compare the number of environments to reports of 7 cm diameter or larger hail in the United States over a comparable period. As was mentioned earlier, there has been a large increase in the number of hail reports in the United States, but the increase in the largest hail has not been so great, although it is still large (Fig. 3.7). There is an inflection in the record at the same time (1973) as the environmental estimate. For the period before that, the hail reports are essentially flat, while the environments decreased by about 1% per year. Since then, the number of reports has increased by 6% per year. If we take the estimate of environments as a baseline, the change in environments is 7% of the change in reports.

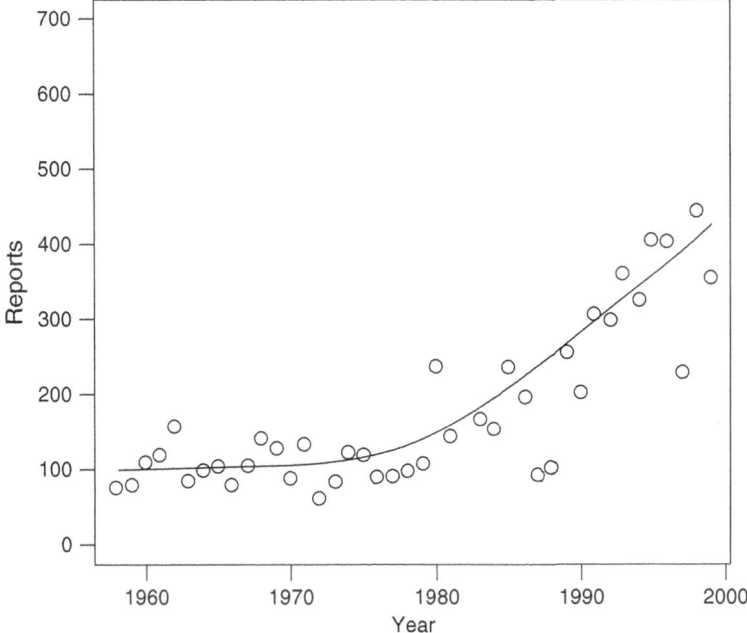

Figure 3.7. Number of reports of hail 7 cm or larger in diameter in the United States per year. The line is the local regression fit to the series.

Other regions show different trends in environments. In particular, a region encompassing the South American maximum shows a decrease of almost 40% over the period of the reanalysis (Fig. 3.8). Caution has to be applied in interpreting this figure for the large data-sparse region associated with the Pacific and the relatively small data-filled region of South America. The changes may be due to analysis or data issues, particularly for the early part of the reanalysis record.

3.5 Conclusions and future needs

Many of the questions raised here could be addressed by better collection of reports. At the very least, that would provide a baseline of the occurrence of events. At a deeper level, large datasets of events would be useful for better determining relationships between environments and events. Those could then be used to refine the reanalysis estimates of favorable environments. To that end, the development of the European Severe Weather Database for the entire continent is an important step. Significant tornadoes do occur in Europe (e.g., Wegener, 1917; Fulks, 1967, 1969; Laun, 1969; Dotzek, 2001, 2003), but the reanalysis appears to underestimate them. This underestimate could

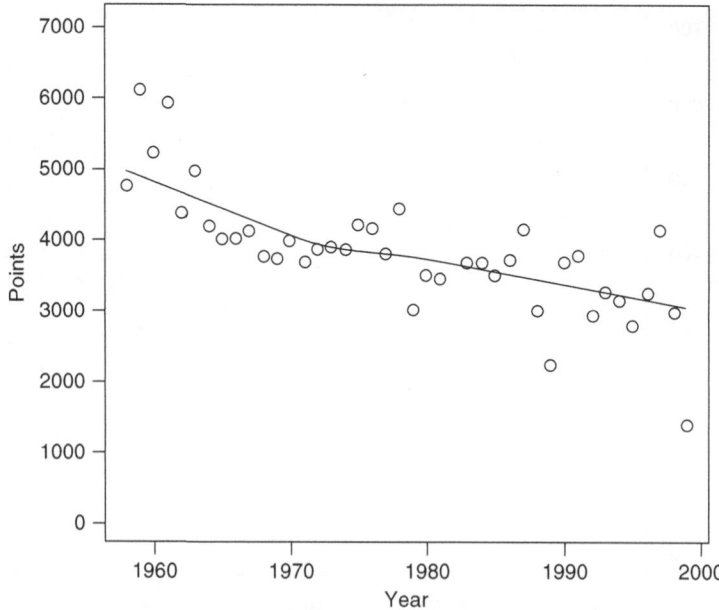

Figure 3.8. Same as Figure 3.6, except for a region of equal size encompassing local maxima in severe thunderstorm counts in South America shown in Figure 3.4.

occur for a variety of reasons, including a failure to identify the environments in which they occur, poor representation of important processes such as orographic forcing in the reanalysis, or differences in the efficiency of the atmosphere in taking an environmental condition and producing a storm.

In order to look at climate change effects, climate model simulations must be analyzed. The challenge in such analysis is that there are important differences compared with the reanalysis problem. The reanalysis has observed events to tie itself to reality. A priori, the ability of the models to represent the future distribution is unknown, and relies strongly on successful verification of present-day simulations. In the model world, the verification can be done by simulating the climate of recent decades, and relating derived quantities (analogous to the reanalysis approach) to observed severe convective storm events. This approach further substantiates the need for reliable, homogeneous, long-term severe weather report databases. Only after such verification, analysis of the distribution of environments in (regional) climate model runs for future scenarios can become feasible, even without having particular severe weather events with which to associate them. It is likely that the analysis will require the development of relationships that are unique to the models. If they can be developed, changes in the distributions in climate change scenarios can be evaluated. The models also have the advantage of producing very large samples of environments, in both space and time, from which the distributions can be studied.

Severe thunderstorms and tornadoes will continue to be a threat. Increases in population and wealth mean that larger losses are possible (Brooks and Doswell, 2001a), even without changes in the meteorological events. Thus, awareness of the threats is important. If, however, changes in the distribution of those threats could be identified, additional preparation for them could be carried out.

Acknowledgments

HB has been supported over the years by a grant from the NOAA Office of Global Programs and by the NCAR Weather and Climate Impact Assessment Science Program. ND has been supported by a DLR *Forschungssemester* grant as a visiting scientist to the NOAA National Severe Storms Laboratory.

References

Beebe, R. G. (1955). Types of airmasses in which tornadoes occur. *Bulletin of the American Meteorological Society*, **36**, 349–50.

Beebe, R. G. (1958). Tornado proximity soundings. *Bulletin of the American Meteorological Society*, **39**, 195–201.

Bissolli, P., Grieser, J., Dotzek, N., and Welsch, M. (2007). Tornadoes in Germany 1950–2003 and their relation to particular weather conditions. *Global and Planetary Change*, **57**(1–2), 124–38, doi:10.1016/j.gloplacha.2006.11.007.

Brooks, H. E. (2004). Tornado warning performance in the past and future: a perspective from signal detection theory. *Bulletin of the American Meteorological Society*, **85**, 837–43.

Brooks, H. E. (2006). A global view of severe thunderstorms: estimating the current distribution and possible future changes. Preprints, *Symposium on the Challenges of Severe Convective Storms*, Atlanta: American Meteorological Society, Conference CD. (Available at http://www.nssl.noaa.gov/users/brooks/public_html/papers/AMS2K6.pdf).

Brooks, H. E., and Craven, J. P. (2002). A database of proximity soundings for significant severe thunderstorms. 1957–1993. Preprints, *Twenty-first Conference on Severe Local Storms*. San Antonio: American Meteorological Society, pp. 639–42.

Brooks, H. E., and Doswell, C. A., III (2001a). Normalized damage from major tornadoes in the United States: 1890–1999. *Weather Forecasting*, **16**, 168–76.

Brooks, H. E., and Doswell, C. A., III (2001b). Some aspects of the international climatology of tornadoes by damage classification. *Atmospheric Research*, **56**, 191–201.

Brooks, H. E., Anderson, A. R., Riemann, K., Ebbers, I., and Flachs, H. (2007). Climatological aspects of convective parameters from the NCAR/NCEP reanalysis. *Atmospheric Research*, **83**, 294–305, doi:10.1016/j.atmosres.2005.08.005.

Brooks, H. E., Doswell, C. A., III, and Cooper, J. (1994). On the environments of tornadic and nontornadic mesocyclones. *Weather Forecasting*, **9**, 606–18.

Brooks, H. E., Doswell, C. A., III, and Kay, M. P. (2003a). Climatological estimates of local daily tornado probability. *Weather Forecasting*, **18**, 626–40.

Brooks, H. E., Lee, J. W., and Craven, J. P. (2003b). The spatial distribution of severe thunderstorm and tornado environments from global reanalysis data. *Atmospheric Research*, **67–68**, 73–94.

Brown, B. G., and Murphy, A. H. (1996). Verification of aircraft icing forecasts: the use of standard measures and meteorological covariates. Preprints, *Thirteenth Conference on Probability and Statistics in the Atmospheric Sciences*. San Francisco, CA: American Meteorological Society, pp. 251–2.

Concannon, P. R., Brooks, H. E., and Doswell, C. A., III (2000). Climatological risk of strong and violent tornadoes in the United States. Preprints, *Second Symposium on Environmental Applications*. Long Beach, CA: American Meteorological Society, pp. 212–19.

Conradsen, K., Nielsen, L. B., and Prahm, L. P. (1984). Review of Weibull statistics for estimation of wind speed distributions. *Journal of Climatology and Applied Meteorology*, **23**, 1173–83.

Craven, J. P., and Brooks, H. E. (2004). Baseline climatology of sounding derived parameters associated with deep, moist convection. *National Weather Digest*, **28**, 13–24.

Doswell, C. A., III (ed.) (2001). *Severe Convective Storms. (Meteorological Monographs, No. 50.)* Boston, MA: American Meteorological Society.

Doswell, C. A., III, Brooks, H. E., and Kay, M. P. (2005). Climatological estimates of daily local nontornadic severe thunderstorm probability for the United States. *Weather Forecasting*, **20**, 577–95.

Dotzek, N. (2001). Tornadoes in Germany. *Atmospheric Research*, **56**, 233–51.

Dotzek, N. (2003). An updated estimate of tornado occurrence in Europe. *Atmospheric Research*, **67–68**, 153–61.

Dotzek, N., Grieser, J., and Brooks, H. E. (2003). Statistical modeling of tornado intensity distributions. *Atmospheric Research*, **67–68**, 163–87.

Dotzek, N., Kurgansky, M. V., Grieser, J., Feuerstein, B., and Névir, P. (2005). Observational evidence for exponential tornado intensity distributions over specific kinetic energy. *Geophysical Research Letters*, **32**, L24813, doi:10.1029/2005GL024583.

Feuerstein, B., Dotzek, N., and Grieser, J. (2005). Assessing a tornado climatology from global tornado intensity distributions. *Journal of Climate*, **18**, 585–96.

Fujita, T. T. (1971). Proposed characterization of tornadoes and hurricanes by area and intensity. SMRP Research Paper 97, University of Chicago.

Fulks, H. W. (1967). Thunderstorms and related severe weather in Europe. *European Theater Weather Orientation* (ETWO), US Air Force Europe, July 1967.

Fulks, H. W. (1969). A synoptic review of the Pforzheim tornado of 10 July 1968. *Technical Bulletin of the Second Weather Wing*, Air Weather Service, US Air Force, April 1969, pp. 26–43.

Gaffen, D. J., and Ross, R J. (1999). Climatology and trends of U.S. surface humidity and temperature. *Journal of Climate*, **12**, 811–28.

Intergovernmental Panel on Climate Change (IPCC) (2001). *Climate Change 2001: The Scientific Basis*. Cambridge, UK: Cambridge University Press.

Intergovernmental Panel on Climate Change (IPCC) (2002). *Workshop Report, IPCC Workshop on Changes in Extreme Weather and Climate Events*, Beijing, China.

Intergovernmental Panel on Climate Change (IPCC) (2007). Climate Change 2007: The Physical Science Basis. (available at http://www.ipcc.ch/ipccreports/ar4-wg1.htm.)

Kelly, D. L., Schaefer, J. T., McNulty, R. P., Doswell, C. A., and Abbey, R. F., Jr. (1978). An augmented tornado climatology. *Monthly Weather Reviews*, **106**, 1172–83.

Kurgansky, M. V. (2000). The statistical distribution of intense moist-convective, spiral vortices in the atmosphere. *Doklady Earth Sciences*, **371**, 408–410. (Available from essl.org/pdf/Kurgansky2000.pdf.)

Laun, W. (1969). An investigation of recent tornadoes over France and Germany. *Technical Bulletin of the Second Weather Wing*, Air Weather Service, US Air Force, April 1969, pp. 3–25.

Lee, J. W. (2002). Tornado proximity soundings from the NCEP/NCAR reanalysis data. M.S. thesis, University of Oklahoma.

Levizzani, V., and Setvák, M. (1996). Multispectral, high-resolution satellite observations of plumes on top of convective storms. *Journal of Atmospheric Science*, **53**, 361–9.

Maddox, R. A., Howard, K. W., and Dempsey, C. L. (1997). Intense convective storms with little or no lightning over central Arizona: a case of inadvertent weather modification? *Journal of Applied Meteorology*, **36**, 302–14.

Rasmussen, E. N., and Blanchard, D. O. (1998). A baseline climatology of sounding-derived supercell and tornado forecast parameters. *Weather Forecasting*, **13**, 1148–64.

Toracinta, E. R., and Zipser, E. J. (2000). Lightning and SSM/I – ice-scattering mesoscale convective systems in the global tropics. *Journal of Applied Meteorology*, **40**, 983–1002.

Trapp, R. J., Wheatley, D. M., Atkins, N. T., Przybylinski, R. W., and Wolf, R. (2006). Buyer beware: some words of caution on the use of severe wind reports in postevent assessment and research. *Weather Forecasting*, **21**, 408–15.

Verbout, S. M., Brooks, H. E., Leslie, L. M., and Schultz, D. M. (2006). Evolution of the U.S. tornado database: 1954–2003. *Weather Forecasting*, **21**, 86–93.

Wegener, A. (1917). *Wind- und Wasserhosen in Europa (Tornadoes in Europe)*. Die *Wissenschaft*, Bd. 60. Braunschweig: Verlag Friedrich Vieweg und Sohn. (In German; available at essl.org.)

4

Regional storm climate and related marine hazards in the Northeast Atlantic

HANS VON STORCH AND RALF WEISSE

4.1 Introduction

Storms represent a major environmental threat. They are associated with abundant rainfall and excessive winds. Windstorms cause different types of damage on land and sea. On land, infrastructure, houses, and other structures may be damaged. In forests, trees may break in large numbers. At sea, wind drags water masses towards the coasts, where the water levels may become dangerously high, overwhelm coastal defenses, and inundate low-lying coastal areas. In addition, the sea surface is affected: wind waves are created that eventually transform into swell. Obviously, wind waves represent a major threat for shipping, offshore activities, and coastal defenses.

With these hazards and threats in mind, we attempt to answer a number of questions related to windstorms in the Northeast Atlantic and northern European region:

1. *How can we determine decadal and longer variations in the storm climate?* The methodological problem is that many variables, which seem to be well suited for this purpose, are available only for too short a period or suffer from inhomogeneities; i.e., their trends are contaminated by signals related to the observation process (e.g., changes in instrumentation, observational practice, or surrounding environmental conditions). Useful indicators of storm climate variation may be derived from a variety of data, including air pressure readings at weather stations and water level readings at tide gauges.
2. *How has the storm climate developed in the past few decades and past few centuries?* Storm activity over the Northeast Atlantic and northern Europe increased for a few decades after the 1960s following an earlier downward trend that started in about 1900. When longer periods are considered (such as by analyzing air pressure readings at stations in Sweden since about 1800), no significant changes are found.
3. *How is storm climate variability linked to hemispheric temperature variations?* Some argue that a general warming would lead to an increase of water vapor in

Climate Extremes and Society, ed. H. F. Diaz and R. J. Murnane. Published by Cambridge University Press. © Cambridge University Press 2008.

the atmosphere, thus providing more "fuel" for the formation of storms. The hypothesized link between large-scale temperature fluctuations and storminess was examined in the framework of a millennium simulation with a state-of-the-art climate model, which was run with reconstructed natural and anthropogenic forcing for the past 500 years, and extended until the year 2100, assuming scenarios for future greenhouse gas emissions. It turns out that for pre-industrial and industrial times (i.e., until about the end of the twentieth century), the hypothesized link could not be detected, even if significant temperature fluctuations were simulated. A joint trend between Northeast Atlantic storm intensity and hemispheric temperature emerged only when future greenhouse gas concentrations were greatly increased.

4. *How did the impact of windstorms on storm surges and ocean waves develop over past decades, and what may happen in the expected course of anthropogenic climate change?* Regionally detailed reconstructions of surface winds since about 1960 have been used to run dynamical models of water levels, currents, and ocean waves in the North Sea. Changes were found to be consistent with the changes of storm activity, namely, a general increase since 1960 to the mid-1990s and thereafter a decline, apart from the southern North Sea, where the upward trend is still going on. Scenarios prepared by a chain of assumed emission scenarios, and global and regional climate models, point to a future of slightly more violent storminess, storm surges, and waves in the North Sea. For the end of the century, an intensification of up to 10% is envisaged, mostly independently of the emission scenario used. When not only the change in windiness but also the thermal expansion of the ocean is considered, increases of 20–30 cm by 2030 and of 50 cm by 2085 appear to be reasonable guesses for future extreme water levels along the German Bight coastline.

There have been many publications that have considered changing storminess in the recent past. Here we focus on changes that have appeared in the past and that may occur in the future over the Northeast Atlantic and northern Europe. This chapter does not aim to present all new material, nor is it meant to be a review covering and evaluating all published material related to the issue of Northeast Atlantic storminess. Instead, its goal is to provide an overview of the issues and to summarize results available from a series of reviewed manuscripts, mostly beginning with the landmark European project Waves and Storms in the North Atlantic (WASA, 1998) paper. Details are available in the referenced literature. Where possible, we cite review papers or manuscripts with extended references that tackle many aspects not explicitly addressed in this chapter and that contain many references not explicitly cited here because of space limitations.

We limit ourselves to the discussion of the four questions outlined above. We do not explicitly discuss other issues, such as changes in the North Atlantic Oscillation (NAO), which has been shown to be closely linked to the variability of North Atlantic storminess and has been discussed in detail elsewhere (e.g., Hurrell et al., 2002). We examine the issue of expected changes in North Atlantic storminess due to anthropogenic climate change only briefly. This topic will be the subject of a detailed account by the forthcoming Fourth Assessment Report of the Intergovernmental Panel on Climate Change (IPCC); Lambert and Fyfe (2006) also discussed the issue in detail. See also Bengtsson et al. (2006). Scenarios of North Atlantic wave climate are offered by Wang et al. (2004). However, we discuss details of the expected change in storms and their impact in terms of storm surges and waves for the case of the North Sea in Section 4.4.

4.2 How can we determine decadal and longer variations in the storm climate?

A major problem with determining changes in windiness concerns the homogeneity, or more precisely the lack of homogeneity, of observed time series. The term *inhomogeneity* refers to the presence of contaminations in a dataset, so that the meteorological data, which are supposed to describe the meteorological conditions and their changes over time, are actually a mix of the sought-after signal and a variety of factors reflecting changing environmental conditions, instrumentation, and observation practices (Karl et al., 1993). For instance, air pressure is generally independent of the specifics of the location (apart from the height of the instrument, which is routinely corrected for) and has been recorded over long periods of time with virtually unchanged instruments, namely, the mercury barometer. Wind measurements represent a rather different case and depend very strongly on the details of the surroundings of the measurement site, in particular, the exposure and obstacles. In addition, instruments and observation practices have changed frequently, particularly so for wind observations and wind estimates over the sea (e.g., Gulev et al., 2003; Gulev and Grigorieva, 2004).

The problem of inhomogeneity is illustrated in Figure 4.1 for a series of examples. A very obvious example is presented in Figure 4.1a, in which the frequency of strong wind events in the city of Hamburg (Germany) by decade is shown. Obviously, a very strong decline took place from the 1940s to the 1950s, the explanation for which is that the wind instrument was moved from the harbor to the airport.

A less obvious example that occasionally has been mistaken as evidence for a worsening of the storm climate in northern Europe is illustrated in Figure 4.1b. It shows the frequency of recorded storm days (with wind speed $\geq 21\,\mathrm{m\,s^{-1}}$) in Kullaberg, southwestern Sweden (after Pruszak and Zawadzka, 2005). Apparently, in recent years the number of storm days was

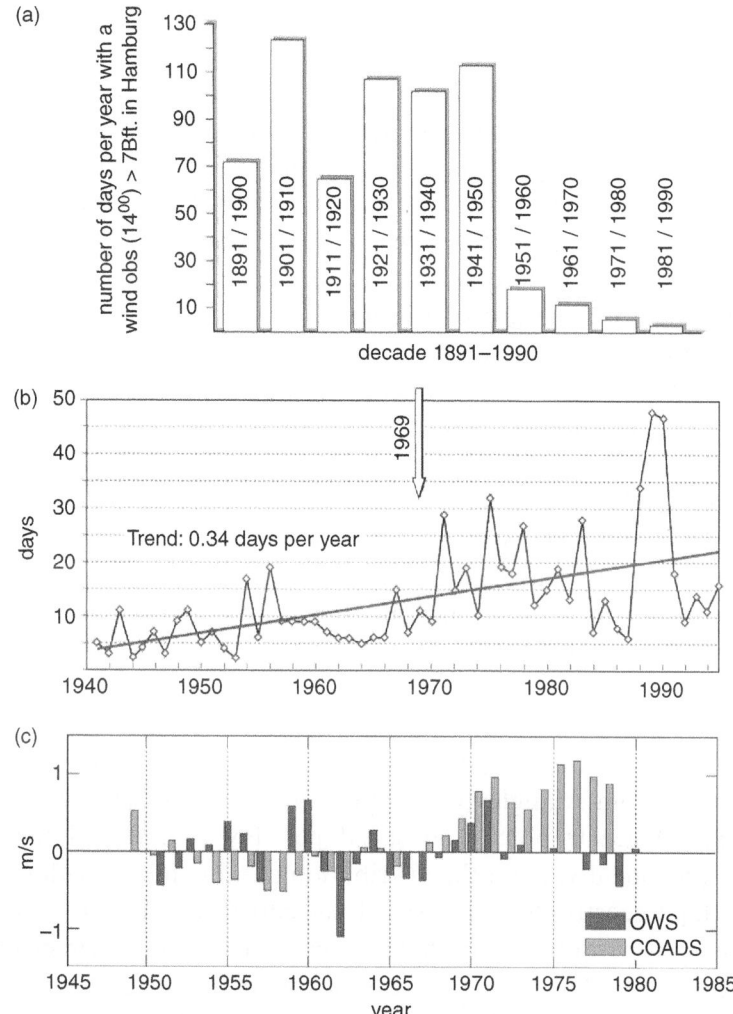

Figure 4.1. (a) Number of days per year with wind speeds of Beaufort Force 7 and greater in Hamburg (after Schmidt, personal communication); (b) frequency of stormy days per year (wind speed $\geq 21\,\mathrm{m\,s^{-1}}$) in Kullaberg, southwestern Sweden (after Pruszak and Zawadzka, 2005); (c) annual mean wind speed anomalies in the North Pacific in the area of ocean weather station OWS P. Data from the ocean weather station are marked as "OWS" (ocean weather station) and those from the ships of opportunity in the vicinity of OWS as "COADS." (After Isemer, personal communication.)

considerably higher than in earlier years. It seems, however, that a severe windstorm damaged the surrounding forest in 1969, so that the locally recorded winds became stronger after the forest windbreak was reduced and reduced surface roughness. We will see later that proxies of storminess indicate no such change in storminess in that area.

Another example of inhomogeneous data records is shown in Figure 4.1c. The example is based on surface marine wind measurements from the Pacific Ocean in the vicinity of the stationary ocean weather ship (OWS) P. While the OWS is expected to take quality-controlled wind measurements, additional wind reports are available from ships traveling near the stationary OWS. These reports enter the Comprehensive Ocean–Atmosphere (COADS) global marine dataset, from which the ship observations can be averaged for each year and compared to the quality-controlled data from the ocean weather station. When this is done, a strong discrepancy emerges: whereas the ship data indicate an upward trend in average wind speed conditions, the OWS P reports variable but by and large stationary conditions. Obviously, at least one of the data products is not homogeneous. These examples suggest that interpreting inhomogenous records may become particularly misleading when long-term changes and trends are analyzed.

The inhomogeneity problem has frequently been overlooked and ignored. Direct wind measurements are hardly ever helpful in assessing changes in storminess for longer periods such as decades. As an alternative, a number of different proxies for storminess in a year or a season have been examined. These proxies are based mainly on air pressure readings and water levels obtained from tide gauge records.

Schmidt and von Storch (1993) have suggested the calculation of geostrophic winds from triangles of air pressure readings. This way, one (or possibly more) geostrophic wind speed per day is obtained for a given location. Subsequently, from the distribution of all numbers within a season or a year, high geostrophic wind speed percentiles are derived that serve as a proxy index for storminess in that season or year. Long-term changes can then be studied based on annual or seasonal proxy index time series. The data in Figure 4.2 demonstrate that such a proxy indeed reflects the observed wind and storm conditions. It shows a comparison between percentiles derived from geostrophic wind speed estimates and local wind observations at five stations that have been known to be quite homogeneous for the 5-year period 1980–84. A remarkably linear relationship is found that suggests that any change in local high wind speed percentiles would be reflected in changes of the geostrophic wind speed index and vice versa (Kaas *et al.*, 1996). Thus time series of the geostrophic wind percentiles can be considered as proxies for changing wind and storm conditions over time (Schmidt and von Storch, 1993; Alexandersson *et al.*, 1998, 2000). Typically, 95th or 99th percentiles are used for examining changes in extremes. Alternatively, one can use the annual frequency of days when the geostrophic wind speed exceeds a threshold of say, $25 \, \mathrm{m \, s^{-1}}$, to examine changes in wind extremes.

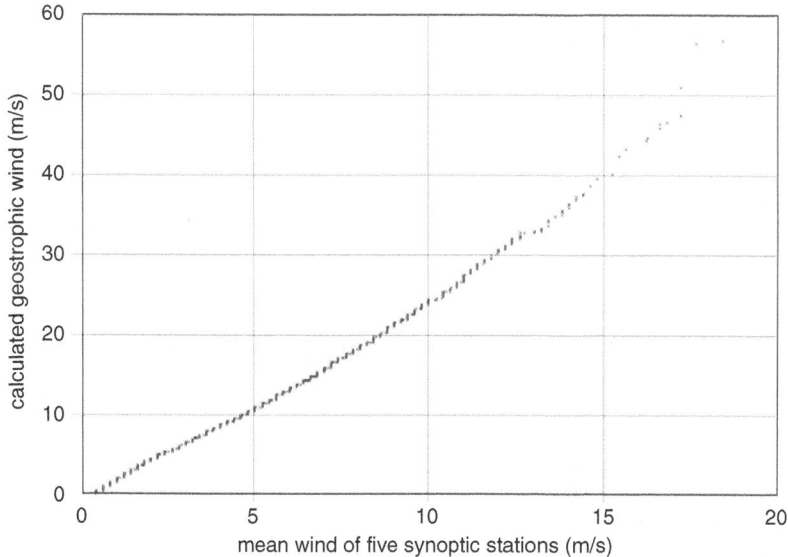

Figure. 4.2. Percentile–percentile plot of observed daily near-surface wind speeds averaged over five synoptic stations in Denmark for which homogeneous wind measurements have been available for 1980–84 and geostrophic wind speeds have been derived from station pressure data. (After Kaas *et al.*, 1996.)

Two alternative proxies are based on local pressure observations, reflecting the experience that stormy weather is usually associated with low air pressure and a rapid fall of the barometer reading (Kaas *et al.*, 1996). These proxies have the advantage that they are available for very long periods of time for some locations (Bärring and von Storch, 2004). The latter point is essential in order to avoid misinterpretation of short-term fluctuations as representing long-term trends. Changes in the statistics of the local proxies may be related to a change in the level of general storm activity or to a change of spatial patterns. Table 4.1 demonstrates that the different indices are mostly consistent among each other, with the exception of the number of deep low-pressure readings. The latter becomes intuitively clear, as low pressure alone is not necessarily sufficient to generate high wind conditions, but strong pressure gradients are required.

Other proxies for storminess may be derived from the variations of water levels at tide gauges as first suggested by John de Ronde, of the Netherlands National Institute for Coastal and Marine Management (RIKZ). While local water level variations at tide gauges are often influenced by local construction works and by slow variations related to global mean sea level rise or geological phenomena such as land subsidence or uplift, some preprocessing is required to derive proxy storm indices from tide gauge data. One option is to first

Table 4.1. *Correlation coefficients between different proxies for storminess*[a]

Correlations	p_{95}	F_{25}	Δ_{16}	N_{980}
p_{99}	0.75	0.90	0.38	0.08
p_{95}		0.64	0.44	0.15
F_{25}			0.35	0.07
Δ_{16}				0.35

[a] p_{95} and p_{99} represent the 95th and the 99th percentiles of seasonal geostrophic wind speeds, F_{25} the seasonal frequency of events with geostrophic wind speeds stronger than $25\,\mathrm{m}\ \mathrm{s}^{-1}$, Δ_{16} the seasonal frequency of air pressure decreasing 16 hPa or more within 24 h, and N_{980} the frequency of barometer readings of 980 hPa and less. Data are from a case study for Denmark. (Reprinted from WASA, 1998.)

determine annual mean high water levels and to subsequently consider variations of the high water levels relative to this annual mean (Pfizenmayer, 1997; von Storch and Reichardt, 1997; Langenberg et al., 1999). Other options for determining storm proxies from tide gauge records are presented in Woodworth and Blackman (2002).

For times when barometers were not yet available, historical accounts help us to assess wind conditions; for example, repair costs of dikes in Holland during the seventeenth century (de Kraker, 1999) or sailing times of supply ships on predetermined routes (e.g., Garcia et al., 2000). However, the homogeneity of historical sources has to be considered with care. Microseismic intensity has been examined to determine whether it may serve as a proxy for regional storm activity (Essen et al., 1999; Grevemeyer et al., 2000), but even if the microseismic records contain signals related to wave activity and thus storminess, a homogeneous long-term record representative for a well-defined region cannot be extracted.

With the proxies discussed above, an assessment of past storminess in northern Europe and the northeast North Atlantic appears possible. In the following section, we will describe how storm activity has evolved in the area.

4.3 How has the storm climate in the Northeast Atlantic and northern Europe developed in the past few decades and past few centuries?

Serious efforts to study changing storminess over the Northeast Atlantic began in the early 1990s when meteorologists noticed a roughening of storm and wave conditions. Wave observations from lighthouses and ships (Carter and Draper, 1988; Cardone *et al.*, 1990; Hogben, 1994) described a roughening since the 1950s, and an analysis of deep pressure systems in operational weather maps indicated a steady increase of such lows since the 1930s (Schinke, 1992). Unfortunately, these analyses all suffered from the problems described above: either an insufficient length of data series or compromised homogeneity. For instance, the skill of describing weather details in weather maps has steadily improved in the course of time, because of more and better data that have been reported to the weather services and improved analysis practices. For global reanalysis, the improvement related to the advent of satellite data on Southern Hemisphere analysis is described by Kistler *et al.* (2001) and Bromwich *et al.* (2007). Another example of the effect of better data coverage is provided by Landsea *et al.* (2004) for a tropical storm.

A breakthrough came when most of the proxies defined in the previous section were introduced, mostly within the European Union (EU) project WASA (WASA, 1998). Alexandersson *et al.* (1998, 2000) assembled homogeneous series of air pressure readings for the period from 1880 for a variety of locations covering most of northern Europe. They calculated 99th percentiles of geostrophic winds from a number of station triangles. After some normalization and averaging, they derived proxy time series for the greater Baltic Sea region and for the greater North Sea region. The time series are shown in Figure 4.3. According to these proxies, the storm activity intensified between 1960 and 1995,[1] but from the beginning of the record until about 1960 there was a long period of declining storminess (Alexandersson *et al.*, 1998, 2000). Since about 1995 the trend has been towards less storminess in most areas of the Northeast Atlantic (Weisse *et al.*, 2005; Matulla *et al.*, 2007).

A similar result was obtained by analyzing the record of high water levels in Den Helder and Esbjerg, two harbors on the Dutch and Danish North Sea coasts (Pfizenmayer, 1997). Figure 4.4 displays two statistics for each of the two tide gauges: annual mean high water levels and the annual 99th percentiles of the deviations of the observed high water levels from the annual mean. While the former, the annual mean, is influenced by a number of non-storm-related processes – such as local construction works, geological changes (land subsidence), and global mean sea level rise – the upper percentiles of the deviations from this mean are expected to be more homogeneous and to better

---

[1] Interestingly, in the early 1990s there were widespread claims in northern Europe (e.g., Berz, 1993; Berz and Conrad, 1994) that there was a significant increase in storminess, which would be consistent with anthropogenic climate change. Following this logic, one would have to assume that the trend would continue into the future, and thus wind-related risks would increase and cause problems for the insurance industry.

H. von Storch and R. Weisse

(a)   British Isles, North Sea, Norwegian Sea, 1881–2002 99-percentile

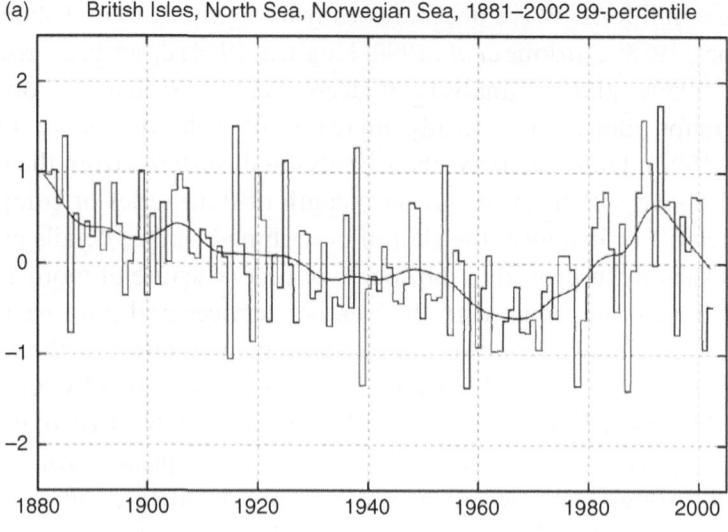

(b)   Scandinavia, Finland, Baltic Sea, 1881–2002 99-percentile

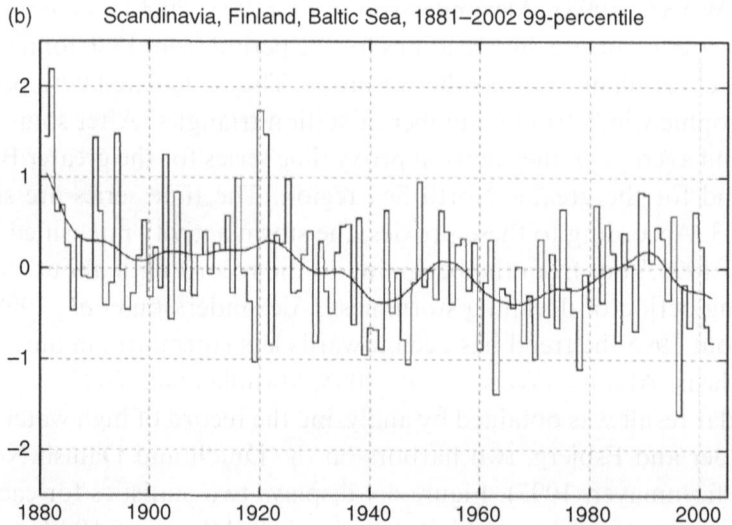

Figure 4.3. Proxy index of storm activity derived from intra-annual percentiles of geostrophic wind speeds derived from air pressure measurements at a series of triangles of stations for the greater North Sea region (a) and the greater Baltic Sea region (b). The index time series were normalized and are thus dimensionless. (Updated version of diagram provided by Alexandersson *et al.*, 2000.)

represent long-term storm-related fluctuations (e.g., von Storch and Reichardt, 1997). Both locations exhibit a marked increase in annual mean high water levels, but the rate of increase is different at the two locations. The latter is likely related to different regional processes related to water works

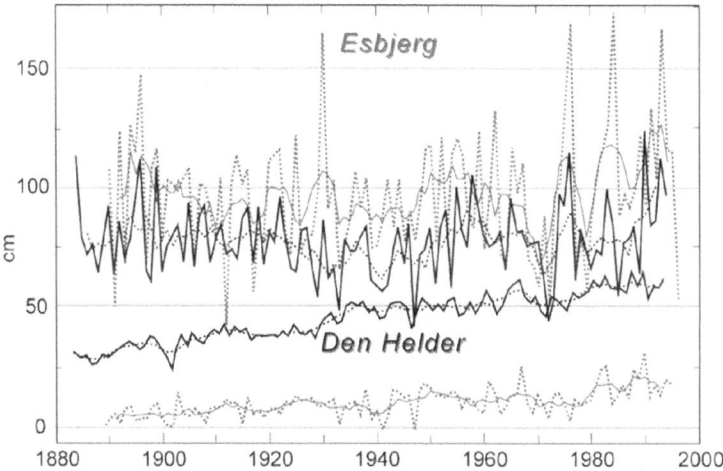

Figure 4.4. Intra-annual statistics of high water levels at Esbjerg (Denmark) and Den Helder (The Netherlands) since the late nineteenth century. The lower two curves display the annual mean high water; the upper two curves represent annual 99th percentiles of the variations around the annual mean. (After Pfizenmayer, 1997.)

and the implementation of coastal defense measures. The 99th percentiles of the deviations from the annual mean high water reveals a somewhat different figure. Again, an increase is found for the period from 1960 through the 1990s, which is, however, not significant compared with the development prior to 1960.

The 1960–95 increase in Northeast Atlantic storminess also does not appear to be dramatic when even longer time windows are considered. Bärring and von Storch (2004) analyzed homogenized local air pressure readings at two locations in Sweden, Lund and Stockholm, which have been recorded since the early 1800s and earlier. The number of deep pressure systems, as well as the number of rapid pressure falls of 16 hPa and more within 12 hours (not shown), has been remarkably stationary since the beginning of the barometer measurements. This is remarkable in view of the notable increase in regional temperatures; e.g., in Denmark (Cappelen, 2005). Using storm indices derived from tide gauge data, other authors have reached similar conclusions. For instance, Woodworth and Blackman (2002) analyzed changes in extreme high waters in Liverpool since 1768. In considering the entire period, they found considerable interannual variability but no clear long-term trend. Bijl *et al.* (1999) considered sea level variations from a number of stations in the coastal zones of northwest Europe over the past 100 years. Similarly, they concluded that there is strong natural variability present in the data but no sign of a significant increase in storm-related water levels.

4.4 How is storm climate variability linked to hemispheric temperature variations?

The link between decadal and centennial variations of mean temperature and storminess has hardly been studied, because of the lack of sufficient data. Specifically, it has been argued that a general warming would be associated with elevated water vapor levels, which in turn would be associated with stronger extratropical storms. Obviously, this argument is to first order symmetric, so a general cooling would be associated with less storminess. The history of climate variability in past centuries is a good framework to test such a hypothesis.

Climate models exposed to time-variable solar, volcanic, and greenhouse gas forcing of past centuries provide good data for the study of historical covariability of temperature and storminess. Such a study was performed by Fischer-Bruns *et al.* (2002, 2005), who counted for each model's grid box the annual frequency of gales in a simulation beginning in 1550 and extending to 2100 (using the IPCC A2 scenario for 2000–2100). They found no obvious link between the levels of storm activity and hemispheric mean temperatures for historical times (not shown). The simulation has a parallel development of storminess and temperature only during the period of anthropogenic climate change in the twenty-first century, and this development is associated mainly with a spatial displacement of the storm track to the northeast and not a major intensification.

The lack of a link between hemispheric mean temperatures and storminess during historical times is demonstrated by Figure 4.5, which shows the spatial patterns of the differences of temperature and of storm frequency (given as number of gale days per year and grid box) between the Late Maunder Minimum (LMM, 1675–1710) and the pre-industrial period of the simulation (1550–1850). The Late Maunder Minimum was the coldest period of the Little Ice Age (LIA), at least in Europe, and the model simulation indicates that this cooling was of almost global extent, affecting all of the Northern Hemisphere. This period was, at least in the model, not associated with reduced storminess in the North Atlantic or in the North Pacific.

Thus neither the admittedly very limited empirical evidence discussed in the previous section nor the modeling study by Fischer-Bruns *et al.* (2002, 2005) support the hypothesis that a general warming would lead, plausibly via increased availability of humidity, to a more severe storm climate.

The parallel development of changes in storminess and temperature in scenario simulations is likely related not to the general increase in temperature but to changes in temperature gradients.

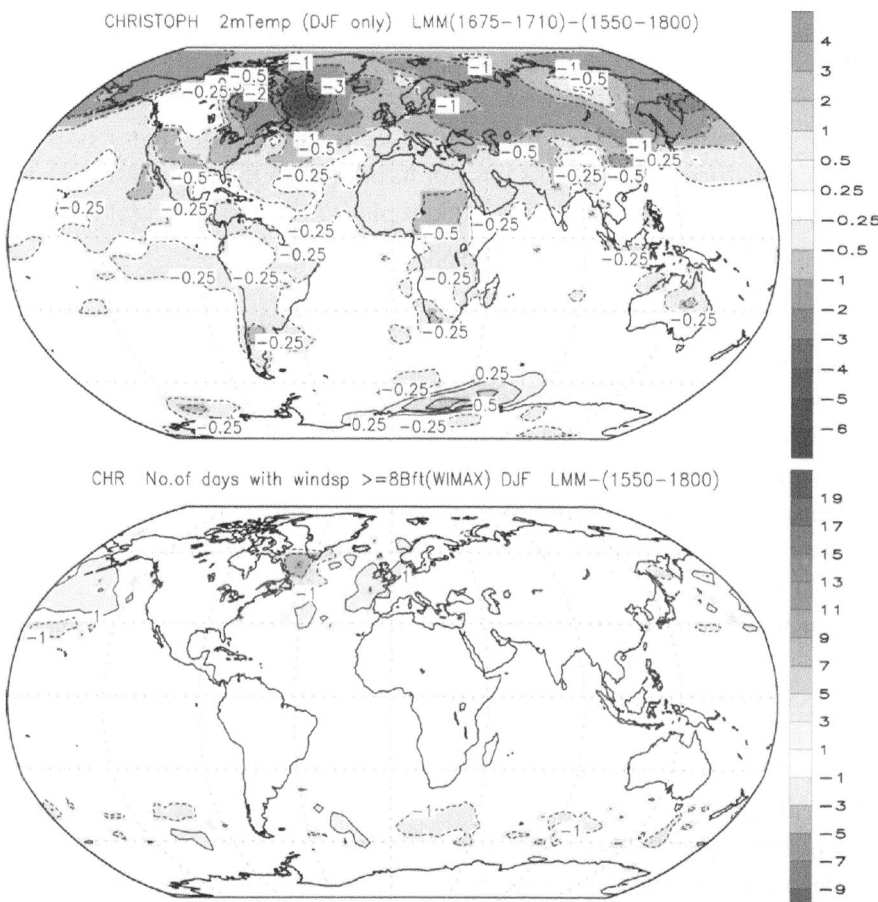

Figure 4.5. Simulated differences in winter between the Late Maunder Minimum (LMM, 1675–1710) and the pre-industrial time (1550–1850), in terms of air temperature (top, K) and of number of gale days (wind speeds of Beaufort Force 8 and more). Note that the LMM is portrayed by the model as particularly cold, but the storm activity shows little change. (Courtesy Irene Fischer-Bruns.) For color version, see plate section.

4.5 How did the impact of windstorms on North Sea storm surges and ocean waves develop over past decades, and what may happen in the expected course of anthropogenic climate change?

Changes in storminess have a significant impact on a variety of relevant socioeconomic activities and risks. An economic segment obviously sensitive to changes in the risk of wind-related damage is the insurance industry (Berz,

1993; Berz and Conrad, 1994).[2] Other relevant aspects are related to ocean waves and storm surges, and their impact on offshore activities, shipping, and coastal protection structures.

Proxy studies, as described in previous sections, indicate that a systematic worsening of storm-related risks has not happened in the past 200 years or so. On the other hand, a worsening has taken place in the past 50 years, and data for that period are good enough to examine the changes of storm surge and ocean wave statistics in more detail.

The availability of good weather analyses – on a global basis, for instance, in the National Centers for Environmental Prediction (NCEP) reanalyses (Kalnay *et al.*, 1996; Kistler *et al.*, 2001), and, for the European region, dynamical downscaling of this reanalysis (Feser *et al.*, 2001) – allows detailed analysis of changing ocean wave and storm surge conditions. To carry out such analysis, 6-hourly (or even more frequent) wind and air pressure fields are used to run ocean wave (Günther *et al.*, 1998; Sterl *et al.*, 1998) and storm surge models (Flather *et al.*, 1998; Langenberg *et al.*, 1999). In this way, homogeneous estimates of changes in the past 50 or so years can be constructed (Weisse and Plüß, 2006). In using the same models, scenarios of expected climate change also can be processed with respect to windstorms, ocean waves, and storm surges (e.g., Flather and Smith, 1998; Kauker, 1998; Lowe *et al.*, 2001; Debernard *et al.*, 2003; Lowe and Gregory, 2005; Woth, 2005; Woth *et al.*, 2006).

Along these lines, the downscaled NCEP reanalyses (Feser *et al.*, 2001) have been used to examine changes in patterns of storminess (Weisse *et al.*, 2005). In most parts of the Northeast Atlantic, storminess – given as annual frequency of gales per grid box – increased until the early 1990s; south of about 50° N there was a decrease (Figure 4.6). This pattern reversed almost completely in the early 1990s apart from the southern North Sea, where the trend towards more storms continued, albeit somewhat decelerated towards the end of the period, at least until 2002. Accordingly, storm surge simulations reveal an increase in high water levels of a few millimeters per year, in the seasonal mean as well as in the high levels relative to the mean (Aspelien, 2006; Weisse and Plüß, 2006), in particular along the German Bight coastline.

---

[2] One should, however, not accept an assertion of the insurance industry as an unbiased and objective description of a situation without careful analysis – overestimating the risks involved does in general not harm the economic interests of an insurance company.

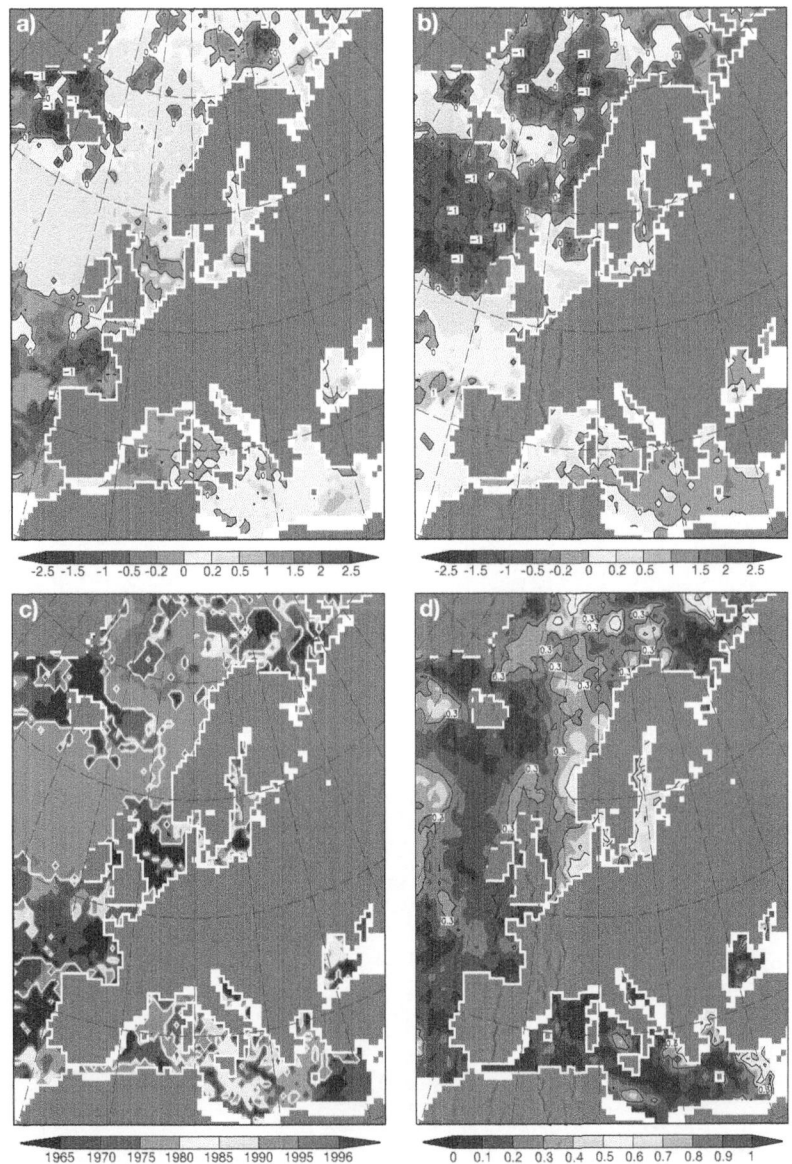

Figure 4.6. Piecewise linear trends before and after a change point T in the total number of storms per year with maximum wind speeds exceeding 17.2 m s^{-1}. Both the trends and the change point are determined by a best fit to the data time series. (a) Trends for the first period 1958–T; (b) trends for the second period T–2002. Units in both cases are number of storms per year. (c) Year T at which a change in trends is indicated by the statistical model; (d) Brier skill score of the bilinear trend fitting the data as compared to using one trend for the entire period. (After Weisse *et al.*, 2005.) For color version, see plate section.

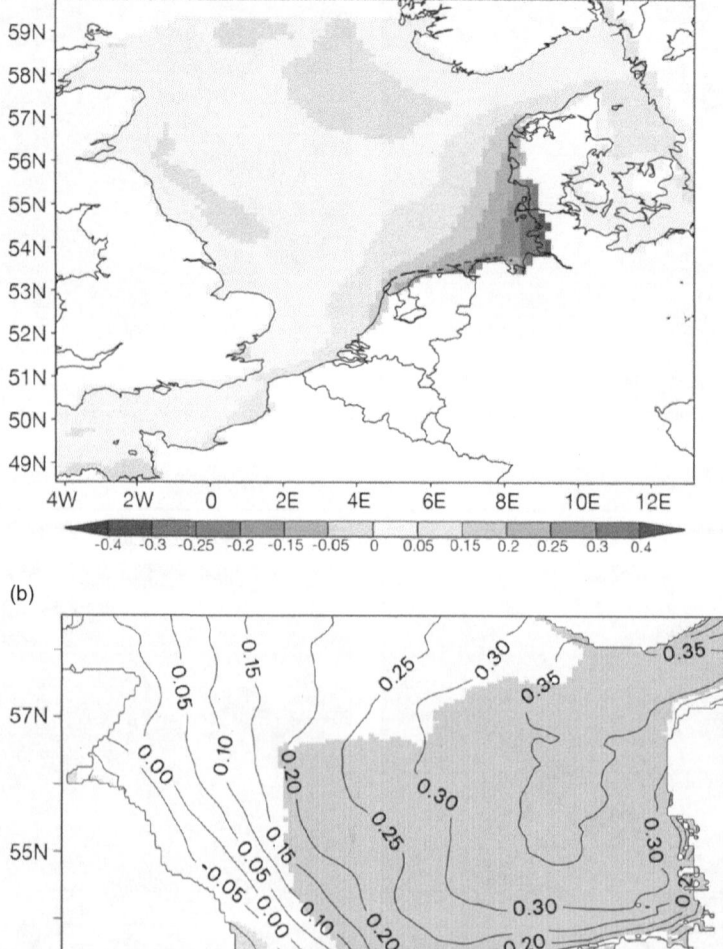

Figure 4.7. Expected changes in wind-related storm surge heights (a) (maximum averaged across many years, Rossby Center Atmosphere Ocean (RCAO) model, emission scenario A2) and ocean wave heights (b) (change of 99th percentile; averaged across a series of simulations using different models and both emission scenarios A2 and B2). Shading in (b) indicates areas where signals from all models and scenarios have the same sign (red, positive; blue, negative) in the North Sea at the end of the twenty-first century. Units are meters. (Courtesy Katja Woth and Iris Grabemann.) For color version, see plate section.

Furthermore, in the HIPOCAS project – Hindcast of Dynamic Processes of the Ocean and Coastal Areas of Europe (Soares *et al.*, 2002), statistics of ocean (surface) waves have been derived. Extreme wave heights increased in the southeastern North Sea during the period 1958–2002 by up to 1.8 cm per year, while for much of the UK coast a decrease was found. The increase in the southeastern North Sea, however, has not been constant in time. The frequency of high wave events increased until about 1985–90 and has remained almost constant since that time (Weisse and Günther, 2007). This development closely follows that of storm activity (Weisse *et al.*, 2005).

Scenarios of future wind conditions have been derived by several groups. The most useful is possibly the set of simulations with the model of the Swedish Rossby Centre, which features not only an atmospheric component but also lakes and a dynamical description of the Baltic Sea (Räisänen *et al.*, 2004). This model was run with boundary conditions provided by two global climate models; also, the effects of two different emission scenarios were simulated. In these simulations, strong westerly wind events are intensified by less than 10% at the end of the twenty-first century (Woth, 2005). A similar result was found by Pryor *et al.* (2006), who empirically downscaled climate change scenarios from ten coupled global climate models and found changes in the mean and 90th percentile wind speeds to be small (less than about 15%) for northern Europe.

These changes of wind speed will have an effect on both North Sea storm surges and wave conditions. For the storm surges along the North Sea coastline, an intensification is expected, which may amount to an increase of 20–30 cm or so, to the end of the century (Figure 4.7a). The mean level has to be added to this wind-related change, so that along the German Bight maximum values of 50 cm are plausible estimates for the increase in water levels during heavy storm surges. In the Elbe estuary, larger values up to 70 cm are derived. These numbers are associated with a wide range of uncertainty ($\pm$ 50 cm) (Grossmann *et al.*, 2007).

Scenarios of future wave conditions show large differences in the spatial patterns and the amplitudes of the climate change signals. There is, however, agreement among models and scenarios that extreme wave heights may increase by up to 30 cm (7% of present values) in the southeastern North Sea by 2085 (Figure 4.7b; Grabemann and Weisse, 2007).

Acknowledgments

We are grateful for the help provided by Irene Fischer-Bruns, Iris Grabemann, and Beate Gardeike.

70 *H. von Storch and R. Weisse*

References

Alexandersson, H., Schmith, T., Iden, K., and Tuomenvirta, H. (1998). Long-term trend variations of the storm climate over northwest Europe. *Global Atmosphere-Ocean System*, **6**, 97–120.

Alexandersson, H., Schmith, T., Iden, K., and Tuomenvirta, H. (2000). Trends of storms in northwest Europe derived from an updated pressure data set. *Climate Research*, **14**, 71–3.

Aspelien, T. (2006). The use of long-term observations in combination with modelling and their effect on the estimation of the North Sea storm surge climate. Ph.D. thesis, Hamburg University, Hamburg, Germany.

Bärring, L., and von Storch, H. (2004). Northern European storminess since about 1800. *Geophysical Research Letters*, **31**, L20202, doi:10.1029/2004GL020441, 1–4.

Bengtsson, L., Hodges, K. I., and Roeckner, E. (2006). Storm tracks and climate change. *Journal of Climate*, **19**, 3518–43.

Berz, G. (1993). Global warming and the insurance industry. *Interdisciplinary Science Review*, **18**(2), 120–5.

Berz, G., and Conrad, K. (1994). Stormy weather: the mounting windstorm risk and consequences for the insurance industry. *Ecodecision*, **12**, 65–8.

Bijl, W., Flather, R., de Ronde, J. G., and Schmith, T. (1999). Changing storminess? An analysis of long-term sea level data sets. *Climate Research*, **11**, 161–72.

Bromwich, D. H., Fogt, R. L., Hodges, K. I., and Walsh, J. E. (2007). Tropospheric assessment of ERA-40, NCEP, and JRA-25 global reanalyses in the polar regions. *Journal of Geophysical Research* (in press).

Cappelen, J. (2005). Sådan var vejret I 2004 – i Danmark, Nuuk i Grønland og i Tórshavn pa Færøerne (The weather of 2004 – in Denmark, Greenland, and Tórshavn in the Faroe Islands). *Vejret*, **102**, 1–10.

Cardone, V. J., Greenwood, J. G., and Cane, M. A. (1990). On trends in historical marine data. *Journal of Climate*, **3**, 113–7.

Carter, D. J. T., and Draper, L. (1988). Has the Northeast Atlantic become rougher? *Nature*, **322**, 494.

Debernard, J., Sætra, Ø., and Røed, L. P. (2003). Future wind, wave and storm surge climate in the northern North Atlantic. *Climate Research*, **23**, 39–49.

de Kraker, A. M. J. (1999). A method to assess the impact of high tides, storms and storm surges as vital elements in climate history. The case of stormy weather and dikes in the northern part of Flanders, 1488–1609. *Climatic Change*, **43**, 287–302.

Essen, J. H., Klussmann, J., Herber, R., and Grevemeyer, I. (1999). Does microseism in Hamburg (Germany) reflect the wave climate in the North Atlantic? *Deutsche Hydrographische Zeitschrift*, **51**, 17–29.

Feser, F., Weisse, R., and von Storch, H. (2001). Multidecadal atmospheric modelling for Europe yields multi-purpose data. *Eos*, **82**, 305–10.

Fischer-Bruns, I., Cubasch, U., von Storch, H., *et al.* (2002). Modelling the Late Maunder Minimum with a three-dimensional OAGCM. *CLIVAR Exchanges*, **7**, 59–61.

Fischer-Bruns, I., von Storch, H., González-Rouco, F., and Zorita, E. (2005). Modelling the variability of midlatitude storm activity on decadal to century time scales. *Climate Dynamics*, **25**, 461–76, doi10.1007/s00382-005-0036-1.

Flather, R. A., and Smith, J. A. (1998). First estimates of changes in extreme storm surge elevation due to doubling CO_2. *Global Atmosphere-Ocean System*, **6**, 193–208.

Flather, R. A., Smith, J. A., Richards, J. D., Bell, C., and Blackman, D. L. (1998). Direct estimates of extreme storm surge elevations from a 40-year numerical model simulation and from observations. *Global Atmosphere-Ocean System*, **6**, 165–76.

García, R., Gimeno, L., Hernández, E., Prieto, R., and Ribera, P. (2000). Reconstruction of North Atlantic atmospheric circulation in the sixteenth, seventeenth and eighteenth centuries from historical sources. *Climate Research*, **14**, 147–51.

Grabemann, I., and Weisse, R. (2007). Climate change impact on extreme wave conditions in the North Sea: an ensemble study. *Ocean Dynamics*, submitted.

Grevemeyer, I., Herber, R., and Essen, H. H. (2000). Microseismological evidence for a changing wave climate in the Northeast Atlantic Ocean. *Nature*, **408**, 349–52.

Grossmann, I., Woth, K., and von Storch, H. (2007). Localization of global climate change: storm surge scenarios for Hamburg in 2030 and 2085. *Die Küste* [*The Coast*], **71**, 169–82.

Gulev, S. K., and Grigorieva, V. (2004). Last century changes in ocean wind wave height from global visual wave data. *Geophysical Research Letters*, **31**, L24601, doi:10.1029/2004GL021032.

Gulev, S. K., Grigorieva, V., Sterl, A., and Woolf, D. (2003). Assessment of the reliability of wave observations from voluntary observing ships: insights from the validation of a global wind wave climatology based on voluntary observing ship data. *Journal of Geophysical Research*, **108**, L3235, doi:10.1029/2002JC001673.

Günther, H., Rosenthal, W., Stawarz, M. *et al.* (1998). The wave climate of the Northeast Atlantic over the period 1955–1994: the WASA wave hindcast. *Global Atmosphere-Ocean System*, **6**, 121–63.

Hogben, N. (1994). Increases in wave heights over the North Atlantic: a review of the evidence and some implications for the naval architect. *Transactions of the Royal Institute of Naval Architects*, **W5**, 93–101.

Hurrell, J. W., Kushnir, Y., Ottersen, G., and Visbeck, M. (2002). An overview of the North Atlantic Oscillation. In *The North Atlantic Oscillation: Climatic Significance and Environmental Impact* (Geophysical Monograph Series 134), ed. J. W. Hurrell, Y. Kushnir, G. Ottersen, and M. Visbeck. Baltimore, MD: American Geophysical Union, pp. 1–35.

Kaas, E., Li, T. -S., and Schmith, T. (1996). Statistical hindcast of wind climatology in the North Atlantic and northwestern European region. *Climate Research*, **7**, 97–110.

Kalnay, E., Kanamitsu, M., Kistler, R., *et al.* (1996). The NCEP/NCAR 40-Year Reanalysis Project. *Bulletin of the American Meteorological Society*, **77**(3), 437–71.

Karl, T. R., Quayle, R. G., and Groisman, P. Y. (1993). Detecting climate variations and change: new challenges for observing and data management systems. *Journal of Climate*, **6**, 1481–94.

Kauker, F. (1998). Regionalization of climate model results for the North Sea. Ph.D. thesis, University of Hamburg, GKSS 99/E/6.

Kistler, R., Kalnay, E., Collins, W., *et al.* (2001). The NCEP/NCAR 50-Year Reanalysis. *Bulletin of the American Meteorological Society*, **82**, 247–67.

Lambert, S. J., and Fyfe, J. C. (2006). Changes in winter cyclone frequencies and strengths simulated in enhanced greenhouse warming experiments: results from the models participating in the IPCC diagnostic exercise. *Climate Dynamics*, **26**, 713–28, doi:10.1007/s00382-006-0110-3.

Landsea, C. W., Anderson, C., Charles, N., *et al.* (2004). The Atlantic hurricane database re-analysis project: documentation for the1851–1910 alterations and additions to the HURDAT database. In *Hurricanes and Typhoons: Past, Present, and Future*, ed. R. J. Murnane and K. -B. Liu. New York: Columbia University Press, pp. 177–221.

Langenberg, H., Pfizenmayer, A., von Storch, H., and Sündermann, J. (1999). Storm related sea level variations along the North Sea coast: natural variability and anthropogenic change. *Continental Shelf Research*, **19**, 821–42.

Lowe, J. A., and Gregory, J. M. (2005). The effects of climate change on storm surges around the United Kingdom. *Philosophical Transactions of the Royal Society*, **A363**, 1313–28, doi:10.1098/rsta.2005.1570.

Lowe, J. A., Gregory, J. M., and Flather, R. A. (2001). Changes in the occurrence of storm surges in the United Kingdom under a future climate scenario using a dynamic storm surge model driven by the Hadley center climate models. *Climate Dynamics*, **18**, 197–88.

Matulla, C., Schöner, W., Alexandersson, H., von Storch, H., and Wang, X.L. (2007). European Storminess: late nineteenth century to present. *Climate Dynamics*, doi: 10.1007/s00382-007-033-y.

Pfizenmayer, A. (1997). Zusammenhang zwischen der niederfrequenten Variabilität in der grossräumigen atmosphärischen Zirkulation und den Extremwasserständen an der Nordseeküste. [The link between low-frequency variability of large-scale circulation and high water level extremes along the German North Sea coast.] Diplomarbeit, Institut für Geographie, Universität Stuttgart.

Pruszak, Z., and Zawadzka, E. (2005). Vulnerability of Poland's coast to sea level rise. *Coastal Engineering Journal*, **47**, 131–55.

Pryor, S. C., Schoof, J. T., and Barthelmie, R. J. (2006). Winds of change? Projections of near-surface winds under climate change scenarios. *Geophysical Research Letters*, **33**, L11702, doi:10.1029/2006/GL026000.

Räisänen, J., Hansson, U., Ullerstig, A., *et al.* (2004). European climate in the late twenty-first century: regional simulations with two driving global models and two forcing scenarios. *Climate Dynamics*, **22**, 13–31, doi:10.1007/s00382-003-0365-x.

Schinke, H. (1992). Zum Auftreten von Zyklonen mit niedrigen Kerndrücken im atlantisch-europäischen Raum von 1930 bis 1991. *Wissenschaftliche Zeitschrift der Humboldt Universität zu Berlin, Mathematisch-Naturwissenschaftliche Reihe*, **41**, 17–28.

Schmidt, H., and von Storch, H. (1993). German Bight storms analyzed. *Nature*, **365**, 791.

Soares, C. D., Weisse, R., Carretero, J. C., and Alvarez, E. (2002). A 40-year hindcast of wind, sea level and waves in European waters. *Proceedings of the 21st International Conference on Offshore Mechanics and Arctic Engineering*, Oslo, Norway, American Society of Mechanical Engineers, OMAE 2002–28604.

Sterl, A., Komen, G., and Cotton, P. D. (1998). Fifteen years of global wave hindcasts using winds from the European Centre for Medium Range Weather reanalysis: validating the reanalyzed winds and assessing the wave climate. *Journal of Geophysical Research*, **103**(C3), 5477–92.

von Storch, H., and Reichardt, H. (1997). A scenario of storm surge statistics for the German Bight at the expected time of doubled atmospheric carbon dioxide concentration. *Journal of Climate*, **10**, 2653–62.

Wang, X. L., Zwiers, F. W., and Swail, V. R. (2004). North Atlantic Ocean wave climate change scenarios for the twenty-first century. *Journal of Climate*, **17**, 2368–83.

WASA (1998). Changing waves and storms in the Northeast Atlantic? *Bulletin of the American Meteorological Society*, **79**, 741–60.

Weisse, R., and Günther, H. (2007). Wave climate and long-term changes for the southern North Sea obtained from a high-resolution hindcast, 1958–2002. *Ocean Dynamics*, **57**(3), 161–72, doi:10.1007/s10236-006-0094-x.

Weisse, R., and Plüß, A. (2006). Storm related sea level variations along the North Sea coast as simulated by a high-resolution model, 1958–2002. *Ocean Dynamics*, **56**, 16–25, doi:10.1007/s10236-005-0037-y.

Weisse, R., von Storch, H., and Feser, F. (2005). Northeast Atlantic and North Sea storminess as simulated by a regional climate model, 1958–2001, and comparison with observations. *Journal of Climate*, **18**, 465–79.

Woodworth, P. L., and Blackman, D. L. (2002). Changes in extreme high waters at Liverpool since 1768. *International Journal of Climatology*, **22**, 697–714.

Woth, K. (2005). Projections of North Sea storm surge extremes in a warmer climate: how important are the RCM driving GCM and the chosen scenario? *Geophysical Research Letters*, **32**, L22708, doi:10.1029/2005GL023762.

Woth, K., Weisse, R., and von Storch, H. (2006). Dynamical modelling of North Sea storm surge extremes under climate change conditions: an ensemble study. *Ocean Dynamics*, **56**, 3–15, doi:10.1007/s10236-005-0024-3.

5

Extensive summer hot and cold extremes under current and possible future climatic conditions: Europe and North America

ALEXANDER GERSHUNOV AND HERVÉ DOUVILLE

Condensed summary

The spatial scale of a heat wave is an important determinant of its impacts. Extensive summer hot and cold spells in Europe and North America are studied through observations and coupled model projections. Recent trends towards more frequent and extensive hot spells as well as rarer and less extensive cold outbreaks follow global warming trends, but they are regionally modulated on decadal timescales. Coupled model projections reflect these natural and anthropogenic influences, with their relative contributions depending on the particular scenarios assumed for global socioeconomic development. Europe appears to have had an early warning in 2003 of conditions that are projected for the second half of the twenty-first century, assuming a "business as usual" emissions scenario. North America, on the other hand, in spite of a general summer warming, has not seen the extent of summer heat that it can potentially experience even if global emissions of carbon dioxide and sulfate aerosols remain fixed at their current levels. Extensive and persistent heat waves naturally occur in association with widespread drought. The recent warming over North America is unusual in that it has occurred without the large-scale encouragement of a dry soil associated with precipitation deficit. Regional precipitation anomalies, together with global anthropogenic influences, can explain the atypical spatial pattern of recent North American summer warming. A decrease of precipitation to more normal amounts over the central and eastern United States is expected to result in a substantial summer warming over that region. Drought has the potential to seriously exacerbate the recent warming over North America to levels more in line with the warmest current model projections. Assuming realistic warming scenarios, a long-term anthropogenic increase (decrease) in the frequency and spatial extent of regional hot (cold) spells is projected to be strong and strongly modulated by decadal-scale variability throughout the twenty-first century.

Climate Extremes and Society, ed. H. F. Diaz and R. J. Murnane. Published by Cambridge University Press. © Cambridge University Press 2008.

5.1 Introduction

Outbreaks of anomalous summer heat occur each year somewhere on Earth. Typically associated with persistent blocking anticyclones, summertime heat waves in the mid-latitudes have a coherent spatial structure that is characterized by lack of rainfall, dry air and soil, and increased fire risk. Disastrous consequences (Macfarlane and Waller, 1976; Sheridan and Kalkstein, 2004) can result from hot spells that are extreme in their duration and spatial extent. The summer of 2003, very likely the hottest in at least 500 years (Luterbacher *et al.*, 2005), brought periods of sustained temperatures exceeding 35 °C over much of western and central Europe. The heat wave was accompanied by an almost complete lack of rainfall (Levinson and Waple, 2004), which resulted in wide-ranging environmental degradation, including severe impacts on agriculture, river flow, mountain glaciers, energy production, and toxicity (e.g., Beniston and Diaz, 2004), as well as wildfires in southwestern Europe, and over 10,000 heat-stress-related deaths in France alone (Levinson and Waple, 2004; Dhainaut *et al.*, 2004).

The Dust Bowl of the 1930s (Schubert *et al.*, 2004), a period of summertime heat and drought affecting large parts of North America – sustained over a decade and punctuated by exceptionally intense and extensive heat outbreaks in 1934, 1936, and 1937 – saw widespread hardship, farmland abandonment, and migration. Even mild hot spells of large spatial extent can cause havoc in the energy sector as demand for air conditioning rises beyond the capacity of power utilities to provide the needed electricity. Power outages during heat waves can lead to still higher human mortality through exposure. However, with adequate infrastructure in place, the health effects of heat waves can be effectively mitigated if the heat wave is anticipated even in the short term (Palecki *et al.*, 2001; Sheridan and Kalkstein, 2004).

Economic hardships (Subak *et al.*, 2000) due to unanticipated and unmitigated heat waves can result from rising power costs, as well as from decreased crop yields and increased livestock mortality. Environmental consequences of hot spells can range from loss of flora and fauna due directly to heat stress and indirectly by fire to depletion of natural water reservoirs and streamflow through related precipitation deficit and increased evaporation. Sustained hot spells can increase the risk of vector-borne and other infectious diseases (Ballester *et al.*, 2003; Zell, 2004). The larger the spatial extent of a heat wave, the more the related hydrologic deficit should be able to exert a positive feedback prolonging the condition.

The spatial scale of a heat wave is an important determinant of its environmental, economic, and health impacts. The scale of effort required to mitigate these impacts also depends in large part on the event's spatial extent. Yet the scale parameter has been largely overlooked in climatological studies of heat waves. Extensive summer cold spells, although not as severe or dangerous in their impacts, can also have important consequences for agriculture and energy demand. They are considered here together with extensive heat waves to provide a more complete picture of variability and trends in summertime temperature extremes over Europe and North America.

We shall see that the regional hot and cold spell indices plainly describe the behavior of regional extreme temperature outbreaks in a way that is complementary to, but fundamentally different from, examinations of temperature magnitudes on local or global scales. One important and robust feature of regional hot and cold spells is their strong low-frequency modulation. Global analyses mask regional decadal variability by averaging over it; local analyses tend to obscure it in higher frequencies. Super-outbreaks of hot and cold air rarely occur counter to prevailing decadal and longer-term trends. Recent trends towards more frequent and extensive hot spells as well as rarer and less extensive cold outbreaks can be explained through a combination of natural multidecadal and anthropogenic influences.

Temperature anomaly magnitude and duration, as well as the spatial extent, of a heat wave all contribute to the severity of its impacts. It is difficult to address all three of these characteristics in one study. Recent studies of heat wave occurrence spurred by the record-breaking 2003 European event focused on the magnitude and duration of *local* temperature anomalies (Beniston, 2004; Beniston and Diaz, 2004; Schar *et al.*, 2004). These studies suggest that the unprecedented temperature anomalies observed at a specific location in connection with the 2003 heat wave were extraordinary with respect to current climate, but were emblematic of expected future conditions. Meehl and Tebaldi (2004), moreover, project that heat waves will become more intense, more frequent, and longer-lasting over Europe and North America in general and specifically at model grid cells around Paris and Chicago. All of these studies examined time slices of several decades in observations and climate models to characterize the effects of anthropogenic climate change projected for an average summer at the end of the twenty-first century. In a rather different study, Stott *et al.* (2004) considered spatially averaged temperatures over the greater Mediterranean region to illustrate that anthropogenic activities have likely increased the risk of an event such as the 2003 European heat wave more than twofold in the current climate, and they projected it to increase 100-fold over the next 40 years. This transition to enhanced heat wave activity over Europe and North America and its dependence on scenarios for political and social action designed to combat global warming, or not, is a major focus of this chapter.

Since large regions (e.g., continents) can and do experience simultaneous subregional hot and cold outbreaks, broad regional temperature averages are not the most appropriate indices for describing variability of regional temperature extremes. In this work, we define an index that *explicitly* reflects the spatial scale of hot and cold outbreaks as well as, although implicitly, their magnitude and duration. The spatial extent of European and North American summertime extreme temperature outbreaks is then considered in the context of decadal and interannual observed variability and coupled model projection of anthropogenic climate change given different scenarios for future emissions and socioeconomic development. Instead of aggregating observed and modeled data in samples of several decades to represent present and future climates as was done in recent studies (Beniston, 2004; Meehl and Tebaldi, 2004; Schar *et al.*, 2004), we present time series of hot and cold spell indices at annual resolution computed for each summer on record – observed and modeled. We then provide a qualitative assessment of the temporal character of spatially extensive temperature extremes over Europe and North America. We consider the extent to which widespread extremes such as the 2003 European event reflect natural climatic variability as opposed to anthropogenic influences. We compare and contrast the recent warming over Europe and North America in the context of their respective regional summer temperature histories and model projections for the future. We discuss the extent to which natural variability is expected to modulate anthropogenic projections of hot and cold extremes over Europe and North America. To better understand the unusual recent summer temperature regime over North America and its likely developments for the near future, we finally focus specifically on the effect of precipitation on regional summer temperatures.

5.2 Regional hot and cold summer indices

We define local heat wave conditions as exhibiting temperatures in the upper 10% of the local climatology over a base period (1950–99). To focus on the spatial extent of heat outbreaks, we construct the regional hot summer index (HSI) by counting the frequency with which each summer (June–July–August, JJA) appears as one of the warmest 10% of summers on the available record at individual locations (stations or grid cells) covering the region of interest. This approach amounts to detecting average summer temperatures warmer than the 90th percentile of the local 1950–99 JJA temperature for all locations over recorded or modeled summers describing a region's climate evolution. The cold summer index (CSI) is constructed similarly for the coldest 10% of summers. Because HSI and CSI (H&CSIs) are computed relative to the local input data, these indices are insensitive to local systematic biases and extremely robust with respect to the nature of the input data used, as long as the data coverage reasonably represents the region of interest. The locally warmest (coldest) summer, by design, does not have a heavier weighting than the second, third, etc., warmest (coldest) summers on a specific record. The indices are, therefore, very robust with respect to outliers as well as to the spatial detail of heat wave patterns, which may be noisy and/or model specific. But the hot and cold summer indices are designed to be highly sensitive to the *spatial scale* of the individual summer's hot and cold air outbreaks. The H&CSIs efficiently detect interannual variability in spatially extensive extreme temperature outbreaks that are long-lived enough to strongly mark local average JJA temperature. Inasmuch as H&CSIs are sensitive to hot and cold summer extreme temperatures, they are shaped by hot and cold spells. Actual hot and cold *spells* can be quantified more precisely from daily data, and this shall be done in future work. All the same, here we regard the summer indices as reflecting strong and persistent extreme temperature outbreaks. This association can fail during summers marked by hot and cold spells following each other locally in time. However, such summers are highly unusual. It is much more common to have hot and cold spells occurring during the same summer and even simultaneously over different parts of a continent (result not shown). A fundamental difference between H&CSIs and standard regional indices constructed by spatially averaging seasonal temperature anomalies can be appreciated by observing the fact that the H&CSIs perform as intended even when different parts of the region experience opposite temperature extremes.

Of course, H&CSIs are sensitive to the spatial scale of the region of interest and to the percentile of the local temperature climatology chosen to define hot and cold extremes. Both regions considered here are large enough to experience significant hot and cold outbreaks in their different subregions in a specific summer, but also compact enough to allow most of their area to be covered by unusually extensive hot or cold extremes. The temporal structure of H&CSI becomes spikier for smaller regions as well as for more extreme temperature thresholds, more saturated for much larger regions and less extreme percentile thresholds. The main conclusions of this study, however, do not change with the choice of, say, a 75% or 98% threshold for HSI. We apply a 90% JJA temperature threshold as a reasonable compromise between spikiness and saturation.

The new and improved Climatic Research Unit (CRU) observational $5° \times 5°$ gridded surface air temperature, CRUTEM2v (i.e., 2 m air temperature, Ta2m), was used to define past heat wave activity. This monthly global land surface temperature record covers the years 1851–2004 and includes variance adjustments due to changing station density within each grid box (Jones *et al.*, 1997, 2001). More information on the CRUTEM2v dataset, hereafter referred to as CRU2, can be found in Jones and Moberg (2003). To avoid using values derived from sparse station records, we used data for 1900 on.

Surface air temperatures from 4 coupled global dynamical climate models (CGCMs) out of the 22 available in the IPCC Fourth Assessment simulations database (www-pcmdi.llnl.gov/ipcc) were analyzed. Because our indices are computed relative to regional climatologies, they downplay individual model biases. And since most models show generally similar features of H&CSI behavior relative to their own climatologies, we show results based on one model that is reasonably close to the average of model projections. The CGCM (Douville *et al.*, 2002) is a fully coupled land–ocean–ice–atmosphere dynamical spectral model developed and run at the Centre National de Recherches Météorologiques (CNRM) of Météo France at the spatial resolution of approximately 2.8°. Four integrations of the CGCM have been analyzed. The "historical" run (1860–1999) was forced with observed greenhouse gas and sulfate aerosol concentrations. The "commit" run (2000–2099) is based on concentrations of these gases fixed at the year-2000 level. The SRES-B1 and A2 projections evolve according to different scenarios for socioeconomic development (Arnell *et al.*, 2004). The B1 is a conservative warming scenario that assumes enlightened action by governments to reduce anthropogenic emissions and population growth, while the A2 is essentially "business as usual."

When truly global observational data are required, we use 2 m air temperatures from the National Centers for Environmental Prediction and National Center for Atmospheric Research (NCEP/NCAR) Reanalysis (Kistler *et al.*, 2001) for the period from 1948 to the present. Although known biases exist in these data (Simmonds *et al.*, 2004), they are the most globally complete and physically consistent data available and are adequate for the purposes of this investigation. Reanalysis has a cold bias in surface air temperature owing to the fact that only upper air temperature observations are assimilated, meaning that land use effects are not incorporated. Also, for the period before the late 1970s, the bias is stronger because fewer observations were available for assimilation. Nonetheless, reanalyzed H&CSIs are well correlated with those derived from CRU2.

The different spatial resolutions of the temperature data resulted in 29, 76, and 115 (70, 186, and 354) grid cells available over Europe (North America) from CRU2, CGCM, and reanalysis data, respectively. The close interannual correspondence between H&CSIs derived from CRU2 and reanalysis,[1] two

---

[1] This correspondence can be seen for Europe in Figure 3.2a.

datasets with vastly different spatial resolutions, provides further evidence of the robustness of our indices with respect to the resolution of the input data.

Near-global gridded station precipitation data (from the Global Historical Climatology Network, GHCN V2) were obtained from the National Climatic Data Center (NCDC); these data consist of observations from 1900 to the present on a 5 × 5 degree grid (www.ncdc.noaa.gov/oa/climate/research/ghcn/ghcngrid_prcp.html#Overview). All gridded data were weighted by cosine of latitude, although this approach does not significantly affect any of the computed indices.

Over North America, where extensive original and homogenized station temperature and precipitation records are available, results derived from the gridded products (i.e., CRU2, GHCN V2, and Reanalysis), were further validated with an extensive station dataset derived from US (Easterling, 2002; NCDC, 2003; Groisman *et al.*, 2004), Mexican (Miranda, 2003), and Canadian (Vincent and Gullett, 1999) networks over North America. In the interest of brevity, we do not show results based on these extensive daily station records here, but we note that the gridded products give essentially the same results. The interesting regional and intraseasonal details that emerge from the station analyses will be presented in future publications.

5.2.1 Europe

Defining Europe by its mid-latitude west-central area as the region situated between 10° W and 25° E, 37° N and 57° N, we first compute the average temperature anomaly from observations and the coupled model relative to their respective 1950–99 climatologies (Figure 5.1a). In contrast to regionally averaged temperatures, which reflect a mixture of magnitude and spatial extent of all seasonal temperature anomalies with coexisting warm and cold anomalies canceling each other, the hot and cold spell indices (H&CSIs, Figure 5.1b) reflect primarily the spatial extent of seasonal warm and cold temperature extremes. Of course, strong coherent anomalies that cover most of the region (i.e., summer 2003) are reflected in both average temperature and in H&CSIs. Regional average temperature, needless to say, is closely correlated with the sum of HSI and CSI, the latter being denoted by negative values as a matter of convention in display. We provide the average temperature anomaly as a reference that shows general temperature tendencies of an entire region, but the H&CSIs provide more detail and will therefore be discussed in greater detail.

The general character of H&CSIs, viewed simultaneously, evolves along with average temperature, but the H&CSI time series are marked with more

(a)

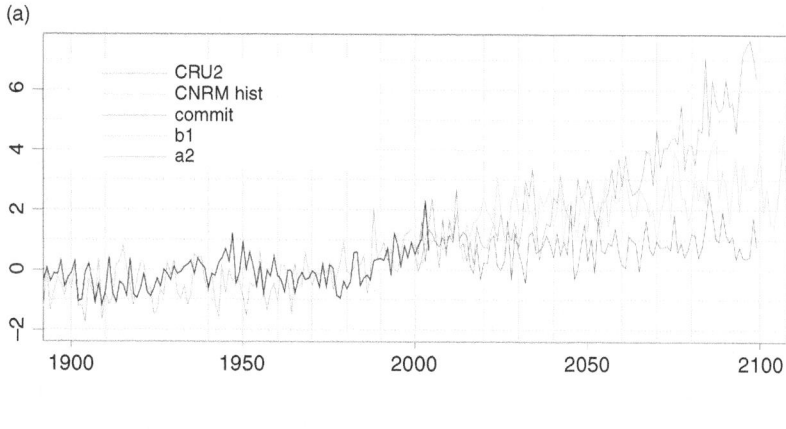

(b)

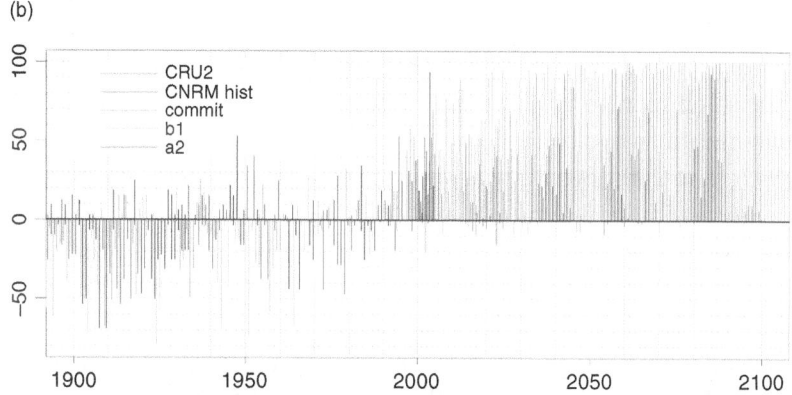

Figure 5.1. (a) European average temperature anomaly relative to the base period 1950–99 from CRU2 observations and the CGCM historical and future scenarios. (b) H&CSIs displayed as percentage of European grid points with summer temperatures above the 90th or below the 10th percentiles of their local summer 1950–99 climatology. HSI (CSI) is displayed in positive (negative) values. By definition, during the base period (delineated by broken orange lines), the mean values are 10% of the area experiencing unusually hot or cold summers. For color version, see plate section.

pronounced multidecadal and interannual variability. Apart from several localized heat waves that covered less than a quarter of the area, Europe was predominantly cold until the early 1940s, with the largest cold extremes, affecting almost 70% of the region, occurring in 1907 and 1909. A general warming ensued and, after a few warmer summers, a heat wave occurred in 1947 that covered more than half of Europe. After that, extremes of both signs became more common. However, Europe continued to experience mostly cold summers in the 1950s, 1960s, and 1970s.

The summer of 1976, when approximately 30% of the area experienced hot spell conditions, stands out as the most extreme heat wave since 1952, while the concurrent cold spell of similar spatial extent seems unremarkable for that period. The 1976 heat wave affected northwestern Europe (Figure 5.2b). It was centered on Great Britain (Green, 1978), where it was the hottest summer on record until 2003 and the cause of much adversity (Subak *et al.*, 2000). However, summer 1976 was actually uncommonly cold over eastern Europe and most of the rest of the northern midlatitudes (Figure 5.2b). The hot and cold spell indices reflect this fact (Figures 5.1b, 5.2a), while the average summer temperature anomaly over the region is close to zero (Figure 5.1a). The case of 1976 accentuates the fact that subregions of a continent or of an entire hemisphere can experience temperature extremes opposite in sign to the prevailing large-scale conditions. It also emphasizes the relative nature of extremes and their impacts viewed in terms of human adaptation to decadal trends. Since the early 1980s, a heat event like that of 1976 was no longer exceptional,[2] with seven hot spells surpassing that of 1976, as well as generally higher mean temperatures. The heat wave of 1994, with over 50% of Europe experiencing heat wave conditions (Figure 5.2a,c), was the most intense event since 1947. The 2003 event[3] was significantly more severe, in terms of both temperature anomaly magnitude and spatial extent (Figures 5.1 and 5.2a,d). Furthermore, no significant cold outbreaks have been experienced since 1993. Viewed in the context of the past century, the heat wave of 2003 is unprecedented. However, it exemplifies the warming trend observed over the last several decades in average summer temperatures as well as in the magnitude and scale of European heat waves (Figures 5.1 and 5.2).

This is part of the warming trend that is manifested globally in the snapshots of Figure 5.2 (b, c, and d), a trend consistent with model projections of anthropogenic warming for Europe and the globe. Even without 2003, European HSIs observed during the past decade indicate the longest and warmest such period on record, a period wholly consistent with the model estimation of warming for anthropogenic forcing fixed at year-2000 levels (the commit run: Figure 5.1a,b).

The coupled model cannot reproduce the observed decadal variability (i.e., the warm late 1940s and early 1950s, the cool late 1970s) – it is not supposed to – but it is able to reproduce the observed warming trend quite well, as do most other coupled models, suggesting an anthropogenic cause to the warming observed since the late 1970s. This being a regional manifestation of a global

---

[2] Except, of course, from a distinctly British viewpoint.
[3] The 2003 summer heat wave consisted of two outbreaks, one in June and a second in August. Most of the adverse impacts occurred during the second outbreak, when hot anomalies rose above the seasonal temperature maximum in August. For the sake of convenience, we refer to the sum effect of these two outbreaks as reflected in JJA average temperature as the summer 2003 heat wave.

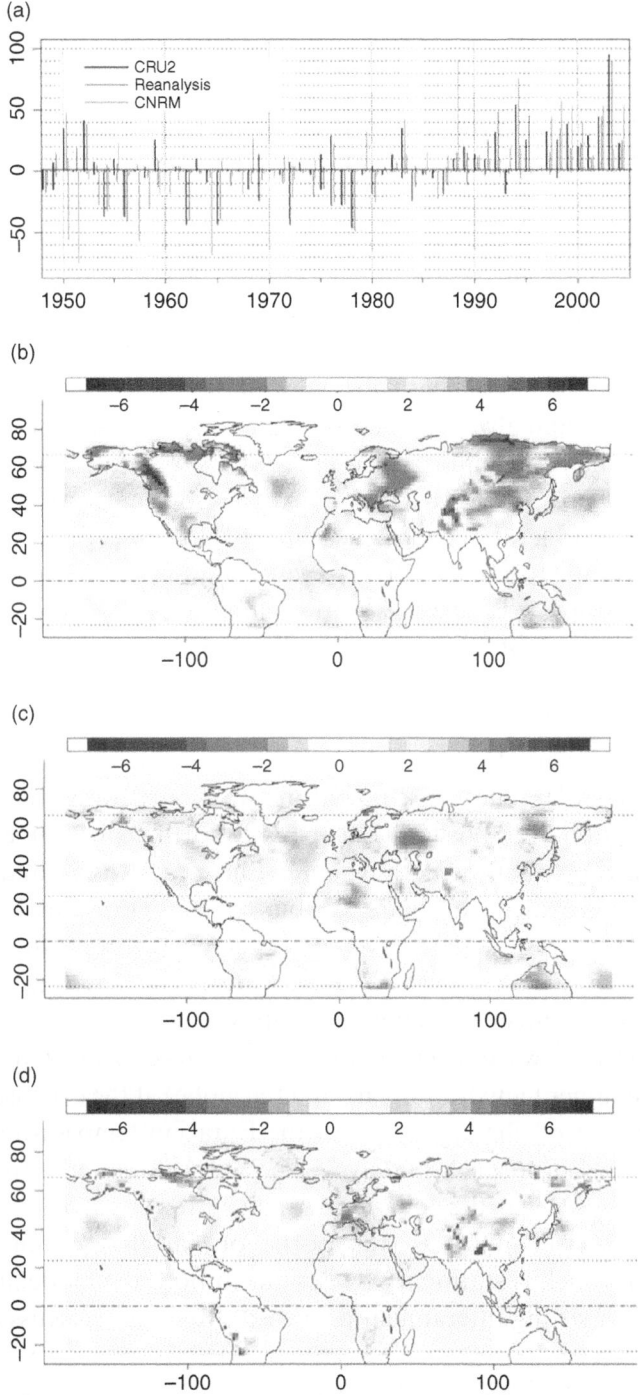

Figure 5.2. (a) H&CSI from CRU2 (black), NCEP/NCAR Reanalysis (red), and CGCM (dashed gray) for the common period 1950–2003. (b) Reanalyzed temperature anomalies for summer 1976, (c) 1994, and (d) 2003. For color version, see plate section.

trend, it is difficult to say to what extent natural decadal variability played a part. The historical model run contains one event of the magnitude of 2003 (model year 1988: Figure 5.1b). The CGCM with forcing fixed at year-2000 levels produced three more events of this general magnitude during the twenty-first century. Aided by natural decadal variability, these events all occur in the model during the same decade, the 2080s. Natural decadal variability masks the difference between the two more realistic scenarios (SRES-A2 and B1) in the first half of the twenty-first century, when a 2003-level average temperature becomes common in both model scenarios. Clear differences in scenarios of anthropogenic forcing become apparent in the second half of the century. After the 2050s, 2003-level heat is exceeded in the milder B1 warming scenario in most summers and always in the A2.[4]

Estimating the probabilities of specific events in various time periods and climate change scenarios for Europe involves several problematic assumptions. We prefer to let the reader qualitatively gauge the danger of extremely hot summers in the future by visually examining Figures 5.1 and 5.2.

To place European heat waves into a global temperature context as well as to validate the CGCM, we correlate the HSI with summer surface air temperatures over the globe derived from the Reanalysis and compare these patterns to those derived from the CGCM's commit (i.e., stationary) run (Figure 5.3). Western European heat waves are seasonally correlated with a hemispheric-scale summer temperature wave structure characterized by in-phase behavior over Western Europe and north-central Siberia and out-of-phase behavior over the North Atlantic, European Russia, and the Russian Far East. This is evident in Figure 5.2, but to see it clearly in Ta2 m correlations with HSI, we should remove the observed warming trend from HSI and Ta2 m. Figure 5.3a shows the correlations with the trend present. This approach has the effect of better illustrating European heat wave activity in the context of a warming planet (notice the mostly positive correlations with boreal summer temperatures around the globe) at the expense of masking especially the out-of-phase portions of the Eurasian temperature wave train associated with European heat waves. The wave train correlation structure becomes obvious when the long-term trend is removed from observations (result not shown), and it is well borne out in the commit model run (Figure 5.3b), which is stationary by design. In both the model and observations, there is a strong interannual propensity for western Russia to be cold

---

[4] According to model results, regional manifestations of global warming should become more evident in the next half century, but differences for varying emissions scenarios (e.g., results of mitigation policies) become discernible only in the second half of the twenty-first century.

(a)

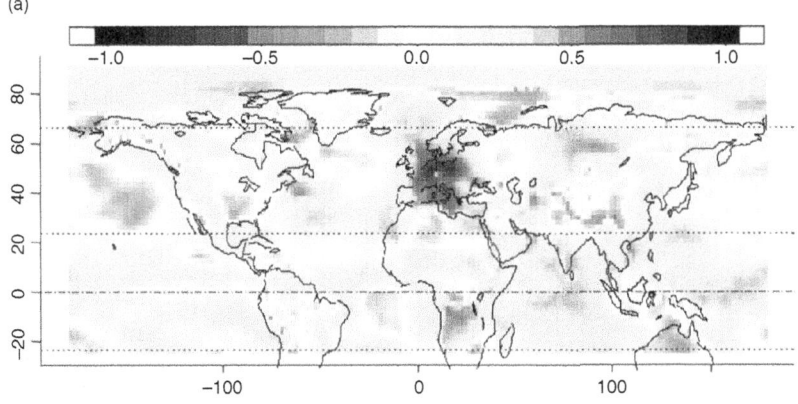

(b)

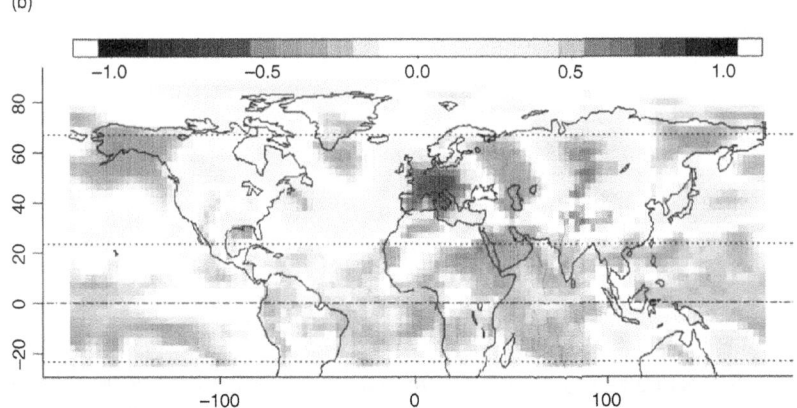

Figure 5.3. Correlation coefficient between Western European HSI and local JJA Ta2 m over the globe in (a) Reanalysis 1948–2003, and (b) the CGCM commit run 2000–99. For color version, see plate section.

during heat wave summers in west-central Europe. Both recent extreme European heat wave summers of 1994 and 2003 were cold in western Russia (e.g., anomalous snowfall occurred in Moscow in June 2003; Levinson and Waple, 2004) and warm over north-central Siberia, thus exhibiting Eurasian summer temperature wave train conditions typical of large European heat waves (Figure 5.2c,d).

5.2.2 *Midlatitude North America*

We limit the study area for North America to the latitudinal band between the Tropic of Cancer and 53° N; north of this band, population and station density becomes sparse, a fact manifested in CRU2 as well as GHCN V2 data gaps

(a)

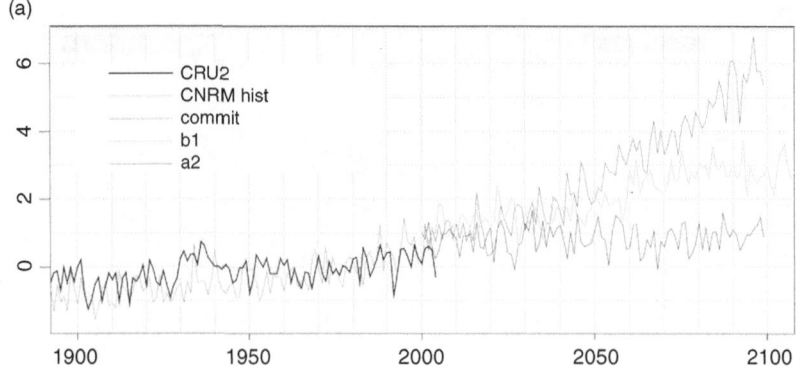

(b)

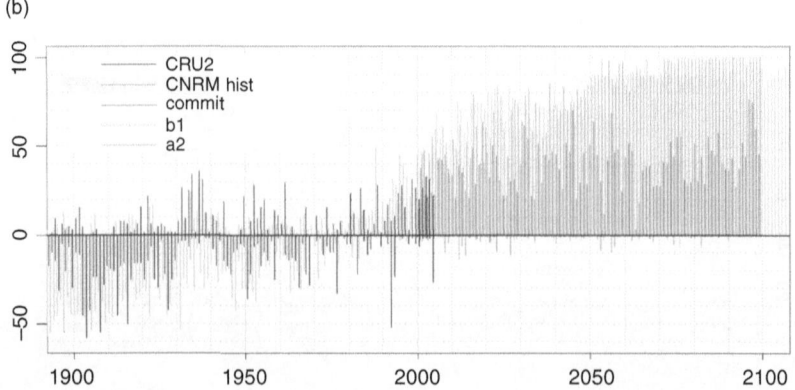

Figure 5.4. Same as Figure 5.1, but for North America; (a) North American average temperature anomaly relative to the base period 1950–99 from CRU2 observations and the CGCM historical and future scenarios. (b) H&CSI displayed as percentage of North American grid points with summer temperatures above the 90th or below the 10th percentiles of their local summer 1950–99 climatology. HSI (CSI) is displayed in positive (negative) values. By definition, during the base period (delineated by broken orange lines), the mean values are 10% of the area experiencing unusually hot or cold summers. For color version, see plate section.

over Canada. North American average temperature and H&CSIs derived from observations and the CGCM are displayed in Figure 5.4.

The first three decades of the twentieth century were the coldest on record, with the summers of 1903, 1907, and 1915 experiencing widespread cold anomalies affecting at least one-half of North America, a scale not reached again until 1992 – the largest summer temperature extreme of the second half of the century. We note that these, as well as smaller-scale recent cold outbreaks (e.g., 1976, 1982, 1993, and 2004), were associated with locally wet summers (result not shown). On the other hand, the 1930s appear to have experienced the warmest and most consistently warm conditions on record, at least until very recently; 7 of 11 (1930–40) summers were more extensively warm (none cool) than expected, and all 11 were warmer than average (see Figure 5.6a for the spatial pattern). The summers of 1934, 1936, and 1937 experienced by far the most severely extensive hot spells of the century (more than 30% of North America was under hot spell conditions). These summers were also exceptionally dry (result not shown for the individual summers; see Figure 5.6b for the 11-year average precipitation anomaly). No summers during the 1930s saw cold outbreaks that even remotely approached their expected spatial extent of 10%.

The 1940s were generally and mildly cool. Temperatures in the 1950s were unusually variable, but the warm summertime temperature anomalies associated with the 1950s drought,[5] although large, were not as spatially extensive, temporally persistent, or exclusive as in the 1930s; they were balanced out by comparably large cold anomalies, a fact reflected in near-zero average temperature anomalies. The 1960s and 1970s saw less variability, with predominantly cool conditions. Temperatures in North America became more variable in the 1980s and 1990s while warming into the start of the twenty-first century. In the midst of general warmth, a cold spell covered 50% of midlatitude North America in 1992, the first time such widespread cold was observed since 1907. The cold summer of 1992 was followed by the mostly cool 1993, both perhaps related to the Mount Pinatubo volcanic eruption[6] (Robock, 2000). But even in the cool summer of 1993 (CSI = 23%), the expected area was hot (HSI = 11%) and the decade that followed (1994–2003) was akin to the 1930s, with 9 out of 10 summers exceeding the mean extent of heat and 3 summers (1998, 2002, and 2003) with hot spell conditions covering more than 30% of the continent for the first time since the 1930s. The last observed summer (2004) was both anomalously warm (HSI = 22%) and cold (CSI = 27%). The recent midwestern heat waves of 1995 and 1999 each resulted in HSI values of under 30%.

The questions as to what global climate conditions are conducive to intense summer heat outbreaks over North America and whether these conditions are reproduced by the model can be addressed by considering maps of correlation coefficients between HSI and local $Ta2\,m$ over the globe (Fig. 5.5). In Reanalysis and the CGCM, spatially extensive hot spells over North America

---

[5] The bulk of the 1950s warmth was coincident with drought, which was mostly concentrated in the fall and winter seasons.

[6] Both these summers, 1992 and 1993, were also anomalously wet over the regions corresponding to the largest observed cooling: Great Plains in 1992, and the northern plains and northwestern United States in 1993 (result not shown). The summer of 1993 was considerably wetter than that of 1992.

(a)

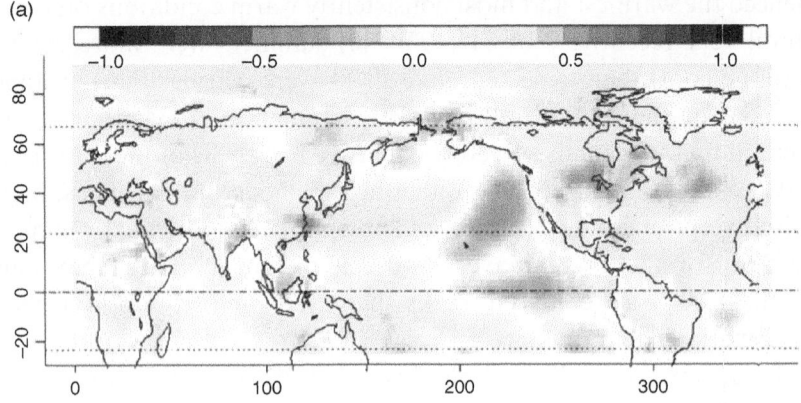

(b)

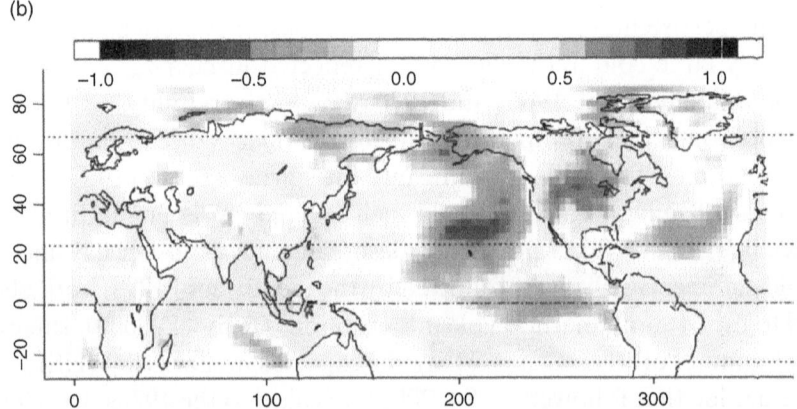

Figure 5.5. Correlation coefficient between North American HSI and local JJA Ta2 m over the globe in (a) Reanalysis, 1948–2003, and (b) the CGCM commit run, 2000–99. For color version, see plate section.

tend to organize preferentially around the midwestern United States,[7] and they tend to be associated with cool temperatures in the northeast midlatitude and tropical eastern Pacific, warm temperatures in the subtropical western Pacific, and a general tendency toward warm temperatures in the tropical and northwestern Atlantic. Other patterns apparent in Reanalysis are characterized by

---

[7] Principal components analysis (PCA) can be used to find coherent patterns of variability that optimally explain the variance of a spatially correlated field; e.g., summer temperature. Principal components analysis applied to CRU2 (as well as to Reanalysis and CGCM) data results in a similar Midwest/Great Plains pattern of summer heat variability (results not shown). Observed H&CSIs are correlated with this main PC mode (correlation = 0.9). This leading mode explains 43% of summer temperature variance (PC2 explains 18%). In CRU2, large peaks (dips) in HSI (CSI) tend to historically line up with those of PC1, indicating that hot (cold) summers tend to have spatially consistent temperature anomalies. The 1930s heat closely followed this pattern. Most other anomalous summers throughout the record did too. Recently, however, the spatial pattern of warming exhibited a rather different structure (see below).

mostly weak positive correlations in patches around the globe, reflecting the general recent global warming. Except for the propensity towards positive correlations dictated by the observed warming trend, the main patterns are reproduced by the model's "commit" climate run, which is stationary by design. The association between HSI and the preferred heat wave location appears to be weakened in Reanalysis, probably because the recent warming over North America did not manifest itself in the preferred heat wave region of the midwestern and north-central United States (see below). Other differences may be due in part to the larger noise level in the shorter reanalyzed climatic sample. As an aside, we have also examined multidecadal changes in global correlation patterns. Modeled correlations computed for consecutive 50-year periods by using the model's historical as well as B1 and A2 runs (figures not shown) plainly resemble the main patterns of Figure 5.5. They also indicate a consistent change in the characteristic pattern of the HSI amounting to a progressive migration of North American heat outbreaks toward the north of the study region. In summary, North American hot summers tend to have a preferred spatial signature that is typically expressed in conjunction with specific sea surface temperature (SST) patterns, and the model appears able to realistically reproduce the observed dynamic structure.

The marine temperature patterns associated with extensive North American summer heat spells are analogous to the anomalous sea surface temperature "forcings" for the 1930s Dust Bowl drought as outlined by Schubert *et al.* (2004). It is also well known that large-scale summer heat can be caused or exacerbated by atmospheric drought through soil moisture deficit (see Alfaro *et al.*, 2006). This was the case in the hot and dry 1930s, when precipitation, especially in summer, was scarce[8] (see Figure 5.6a, b). Although more extreme and persistent than other hot summers on record, the spatial structure of the 1930s heat followed the typical pattern of hot summers over North America (result of PCA not shown). However, the most recent period of summer heat observed over North America, although on the scale approaching that of the 1930s, was not associated either with prolonged large-scale drought or with an anomalously cold tropical Pacific.

In fact, the most recent warm period, 1994–2004, was very much unlike the 1930s (Figure 5.6c). The central and eastern United States were mostly wet over this period[9] (Figure 5.6d). Despite this recent wetness, the central and eastern United States have not cooled as would be naturally expected. Such expectation

---

[8] Drought reconstructions show that the 1930s Dust Bowl drought was the most severe and widespread such event to strike the United States since 1700 (Cook *et al.*, 1999).

[9] In fact, the entire country has been generally wet over the last quarter century.

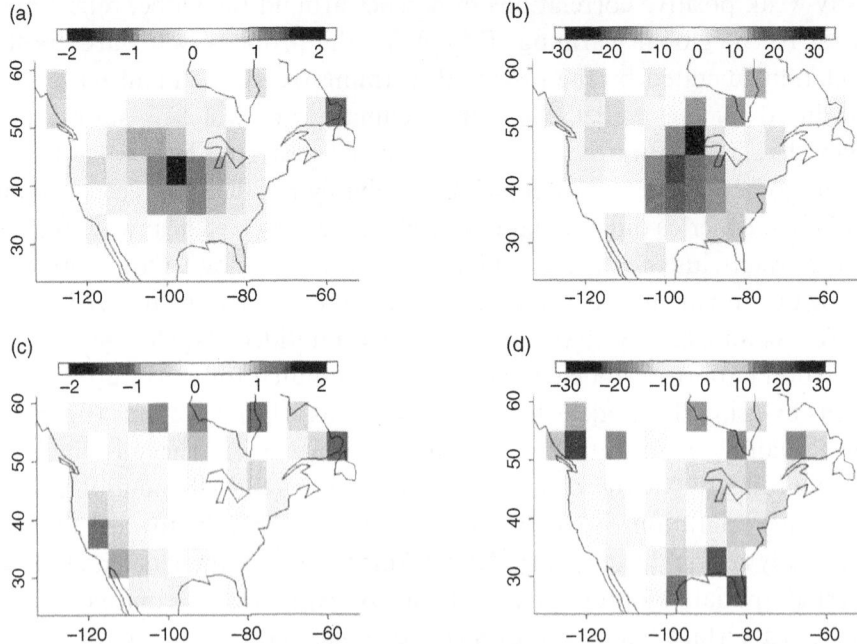

Figure 5.6. Summer temperature (a, c) and precipitation (b, d) anomalies for two 11-year periods: 1930–1940 (a, b) and the most recent period of record, 1994–2004 (c, d). Notice the reversed color scale on the precipitation plots. Canadian precipitation records suffer from missing data at the end of the record, accounting for the visible discontinuity along the US–Canada border in (d). For color version, see plate section.

is supported by Figure 5.7a, which shows observed local correlations between summer temperatures and precipitation. Meanwhile, much of the recent warming over midlatitude North America has been due to warmer summers all around this central and eastern US wet spot (Figure 5.6c,d). Recent summer warmth over California and the rest of the mountainous West is probably related to the strong warming observed in the West during winter and spring, which resulted in changes in the snow-to-rain ratio (Knowles *et al.*, 2006) as well as the spring's earlier arrival and related changes in surface hydrology (Cayan *et al.*, 2001), drying the soil in summer without appreciable changes in precipitation amounts. These same changes in summer temperatures and hydrology have also resulted in increased wildfire activity over the western mountains (Westerling *et al.*, 2006). The western summer warming (Figure 5.6c) is reflected in increasing values of the North American HSI, but this pattern is generally atypical of extensive North American summer heat.

Strong and extensive Canadian warming, apparent in Figure 5.6c, did not strongly affect the earlier results because only far southern Canadian data, for

(a)

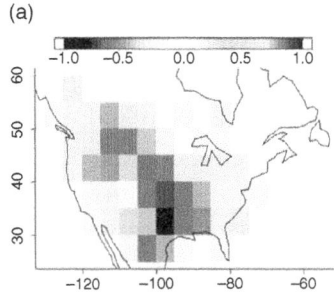

(b)

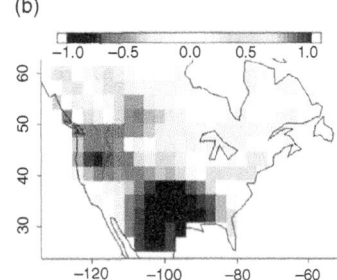

Figure 5.7. Correlation coefficient between local June–August average temperature and precipitation in (a) observations (GHCN V2 and CRU2 evaluated over the period of record, 1900–2004, or slightly shorter based on data availability for the individual grids) and (b) the model (the commit run, 100 years). For color version, see plate section.

reasons of data quality given above, were included in the North American index calculation. Had more extensive Canadian data been included, the recent warming would have appeared larger. This observation is consistent with the model result projecting heat waves to spread progressively towards the north (not shown).

5.3 Role of precipitation

There is a well-known interaction between soil moisture and surface level temperature. On the one hand, anomalous prolonged warmth can desiccate the soil; on the other, the drier the soil, the less energy is required to heat it. Precipitation deficit can, therefore, enhance regional heat wave activity and partially account for multidecadal timescales in extreme summer heat outbreaks that can temporarily either promote or dampen the effects of global warming. Summer heat and precipitation are, of course, dynamically linked, for the anticyclonic circulation that produces heat waves also precludes precipitation. In regions that typically receive much summer precipitation, it is therefore difficult to isolate the effect of contemporaneous (summer) precipitation on temperature. The effect of antecedent precipitation on summer temperature through soil moisture accumulation is typically small and may influence heat waves, primarily in the early summer. In regions where summer precipitation amounts are small, and especially where winter precipitation is stored as snowpack (e.g., California and the mountainous Southwest), the delayed precipitation–soil moisture effect can be much stronger.

We have examined the contemporaneous and lagged relationships between local summer temperature and precipitation. Figure 5.7 shows that the two are strongly anticorrelated over North America,[10] particularly along the Front Range of the Rockies and to the southeast over the Great Plains and over a great part of the Midwest and the Gulf Coast, the Mexican Plateau, and the mountainous northwestern United States. We have also regressed precipitation out of local summer temperature and compared resulting regionally averaged temperatures as well as H&CSI with raw values discussed above over both Europe and North America. These results are not shown, but they indicate that precipitation deficit appears to have played a critical role in enhancing the severity of Europe's 2003 heat wave as well as the Dust Bowl and other observed extremes. However, the bulk of the recent observed warming trend over Europe is not associated with drying.[11] Moreover, the area that traditionally experiences large-scale heat waves, the Midwest and Great Plains region of the United States, has remained close to normal (Figure 5.6c). In this area and to the south and west of it, summer temperature is strongly anticorrelated with precipitation (Figure 5.7a). It is important to emphasize that this has been true historically on interannual as well as longer timescales, but lately, the local anticorrelation on decadal time-scales has disappeared. Although the central/eastern United States has been unusually wet in recent decades, no significant long-term cooling has occurred there. Rather, normal conditions have prevailed in this region recently. The laws of thermodynamics require us to assume that there has been a concurrent warming that has offset the cooling expected in the central and eastern United States in response to increased wetness. Giving further support to this inference, the rest of North America has most certainly warmed. If this background warming persists or continues, as we can expect from model results (see Figure 5.4 as well as results of numerous other modeling studies), we can reasonably expect that the end of the current wet spell should be accompanied by stronger and more extensive summer warmth over the central and eastern United States.

As for the model, the regions of largest projected midlatitude continental summer warming in both the A2 and B1 scenarios are those with projected regional decreases in precipitation (result not shown). This is not surprising. The model is certainly able to reproduce the spatial signature of the local summertime precipitation–temperature relationship, albeit with somewhat stronger than observed coupling (Figure 5.7b).

---

[10] Europe presents a similar picture, but the anticorrelation is somewhat weaker than it is for North America.

[11] Rather, it appears to be associated with the enhanced greenhouse effect of water vapor (Philipona *et al.*, 2005).

5.4 Summary and conclusions

Examination of hot and cold summer indices over continental-scale regions suggests that in practically every summer there is a temperature extreme somewhere (Figures 5.1b and 5.4b). The European HSI shows the 2003 event to be the most intense and widespread since the beginning of record. According to our definition, more than 90% of western and central Europe experienced hot spell conditions in 2003. At many European locations, 2003 was the hottest summer on record, probably the hottest in over 500 years (Luterbacher *et al.*, 2004), with average summer temperature anomalies reaching 5 °C (Figure 5.2d). This intensely and extensively hot summer was consistent with the multidecadal warming trend towards more frequent and extensive European heat waves apparent since the late 1970s (Figure 5.1b). Coupled model projections show a combination of natural and anthropogenic influences, with their relative magnitudes being a function of the particular scenario assumed for global socioeconomic development: i.e., resolute action or inaction on the part of nations. For now, Europe appears to have had an early warning in 2003 of conditions projected for the second half of the twenty-first century, assuming inaction. This conclusion is consistent with those of other recent studies (Beniston, 2004; Beniston and Diaz, 2004; Meehl and Tebaldi, 2004; Schar *et al.*, 2004; Stott *et al.*, 2004). Drought certainly enhanced the severity of the 2003 event (Levinson and Waple, 2004). Apart from this, drought does not appear to have been a general factor in the recently observed warming trend. Fresh observational evidence suggests that much of the recent warming over Europe, at least around the Alps, has been caused by the enhanced greenhouse effect due to water vapor feedback (Philipona *et al.*, 2005) – water vapor being the most important and variable greenhouse gas. For now, Europe is warming at least as much and as fast as predicted by the climate model.

North America, while also warming, has not yet felt the feasible level of heat projected for the current stage of anthropogenic climate change. Interestingly, the current warm period, although at least in spatial extent comparable with the 1930s Dust Bowl, does not involve the same region of North America and is not associated with severe drought; it is, therefore, very much unlike the similar-scale 1930s warming (and other more common warm events). The recent summertime warming over North America has occurred notably in the mountainous West, where it is consistent with observed hydrological changes initiated by warming trends in the winter and spring. Specifically, the western summertime warming occurred alongside a strong observed trend towards warmer winters and springs and earlier snowmelt (Cayan *et al.*, 2001), as well as decreasing snow/rainfall ratios (Knowles *et al.*, 2006). The hydrological changes initiated by these trends in winter and spring result in drying of the soil into the summertime and can partially explain the western summertime warming. It is possible that European-Alpine-type water vapor feedback processes (see Philipona *et al.*, 2005) may also explain a part of this western summertime warming. Supporting such a possibility, as well as giving further verification of a large-scale environmental change, is the fact that North American warming also has a broad Canadian footprint (Figure 5.6c). However, Canadian data were mostly excluded from index calculations here.

The only portion of North America that has not warmed lately is the central and eastern United States, a preferred area for heat wave occurrence and an area where summer temperature and precipitation are strongly anticorrelated; i.e., this region has high drought-related heat wave potential. The anomalous recent wetness over this region would be normally associated with cool summers, but the region has experienced average temperatures recently. These observations point to global warming as a likely cause for the *lack* of cooling over the central and eastern United States over the last couple of decades. In other words, natural decadal variability associated with precipitation appears to be at work taking the edge off the anthropogenic warming over the central and eastern United States. This means that the next large-scale Great Plains summer drought will likely be associated with warming exceeding that of the 1930s in both magnitude and spatial extent. For the same reason, a megadrought of 1930s intensity and spatial extent no longer appears to be required to produce heat wave activity on the scale of the Dust Bowl. A more pedestrian drought will do.

The CGCM driven with anthropogenic forcing reproduces well the warming observed over Europe and North America. It also reproduces well the coupling of summertime temperature and precipitation. Model results interpreted in light of the observations suggest that, even at anthropogenic output fixed at current levels, we can expect much stronger and more widespread heat waves than have been observed up to now in North America as soon as natural decadal variability (e.g., precipitation) turns to conspire with, or at least stops counteracting, the anthropogenic signal. The model also provides variants of likely further evolution of summer heat. The A2 and B1 warming scenarios are indistinguishable from each other in the next several decades mostly because each model run is strongly modulated by its own natural decadal variability. However, by the middle of the century, the two scenarios clearly diverge, with the B1 stabilizing at a temperature anomaly of about 3 °C and HSI between 75% and 98%, while the A2 saturates at HSI = 100% by about 2070 with temperatures continuing to rise.

Our results agree with recent studies in general, and they complement these studies by providing a time-evolving view of regional heat wave activity in individual summers that naturally emphasizes higher-frequency variability on top of the anthropogenic trend. Moreover, our explicit focus on the heat wave's spatial extent emphasizes the European 2003 heat wave type – as well as that of 1934 and 1936, the Dust Bowl's defining years – over much smaller-scale heat waves (such as the more recent Midwestern heat waves of 1995 and 1999). Our results further identify the central, midwestern, and eastern United States as the regions most at risk of "surprise" intensification of summer heat wave activity. These regions are traditionally prone to heat waves, but they have been cooled off recently by unusually wet conditions, which cannot be expected to persist much longer.

Over North America, a spatially extensive heat wave of the magnitude of Europe 2003 has not occurred in observed history. However, the probability of such an event may be significant and increasing. The observed current level of warming agrees with the cooler decades projected by the model run with anthropogenic forcing fixed at current levels. However, even with this fixed forcing, 22% of the projected heat waves cover over half of North America, a level heretofore not reached in observations. Realistically, this level and spatial extent of summer heat, were it to occur in the near future and in its preferred location, will be coupled with drought. It is clear that if such extensively hot and dry summers were to occur in reality, especially if they were unanticipated, they would produce adverse consequences for North America. Observational and model results both suggest, furthermore, that such extremes exhibit natural cycles and tend to congregate in decadal sequences of hot summers.

Public perception of climate variability and change is strongly influenced by seasonal extremes. For example, the summers of 1976–78 experienced extremely cold conditions over widespread northern midlatitude regions even when considered in the context of the 1960s and 1970s, two consistently cold decades. It is not surprising, therefore, that three decades ago, the actual concern was global cooling (Kukla *et al.*, 1977), although no plausible mechanisms for such cooling were identified. Timescales are easily confused, however, and, at least in nonscientific literature, this strong decadal variation was at the time widely taken for the beginning of an ice age (Ponte, 1976). Conversely, the last two decades of the twentieth century experienced a mean warming trend globally and over the midlatitude Northern Hemisphere unprecedented in recent centuries (Mann *et al.*, 1998; Moberg *et al.*, 2005), a trend that continues strongly up to the present (Hansen *et al.*, 2006; Jones and Palutikof, 2006). This recent warming was punctuated by increasing frequencies of large-scale regional hot summer outbreaks and a decrease in the frequencies of cold summers. Plausible mechanisms for such a trend do exist and are exemplified by the IPCC socioeconomic and emissions scenarios (Arnell *et al.*, 2004). Further anthropogenic warming accompanied by more frequent, intense, and extensive heat waves is projected, but, owing to natural decadal-scale climate variability, it appears that regional effects of global policy action (e.g., "business as usual" vs. "enlightened management" scenarios) may not be detectable for several decades ahead. Of course, although the probability of regional-scale cold outbreaks is expected to diminish over this century, they are still possible and will occur with a generally decreasing frequency and spatial extent. However, cold outbreaks will certainly occur frequently enough to spur outbreaks of regional skepticism regarding the nature and causes of climatic change for several decades to come.

Acknowledgments

The original ideas and research presented here were initiated during Gershunov's visit to CNRM in the summer of 2004. This visit was supported by Météo France through the visiting scholar program. We thank Sophie Tyteca and Mary Tyree for help with data acquisition and processing. Partial funding was provided by the California Climate Change Center, sponsored by the California Energy Commission's Public Interest Energy Research Program and by the NOAA Office of Global Programs, under the California Applications Program.

References

Alfaro, E., Gershunov, A., and Cayan, D. R. (2006). Prediction of summer maximum and minimum temperature over the central and western United States: the role of soil moisture and sea surface temperature. *Journal of Climate*, **19**, 1407–27.
Arnell, N. W., Livermore, M. J. L., Kovats, S., *et al.* (2004). Climate and socio-economic scenarios for global-scale climate change impacts assessments: characterising the SRES storylines. *Global Environmental Change: Human and Policy Dimensions*, **14**, 3–20.
Ballester, F., Michelozzi, P., and Iniguez, C. (2003). Weather, climate, and public health. *Journal of Epidemiology and Community Health*, **57**, 759–60.
Beniston, M. (2004). The 2003 heat wave in Europe: a shape of things to come? An analysis based on Swiss climatological data and model simulations. *Geophysical Research Letters*, **31**, L02202, doi:10.1029/2003GL018857.
Beniston, M., and Diaz, H. F. (2004). The 2003 heat wave as an example of summers in a greenhouse climate? Observations and climate model simulations for Basel, Switzerland. *Global and Planetary Change*, **44**, 73–81.
Cayan, D. R., Kammerdiener, S. A., Dettinger, M. D., Caprio, J. M., and Peterson, D.H. (2001). Changes in the onset of spring in the western United States. *Bulletin of the American Meteorological Society*, **82**, 399–415.

Cook, E. R., Meko, D. M., Stahle, D. W., *et al.* (1999). Drought reconstructions for the continental United States. *Journal of Climate*, **12**, 1145–62.

Dhainaut, J. F., Claessens, Y. E., Ginsburg, C., and Riou, B. (2004). Unprecedented heat-related deaths during the 2003 heat wave in Paris: consequences on emergency departments. *Critical Care*, **8**, 1–2.

Douville, H., Chauvin, F., Royer, J. -F., Salas-Mélia, S., and Tyteca, S. (2002). Sensitivity of the hydrological cycle to increasing amounts of greenhouse gases and aerosols. *Climate Dynamics*, **20**, 45–68.

Easterling, D. R. (2002). Recent changes in frost days and the frost-free season in the United States. *Bulletin of the American Meteorological Society*, **83**, 1327–32.

Green, F. H. W. (1978). Exceptional heat-wave of 23 June to 8 July 1976. *Meteorological Magazine*, **107**, 99–100.

Groisman, P. Y., Knight, R. W., Karl, T. R., *et al.* (2004). Contemporary changes of the hydrological cycle over the contiguous United States: trends derived from in-situ observations. *Journal of Hydrometeorology*, **5**, 64–85.

Hansen, J., Ruedy, R., Sato, M., and Lo, K. (2006). Global temperature trends: 2005 summation. (http://data.giss.nasa.gov/gistemp/2005/).

Jones, P. D., and Moberg, A. (2003). Hemispheric and large-scale surface air temperature variations: an extensive revision and an update to 2001. *Journal of Climate*, **16**, 206–23.

Jones, P. D., and Palutikof, J. (2006). Global temperature record. (www.cru.uea.ac.uk/cru/info/warming/).

Jones, P.D., Osborn, T.J. and Briffa, K.R. (1997). Estimating sampling errors in large-scale temperature averages. *Journal of Climate*, **10**, 2548–68.

Jones, P. D., Osborn, T. J., Briffa, K. R., *et al.* (2001). Adjusting for sampling density in grid-box land and ocean surface temperature time series. *Journal of Geophysical Research*, **106**, 3371–80.

Kistler, R., Kalnay, E., Collins, W., *et al.* (2001). The NCEP-NCAR 50-year reanalysis: monthly means CD-ROM and documentation. *Bulletin of the American Meteorological Society*, **82**, 247–67.

Knowles, N., Dettinger, M. D., and Cayan, D. R. (2006). Trends in snowfall versus rainfall in the western United States. *Journal of Climate*, **19**, 4545–59.

Kukla, G. J., Angell, J. K., Korshover, J., *et al.* (1977). New data on climatic trends. *Nature*, **270**, 573–80.

Levinson, D. H., and Waple, A. M. (2004). State of climate in 2003. *Bulletin of the American Meteorological Society*, **85**, 1–72.

Luterbacher, J., Dietrich, D., Xoplaki, E., Grosjean, M., and Wanner, H. (2004). European seasonal and annual temperature variability, trends, and extremes since 1500. *Science*, **303**, 1499–503.

Macfarlane, A., and Waller, R.E. (1976). Short-term increases in mortality during heat waves. *Nature*, **264**, 434–6.

Mann, M. E., Bradley, R. S., and Hughes, M. K. (1998). Global-scale temperature patterns and climate forcing over the past six centuries. *Nature*, **392**, 779–87.

Meehl, G. A., and Tebaldi, C. (2004). More intense, more frequent, and longer lasting heat waves in the twenty-first century. *Science*, **305**, 994–7.

Miranda, S. (2003). Actualizacion de la base de datos ERIC II. Final report of the project TH-0226, Instituto Mexicano de Tecnología y Agua (IMTA) internal reports.

Moberg A., Sonechkin, D. M., Holmgren, K., Datsenko, N. M., and Karlén, W. (2005). Highly variable Northern Hemisphere temperatures reconstructed from low- and high-resolution proxy data. *Nature*, **433**, 613–17.

National Climatic Data Center (NCDC). (2003). Data documentation for data set 3200 (DSI-3200), Surface land daily cooperative summary of the day. Asheville, North Carolina: National Climatic Data Center, 36 pp. (Available online at www.ncdc.noaa.gov/pub/data/documentlibrary/tddoc/td3200.pdf.)

Palecki, M. A., Changnon, S. A., and Kunkel, K. E. (2001). The nature and impacts of the July 1999 heat wave in the midwestern United States: learning from the lessons of 1995. *Bulletin of the American Meteorological Society*, **82**, 1353–67.

Philipona, R., Dürr, B., Ohmura, A., and Ruckstuhl, C. (2005). Anthropogenic greenhouse forcing and strong water vapor feedback increase temperature in Europe. *Geophysical Research Letters*, **32**, L19809, doi:10.1029/2005GL023624.

Ponte, L. (1976). *The Cooling*. Englewood Cliffs, NJ: Prentice-Hall.

Robock, A. (2000). Volcanic eruptions and climate. *Reviews of Geophysics*, **38**, 191–219.

Schar, C., Vidale, P. L., Luthi, D., *et al.* (2004). The role of increasing temperature variability in European summer heat waves. *Nature*, **427**, 332–6.

Schubert, S. D., Suarez, M. J., Pegion, P. J., Koster, R. D., and Bacmeister, J. T. (2004). On the cause of the 1930s Dust Bowl. *Science*, **303**, 1855–9.

Sheridan, S. C., and Kalkstein, L. S. (2004). Progress in Heat Watch–Warning System technology. *Bulletin of the American Meteorological Society*, **85**, 1931–41.

Simmonds, A. J., Jones, P. D. da Costa Bechtold, V., *et al.* (2004). Comparison of trends and variability in CRU, RTA-40 and NCEP/NCAR analyses of monthly-mean surface air temperature. ERA-40 Project Report Series No. 18.

Stott, P. A., Stone, D. A., and Allen, M. R. (2004). Human contribution to the European heat wave of 2003. *Nature*, **432**, 610–13.

Subak, S., Palutikof, J. P., Agnew, M. D., *et al.* (2000). The impact of the anomalous weather of 1995 on the U.K. economy. *Climatic Change*, **44**, 1–26.

Vincent, L. A., and Gullett, D. W. (1999). Canadian historical and homogeneous temperature datasets for climate change analyses. *International Journal of Climatology*, **19**, 1375–88.

Westerling, A. L., Hidalgo, H. G., Cayan, D. R., and Swetnam, T. W. (2006). Warming and earlier spring increase western U.S. forest wildfire activity. *Science*, **313**, 940–3.

Zell, R. (2004). Global climate change and the emergence/re-emergence of infectious diseases. *International Journal of Medical Microbiology*, **293**, 16–26.

6

Beyond mean climate change: what climate models tell us about future climate extremes

CLAUDIA TEBALDI AND GERALD A. MEEHL

Condensed summary

Atmosphere–ocean general circulation models (AOGCMs) necessarily have a limited ability to simulate extreme phenomena, due to their finite – and currently still relatively coarse – resolution. Nevertheless, recent studies have analyzed output from AOGCMs in order to assess their ability to simulate current extremes, mostly temperature-related, in order to infer what the future may bring, under scenarios of continuing and increasing greenhouse gas emissions.

We present results of analyzing heat wave changes as projected by the Parallel Climate Model (PCM; National Center for Atmospheric Research/US Department of Energy [NCAR/DOE]) under a "business as usual" emission scenario, pointing at future increases in frequency, duration, and intensity of heat waves. We also describe a broader study of extreme indicators from multiple AOGCMs that contributed their output as part of the activities Fourth Assessment Report of the Intergovernmental Panel on Climate Change (IPCC).

We looked at ten different indicators, describing aspects of extreme temperature and precipitation events, for current climate and for the rest of the twenty-first century, under a range of emissions scenarios. We find strong and consistent signals of changes towards warmer temperature extremes, and an overall agreement of the different models on the intensification of precipitation, particularly in the high latitudes of the Northern Hemisphere.

6.1 Introduction

Atmosphere–ocean general circulation models (AOGCMs) are the principal tools at our disposal when we seek to characterize projections of future

Climate Extremes and Society, ed. H. F. Diaz and R. J. Murnane. Published by Cambridge University Press. © Cambridge University Press 2008.

climate. Many studies in the past decade have addressed projected changes in global average temperature and precipitation, and other general climate indicators. In the past few years, increasingly accurate and more reliable climate models (Duffy *et al.*, 2003; Govindasamy *et al.*, 2003) have let us ask, and answer, more specific questions with a more direct relevance for impacts on society and ecosystems at regional scales. In particular, extreme events have of late received much attention, and rightly so, as they carry the most devastating consequences (Kunkel *et al.*, 1999; Easterling *et al.*, 2000; Meehl *et al.*, 2000), as the heat waves in Chicago in July 1995 and in Europe in August 2003, and the floods and hurricanes in 2005 have clearly demonstrated.

In this chapter we summarize our recent work that has used AOGCM projections to infer changes in extremes under future scenarios of continuing or increasing greenhouse gas emissions. Analyses of future changes in heat wave intensity, duration, and frequency; in precipitation intensity; and in several other indices quantifying extreme behavior of temperature and precipitation will be presented.

Two approaches to the analysis of AOGCM projections of extremes are used in this work. For the heat wave analysis, we use an ensemble of simulations from a single model, the Parallel Climate Model (PCM; National Center for Atmospheric Research/US Department of Energy [NCAR/DOE]; Washington *et al.*, 2000), run under historical natural and anthropogenic forcings. These forcings are meant to replicate at best the observed climate characteristics in the twentieth century, and a "business as usual" (BAU) scenario for the twenty-first century, assuming no substantial change in policies regulating anthropogenic greenhouse gas emissions. For this analysis, we have performed an in-depth verification of the model's performance in replicating not only the observed characteristics of heat waves, but also the atmospheric dynamics conducive to their onset and duration.

For precipitation intensity and the other indices, we perform an inter-model comparison of current global average tendencies and future global average and geographical tendencies and their statistical significance on the basis of a multi-model ensemble and three different scenarios for the future. Even in the absence of an in-depth model validation exercise, we consider it informative to compare different models and summarize their agreement and disagreement to portray a multifaceted picture of the behavior of future extremes.

The work presented in this chapter is based on Meehl and Tebaldi (2004), Meehl *et al.* (2005), and Tebaldi *et al.* (2007), but a good portion of the detailed results shown here has not been published before.

6.2 Heat waves will become more intense, longer, and more frequent

6.2.1 Defining a heat wave

In lieu of a standard definition of heat wave, we chose to use two real episodes as templates in order to extract relevant characteristics that have been proved to be associated with large negative impacts on human health, the economy, and the environment. Thus we considered two devastating episodes: Chicago in July 1995, associated with at least 700 deaths, and Paris in August 2003, when thousands of deaths were attributed to the extreme heat that afflicted most of Western Europe. The most serious impacts of the Chicago heat wave were identified with a stretch of three days during which the minimum temperature failed to drop significantly at all, thus depriving the population of the city of any of the relief usually found in the nighttime hours. During the heat wave that hit Europe in the summer of 2003, the temperature stayed above extreme thresholds, on average, for over 2 weeks. Accordingly, we used two definitions of heat wave from the literature that fit the observed characteristics:

- From Karl and Knight (1997), we define a heat wave as the hottest spell of three consecutive days every summer, in terms of average minimum temperature.
- From Huth *et al.* (2000), we fix two location-specific thresholds, defined as the 81.5 and the 97.5 percentiles of the climatology of maximum temperature at that location (e.g., Chicago or Paris). Then we define a heat wave as the longest spell of consecutive days satisfying three conditions:
  - Every day, the maximum temperature must be above the lower threshold.
  - At least three days within the period have a maximum temperature above the higher threshold.
  - The average maximum temperature of the entire period is above the higher threshold.

Both definitions are applied to daily data for the warm season, defined as May through October in the Northern Hemisphere and as November through April in the Southern Hemisphere. Figure 6.1 provides an example of a time series of maximum temperature satisfying these three conditions. In the figure, two black lines indicate the two thresholds. The average temperature of the time series is nearly the same as the higher threshold. All the daily values (indicated by small open circles, and connected for display purposes by the solid line) are above the lower threshold and at least three (in this case more) are above the higher line. The average value coincides with the higher threshold (it could be higher in certain cases).

Max Temperature

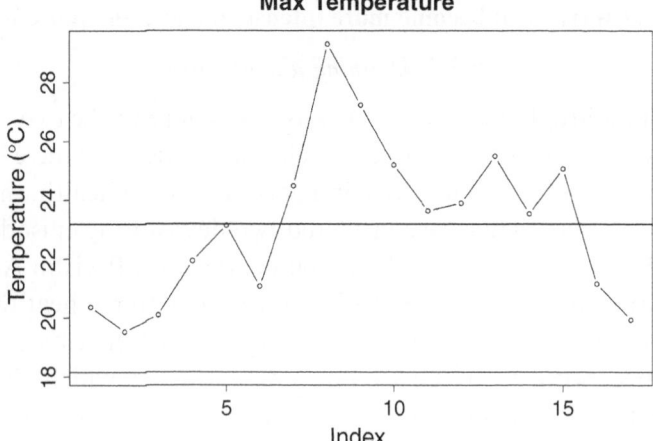

Figure 6.1. Example of a time series satisfying the three conditions defining the second type of heat wave (from Huth *et al.*, 2000).

Clearly, by the first definition, we are fixing the duration as three days and the frequency as once per year (i.e., summer), but we will be comparing the intensity of this first type of heat wave over large geographical regions at different times. We define intensity of the first type of heat wave as the average minimum temperature in those three days (according to the first definition, which under-lines the harmful effects of high nighttime temperatures). On the other hand, the second definition leaves open the possibility of recording more than one heat wave each year, and of durations much longer than three days (which represents only the minimum duration for this second type of heat wave).

6.2.2 Worst three-day events

In Meehl and Tebaldi (2004), we presented a detailed analysis of the PCM simulations of heat waves, defined as the warmest three-day temperature events of each summer (as a reminder, defined as May–October), for two large geographical areas, North America and Europe. After computing aver-age intensity over the areas for the period 1961–90 from the NCEP/NCAR reanalysis record and from the PCM historical ensemble, we found that the model was able to reproduce not only the geographical pattern of the observed events, but also their intensities, in terms of absolute values of average mini-mum temperatures during such events, over the 30 years used as a baseline. We then went on to show average changes in intensity (i.e., in the mean values of minimum temperature over the worst three days of the summer) between the last 20 years of the twenty-first century and the baseline period of 1961–90, as

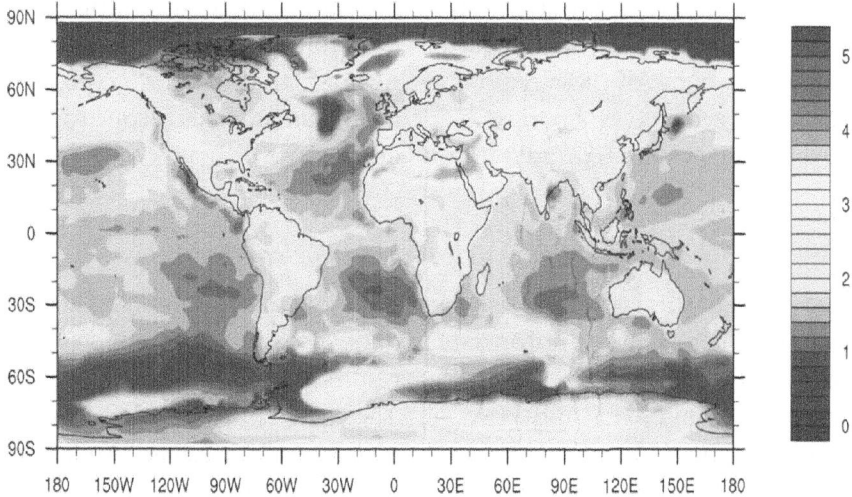

Figure 6.2. How much more intense will heat waves become, on average, in a future with no curbing of greenhouse gas emissions? The answer is in this figure. At each point on the map, the color corresponds to a value in degrees Celsius (see legend on the right of the figure). This value is the projected magnitude of the increase in temperature during the three warmest consecutive nights of the summer, in the climate of the end of the twenty-first century (the last two decades, to be precise), compared with the climate of the last three decades of the twentieth century. Summer is defined "loosely" as the 6-month period from May through October in the Northern Hemisphere and from November through April in the Southern Hemisphere. For color version, see plate section.

geographical patterns of the expected increases in absolute values. Interesting regional differences surfaced, with the West Coast and the Southeast of the United States projected to experience the largest absolute intensification of the heat events, and the Mediterranean basin, especially the Balkans region, representing the hottest spots in the future climate of Europe.

In Figure 6.2 we show a global map of projections of intensification of these heat events. For example, the Mediterranean region will experience the three hottest nights of the summer becoming warmer still, by on average 3–4 °C with respect to what occurs now in this region. Similarly, in the West and Southwest of the United States the three warmest nights of the summer will be warmer by 2.5–3.5 °C. Note that the color scale over inhabited regions covers only positive values, signifying that every populated region of the world will see warmer nights. However, the findings discussed in detail for the two regions in Meehl and Tebaldi (2004) remain true at a global level: some regions will see more substantial warming than others. It is not simply a zonal gradient that we see in this map, but interesting geographical differences surface, linked to changes in the general circulation patterns that may be associated with anthropogenic forcings. What is relevant with regard to probable societal and environmental impacts is that some of these regions are less prepared than others to face the discomfort of heat waves, because their current climate is less prone to extreme heat events, suggesting that adaptation will be more difficult.

6.2.3 Spells of days above climatological thresholds

The second definition of a heat wave necessitates a more complicated computation, and it is thus more appropriate to apply it only to selected locations. In Meehl and Tebaldi (2004), we showed results for time series of maximum temperature (from NCEP and PCM runs) extracted at two grid points, representing Chicago and Paris. We focused on average duration and frequency of such heat events, first verifying that the two characteristics simulated by PCM in the historical runs are consistent with the NCEP reanalysis observations, and then analyzing changes in these characteristics in the future climate. Both duration and frequency at both locations showed a large shift towards higher values, no longer encompassing the observed values from NCEP.

The same statistics were computed for several locations around the world, in both hemispheres and representative of very heterogeneous climates. Figures 6.3 and 6.4 show similar results for the changes in duration and frequency of heat waves at grid points located closest to the following cities: Seattle, Austin, and St. Louis for the United States; Melbourne, Australia; and Santiago, Chile (representative of Southern Hemisphere climate); and other Northern Hemisphere cities – Tokyo, Japan, and London, United Kingdom. In the figures, the ensemble members of the historical and future simulations are represented, respectively, by blue and red points, and the range of their simulations is highlighted by line segments and by dotted vertical lines of the same color. The values from the observations are shown as black marks labeled "NCEP." The figures show that there are considerable shifts toward larger average numbers of heat waves per year (top panels) at all locations except Melbourne, with London being the only location where the ranges of the two experiments partially overlap. The percentage increase ranges from 6% and 18% for Melbourne and London, respectively, to 40%–60% for the other locations. The NCEP estimate is within the range of the model historical runs for all locations except Santiago, where the model consistently overestimates it by 0.25 heat waves per year. For duration, London and St. Louis are the only two cases where the range of current and future simulations overlap in part, still showing a shift in the ensemble average projections of 53% and 39%, respectively.

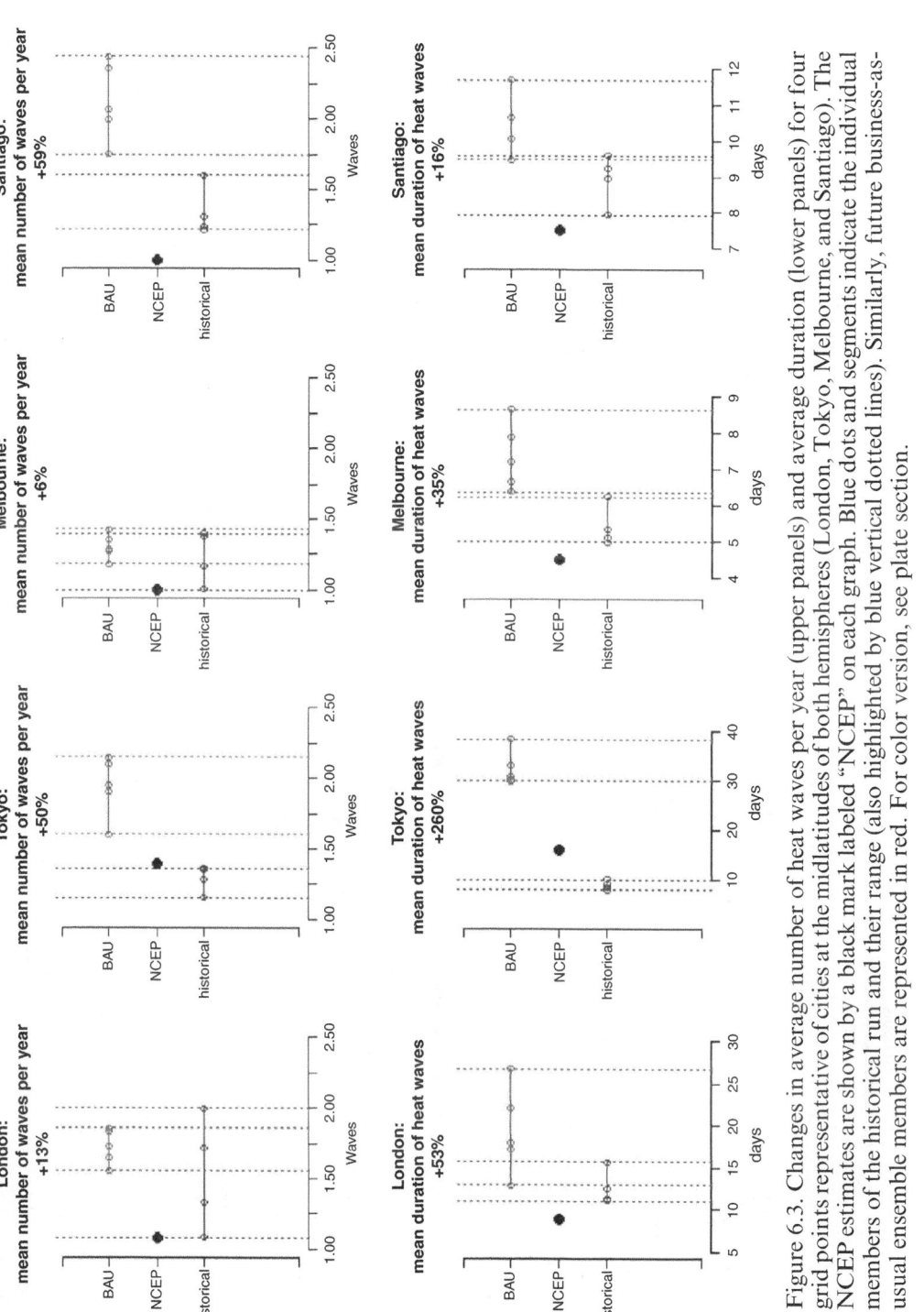

Figure 6.3. Changes in average number of heat waves per year (upper panels) and average duration (lower panels) for four grid points representative of cities at the midlatitudes of both hemispheres (London, Tokyo, Melbourne, and Santiago). The NCEP estimates are shown by a black mark labeled "NCEP" on each graph. Blue dots and segments indicate the individual members of the historical run and their range (also highlighted by blue vertical dotted lines). Similarly, future business-as-usual ensemble members are represented in red. For color version, see plate section.

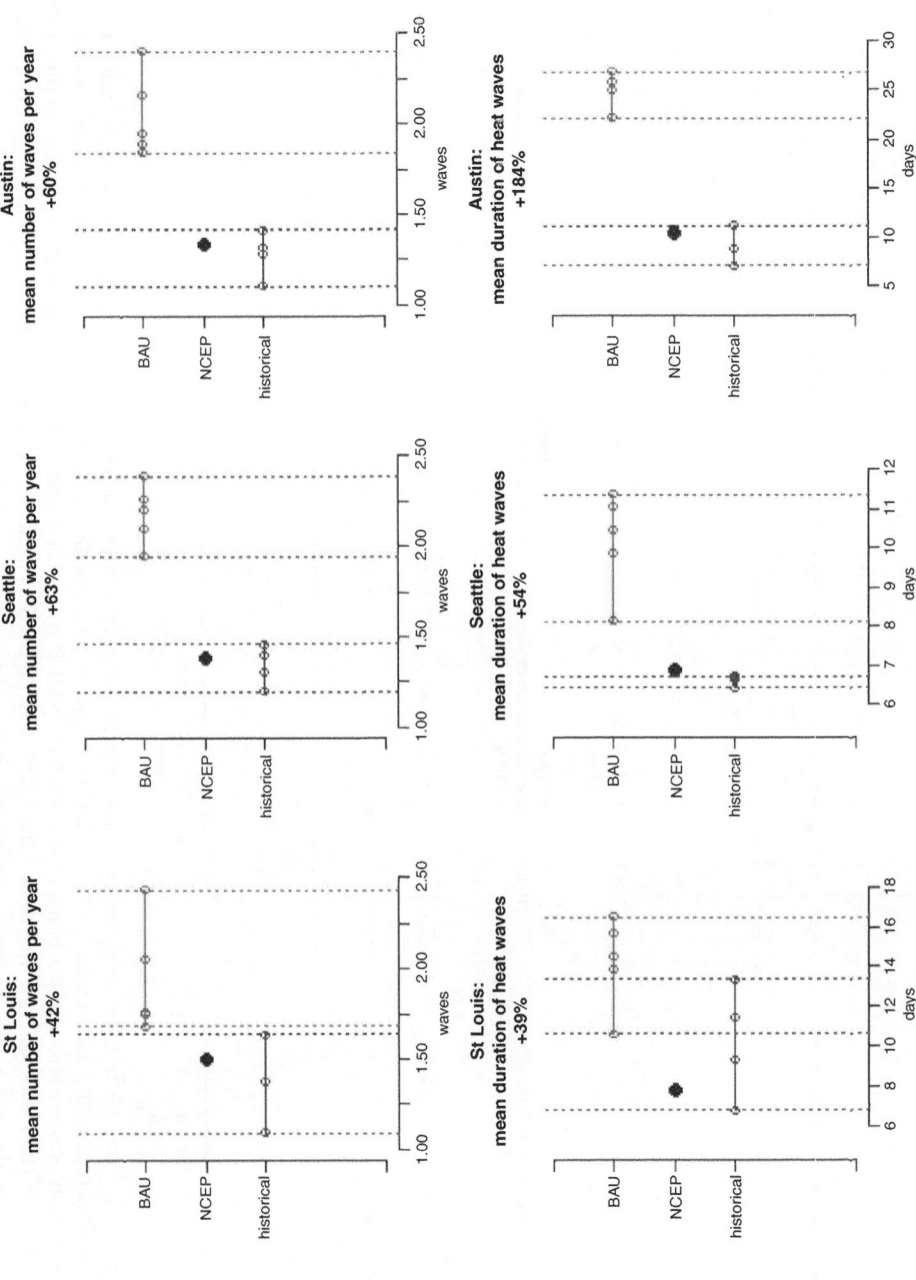

Figure 6.4. As Figure 6.3, for three grid points representative of three US cities (Seattle, Austin, and St. Louis). For color version, see plate section.

Tokyo and Austin see more than 100% increases in duration. The NCEP estimate is in three cases (London, Melbourne, and Santiago) just under the range of the model historical runs, and in three cases (the US locations) well within them. The only case where the PCM historical runs underestimate the duration of current heat waves is Tokyo.

We do not claim that the grid point results should be interpreted literally as projections for the seven cities. Rather, we use these results as a sample of different climate conditions in the world, all of which are consistently showing dramatic changes in heat wave characteristics.

Duration and frequency were computed after hot spells satisfying the three conditions were isolated from the hot season records (defined here as well as the days between May 1 and October 31 in the Northern Hemisphere, and November 1 through April 30 for the Southern Hemisphere). It is important to note that the thresholds of the definition represent the percentiles of current-day climatology, obtained by using NCEP and PCM historical time series at the seven locations for the period 1961–90. It is thus to be expected that a warming climate will produce a worsening of the conditions associated with such events.

But it is not just the change in the mean conditions that causes changes in heat wave characteristics. As was demonstrated in many studies (ranging from Katz and Brown [1994], where the sensitivity of extreme events to changes in variance of the climatological distributions was studied, to Schär *et al.* [2004] and Scherrer *et al.* [2005], where specific expected changes in the variance of temperature distribution over Europe are associated with future heat extremes similar to the heat wave of 2003), changes in the standard deviation of temperature in the future may play an additional important role in explaining changes in future extreme heat.

We show in Figure 6.5 two global maps of the ratio of the variances of minimum and maximum temperature. To limit the effects of the changes in variance due to the seasonal cycle, the variances were computed over only the three hottest months of the year (June–July–August, JJA, in the Northern Hemisphere; December–January–February, DJF, in the Southern Hemisphere) after detrending the time series at each grid point. A comparison of these maps with Figure 6.2, indicating changes in the intensity of heat waves, shows a large degree of correlation in the spatial patterns, especially for minimum temperature (top panel in Figure 6.5), where the areas with values larger than one (indicating areas where the variance of the distribution is larger in the future than in the present) are very similar to the areas of larger changes in three-day warmest temperature events (Alaska, the South of the United States, the Mediterranean basin, and central and northern Asia).

(a)

**Ratio of variances of Min Temp, future vs. present,
June–August in NH, Dec–Feb in SH**

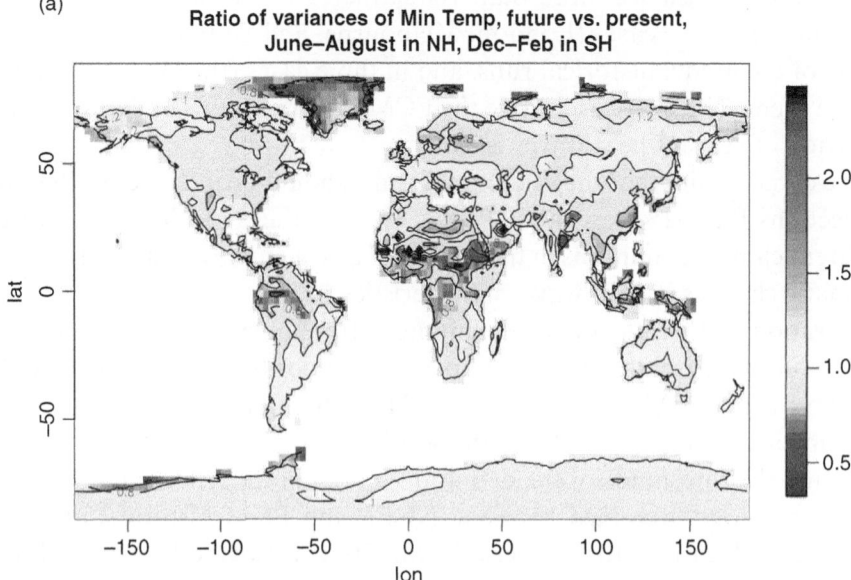

(b)

**Ratio of variances of Max Temp, future vs. present,
June–August in NH, Dec–Feb in SH**

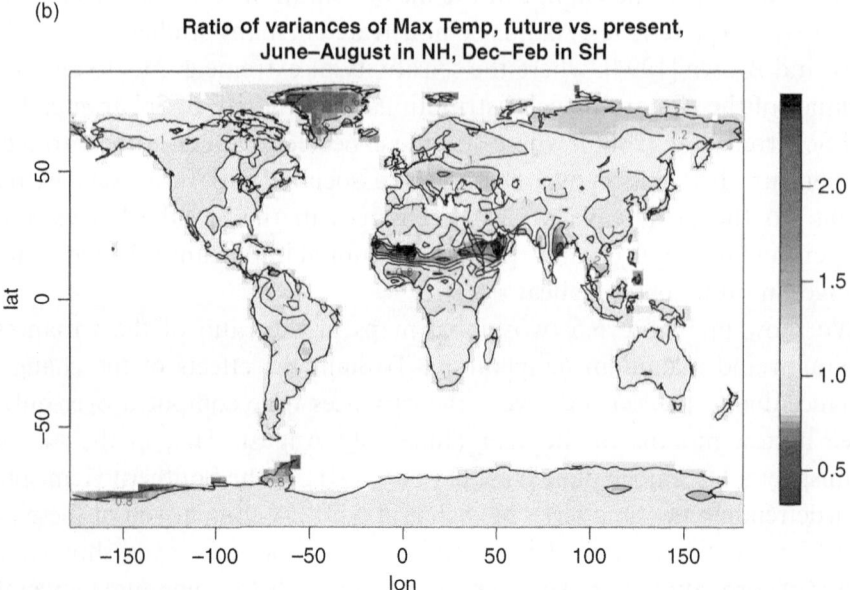

Figure 6.5. Projected changes in standard deviation of maximum (b) and minimum (a) temperature under a business-as-usual scenario, for the end of the twenty-first century compared to the baseline climatology of 1961–90. For color version, see plate section.

6.2.4 *Validation of the climate model*

Is the PCM correctly reproducing heat wave characteristics for the right reasons? In other words, are the dynamical mechanisms responsible for the onset and duration of heat waves the same in the model as in reality? After we identify, during each summer, days within a heat wave and regular days, it is possible to compute large-scale patterns of pressure anomalies associated with heat wave occurrence at a specific location. In Meehl and Tebaldi (2004), the validation was conducted with respect to the two observed episodes in Chicago and Paris, and the PCM-simulated pressure anomalies were consistent with the observed anomalies computed from the NCEP record. Thus we concluded that the climate model is reproducing observed characteristics of heat waves for the right dynamical reasons, and, accordingly, we trust its projections for the future.

As an additional validation step, we compared average changes in global pressure patterns under a scenario of 1% CO_2 increase as obtained by an ensemble average of 12 AOGCMs participating in the Climate Model Intercomparison Project-2 (CMIP2) experiment (Covey *et al.*, 2003) to the changes projected by the PCM, finding acceptable agreement across models. Attributing reliability to future projections is of course always a difficult issue, but inter-model agreement and a process-based validation lend strong support to the PCM projections described here.

6.3 Ten indices of climate extremes: model projected changes during the twenty-first century

Ten indices were chosen in Frich *et al.* (2002) as representative of a wide spectrum of extreme temperature and precipitation conditions, thought to be relevant for societal and environmental impacts. The definitions were intended for adoption by observing stations and thus needed to be easy to compute and store. Monitoring past trends and current characteristics of these indices has been the subject of the original paper and a suite of additional studies (Kiktev *et al.*, 2003; Zhang *et al.*, 2005; Alexander *et al.*, 2006) as increasingly richer datasets have been gathered, quality checked, and regularized by interpolating them to global grids.

In Tebaldi *et al.* (2007), we analyzed changes in the ten indices as simulated by a multi-model ensemble under three different SRES scenarios. The study did not tackle model validation in depth, but was content with deriving common tendencies across the ensemble of models. This was done for global average scales for the entire twenty-first century and in terms of the geographical patterns of changes, comparing the last two decades of the twentieth and twenty-first centuries. We summarize the study results here, after we list the definitions of the ten indices.

6.3.1 Definitions of climate extreme indices

The ten climate extreme indices are annual summaries of daily temperature or precipitation conditions. They are computed separately at each model's grid point. Five indices describe temperature extremes:

- Total number of frost days, defined as the annual total number of days with absolute minimum temperature below 0°C (*frost days*).
- Intra-annual extreme temperature range, defined as the difference between the highest temperature of the year and the lowest (*Xtemp range*).
- Growing season length, defined as the length of the period between the first spell of 5 consecutive days with mean temperature above 5 °C and the last such spell of the year (*growing season*).
- Heat wave duration index, defined as the maximum period of at least 5 consecutive days with maximum temperatures higher by at least 5 °C than the climatological norm for the same calendar day (*heat waves*).
- Warm nights, defined as the percentage of times in the year when minimum temperature is above the 90th percentile of the climatological distribution for that calendar day (*warm nights*).

Five indices describe precipitation extremes:

- Number of days with precipitation greater than 10 mm (*precip>10*).
- Maximum number of consecutive dry days, defined as days with precipitation less than 1 mm (*dry days*).
- Maximum 5-day precipitation total (*5day precip*).
- Simple daily intensity index, defined as the annual total precipitation divided by the number of wet days (*precip intensity*).
- Fraction of total precipitation due to events exceeding the 95th percentile of the climatological distribution for wet day amounts (*precip>95th*).

Note that all precipitation indices except dry days refer to intensity characteristics of daily rainfall.

6.3.2 Extreme indices and climate model output

Clearly, we are not dealing with definitions of extremely rare events. These indices strike a balance between characterizing the tails of the distributions of temperature and precipitation on one hand and still providing enough data points for the statistical behavior to be analyzed by traditional approaches, rather than having to resort to extreme value theory (e.g., Coles, 2001). The definitions, therefore, also seem to be particularly appropriate for indices computed from model output, because the resolution of AOGCMs is still too coarse to expect an accurate representation of the most intense, rarest episodes of extreme temperature and precipitation and the processes behind them. The grid resolution for the nine models of the ensemble varies from the coarsest, $5° \times 5°$, to the finest, $1.25° \times 1.25°$, with the T42 resolution ($2.75° \times 2.75°$) being representative of the whole set, and used as a common grid after bilinear interpolation.

As part of the activities leading to the publication in 2007 of the fourth assessment report by the Intergovernmental Panel on Climate Change (IPCC), up to 23 state-of-the-art general circulation models have been run under historical forcing conditions and three SRES scenarios (Nakicenovic and Swart, 2000): a high-emission scenario (A2), a low-emission scenario (B1), and an intermediate scenario (A1B). Of these modeling groups, nine have computed the ten indicators and have made their annual summaries available on the Program for Climate Model Diagnosis and Intercomparison (PCMDI) web site that collects all the experimental results (www-pcmdi.llnl.gov).

We analyze the nine model simulations for historical and future changes in the extremes indices. As was mentioned in the Introduction, we choose to remain in "model world," leaving any quantitative model validation exercise for future work, which will take advantage of a complete gridded dataset of observed indices available through the work of Alexander *et al.* (2006). Accordingly, we focus on the signs of changes (within the historical climate and for the future) and their statistical significance, for each model and within the multi-model ensemble.

In order to do so, we first take single-model ensemble averages, when more than one run under the same scenario is available for a single model. Then we center each model's output around its 1980–99 average, and we scale it by the standard deviation computed (after detrending) over the entire period 1960–2099, covering both historical and future run lengths. This procedure is applied to global average time series and hemispheric average time series, and to each grid point of the two bidecadal means used to produce geographical patterns of change.

6.3.3 Trends in extreme indices during the twentieth century

As a means of validating the AOGCMs within the terms of our qualitative analysis, we compared observed global average trends as reported in Frich *et al.* (2002) and Kiktev *et al.* (2003) to the trends produced by the historical runs of the AOGCMs. We obtained the following results:

- For the five temperature indices, all the statistically significant trends observed are reproduced by the models. These are decreasing trends for *frost days* and *Xtemp range* and increasing trends for *growing season* and *warm nights*. They are all consistent with a warming climate and seem to hint at particularly substantial increases in minimum (nighttime) temperatures, a result that has been discussed in many studies already (Meehl *et al.*, 2004). The index describing *heat waves* is simulated with an increasing trend in the historical runs by all models (of which four deem it statistically significant), while no detectable trend is found in the observations.
- For the five precipitation indices, all models simulate increasing trends, but these are seldom statistically significant, a result of high intermodel and interannual variability. The observed trends are statistically significant and increasing for *precip>10* and *5day precip*, statistically significant and decreasing for *dry days*, and not significant for the remaining two indices, *precip intensity* and *precip>95th*. It is difficult to draw general conclusions for the historical behavior of precipitation indices, recognizing that it will be relatively more difficult to determine statistical significance and consistency of changes for the precipitation indices for the future climate. This difficulty can be attributed – on the modeling side – to the more difficult nature of the simulation of precipitation in climate models, where the resolved scales are still too large to represent the processes behind precipitation patterns, which are local in scale and short-lived, especially when it comes to extreme/intense rain. From the observational point of view, precipitation statistics are highly variable year-to-year and the quality of the observed record is sometimes poor. Also, the coverage of the observed network is far from exhaustive, especially in the low latitudes, developing countries, and Southern Hemisphere.

We conclude that the AOGCMs agree well with observations indicating that there has already been a trend in the direction of more heat-related extremes, a finding that is consistent with a warming climate. Models also agree with observations that the historical period has seen an intensification of precipitation, albeit with a higher degree of spatial and interannual variability compared to the trend observed within temperature extremes.

6.3.4 *Changes in extreme indices*

The analysis of global trends (shown in Figure 6.6 for four representative indices) and geographical patterns of changes (shown in Figure 6.7 for the same four indices and for the A1B scenario, which is representative of the other two scenarios that envelop it) brings us to the following conclusions:

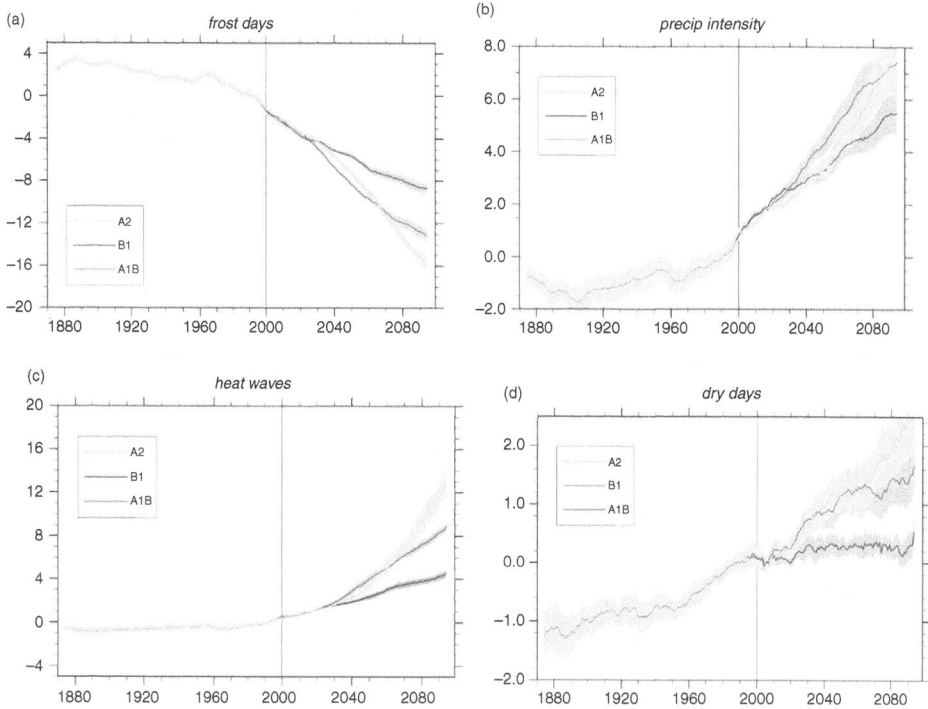

Figure 6.6. Multi-model ensemble means of global average time series (after standardization) for two of the temperature indices (*frost days* and *heat waves*) (a, c) and two of the precipitation indices (*precipitation intensity* and *dry days*) (b, d). The three scenarios are shown in different colors. The shading represents one standard deviation of the ensemble mean, as a measure of inter-model variability. First published in Tebaldi *et al.* (2007). For color version, see plate section.

- For temperature-related indices, the trends already observed in the last part of the twentieth century continue, amplified by the accumulation of greenhouse gas concentrations. Geographical patterns of changes are consistent across models and across scenarios, highlighting several regions of the world as likely hot spots in the future climate: High latitudes of the Northern Hemisphere for all five indices, especially *Xtemp range*; northwest of North America for *frost days* and *growing season*; southwest parts of North America for *heat waves*; Eastern Europe for *frost days* and *growing season*; northern Australia for *heat waves*; and southeast Australia for *frost days* and *growing season*.
- For precipitation-related indices, the geographical patterns are less stable across models and scenarios, but some common features appear nonetheless: High latitudes of the Northern Hemisphere show the most coherent regional patterns of change towards increased intensity of precipitation. *Dry days* is the index showing the largest inter-model geographical variability, making regional statements difficult besides the general prevalence of positive changes at low latitudes.

(a)   *frost days*

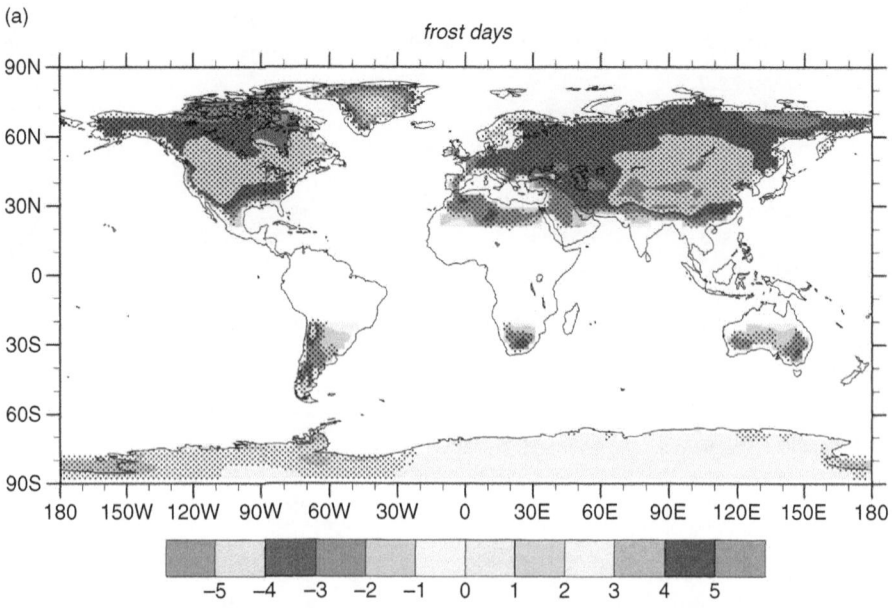

(c)   *heat waves*

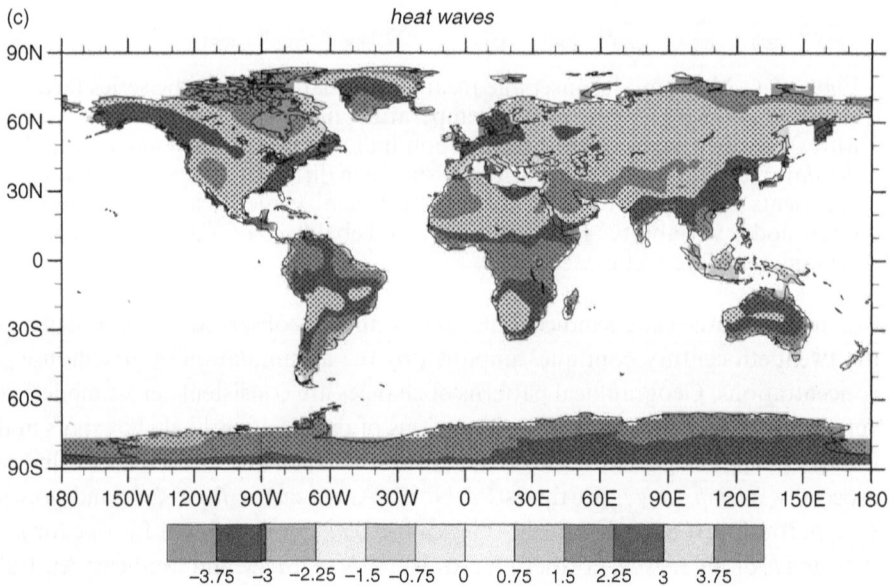

Figure 6.7. Multi-model ensemble means of spatial patterns of change (after standardization) under the A1B scenario for the same four indices as in Figure 6.6. Shown are the differences between two 20-year averages (2080–99 vs. 1980–99) for the temperature indices (a, c) and the precipitation indices (b, d). Grid points over the oceans were not included in the analysis and are thus left blank. Subtropical regions are also not included in the analysis of *frost days*. Stippling indicates that a majority of the models agree on the statistical significance of the change. First published in Tebaldi *et al.* (2007). For color version, see plate section.

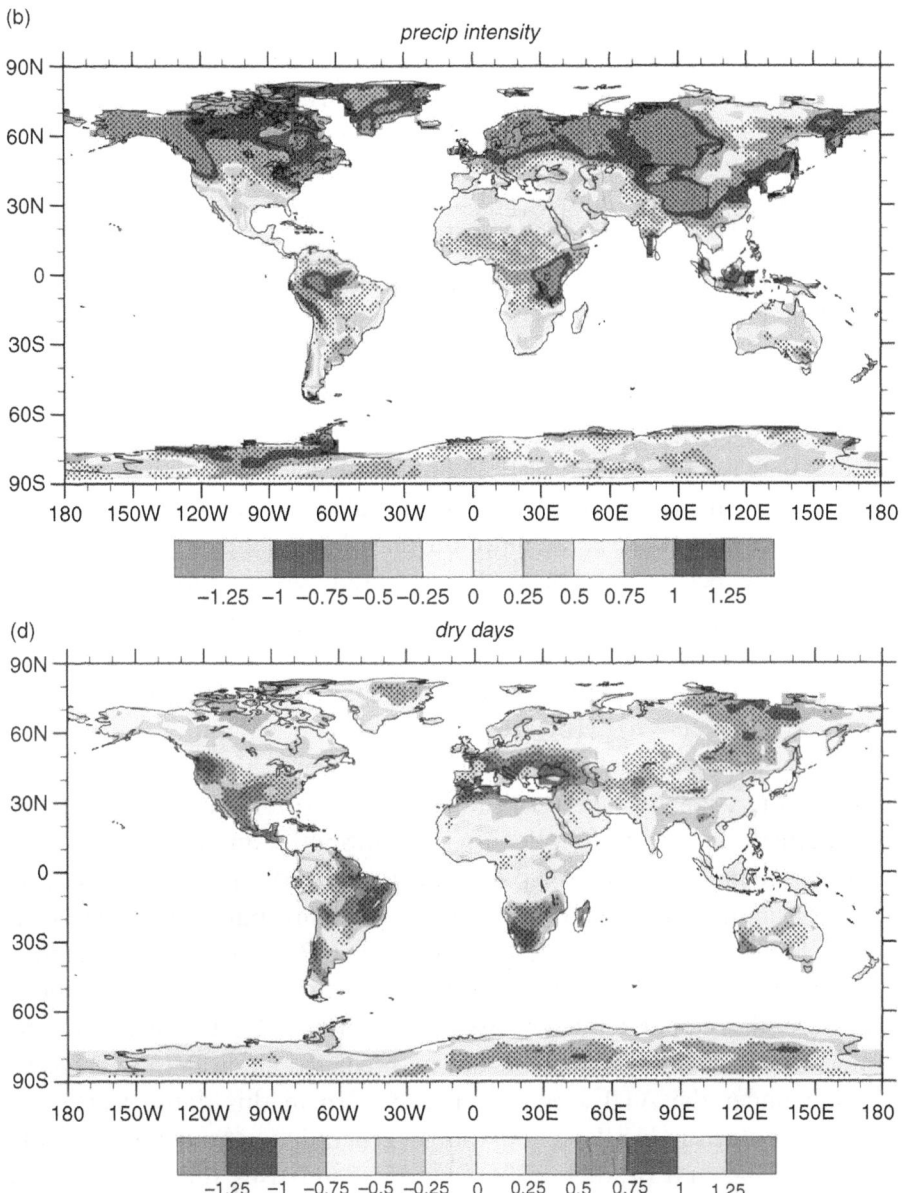

Figure 6.7. (cont.)

Also for this index, however, an increasing trend of the global average time series appears for the higher-emission scenarios, A1B and A2. Taken together, the five indices suggest that future climate may see a simultaneous increase in dry spells coupled to heavy rainfall events, a combination that may be conducive to flooding.

6.3.5 More about precipitation intensity

Meehl *et al.* (2005) took a closer look at changes in the intensity of rainfall through an in-depth analysis of the *precip intensity* index, which is readily interpretable; in addition, similar behavior is found in the other three indices that quantify changes in the intensity of rainfall by considering the changing characteristics of precipitation events whose intensity is above certain thresholds (10 mm and the 95th percentile of climatology). The tendency of climate models to produce frequent small precipitation events in their simulations (documented, for example, in Semenov and Bengtsson, 2002) may make the interpretation of the *precip intensity* index alone problematic, but the similarity of its behavior to that of the other indices that define changes in intensity in a stricter way (using exceedances over high thresholds) made us more comfortable in taking its changes seriously. Also, we are always looking at "changes" in the model statistics, rather than absolute values, another way of adjusting for model biases, under the assumption that they remain constant over the length of the experiments. The goal of the study was to analyze the spatial patterns of change in this index, their statistical significance, and the processes responsible for these changes.

Almost all areas of the globe see an increase in precipitation intensity, as is evident in Figure 6.7(b). However, increases in precipitation intensity are greatest in the tropics, as well as over northern Europe, northern Asia, the east coast of Asia, northwestern and northeastern North America, southwestern Australia, and parts of south-central South America. In a warming climate, more precipitation is expected on average not only as a consequence of higher evaporation rates from the warmer oceans, but also because of the larger holding capacity of relatively warmer air. Possibly, for local climates and brief timescales, more precipitation may actually have some cooling effects, but the global tendency over the century-long timescales is for a warmer, wetter climate. Figure 6.8 shows predicted changes in average precipitation under the A1B scenario. The pattern in this figure suggests an intensification of the Hadley circulation – generally wetter tropics and drier subtropics (Diaz and Bradley, 2005) – and in comparing this figure with the patterns of change in precipitation intensity of Figure 6.7, as can be expected, some of the areas with higher precipitation intensity coincide with areas of larger average precipitation. However, for some areas, the models project large increases in intensity together with lower average precipitation. That is to say, these areas could experience fewer rainy days, but heavier rain amounts when those days occur. These are in fact those areas where concurrently longer spells of dry days are projected.

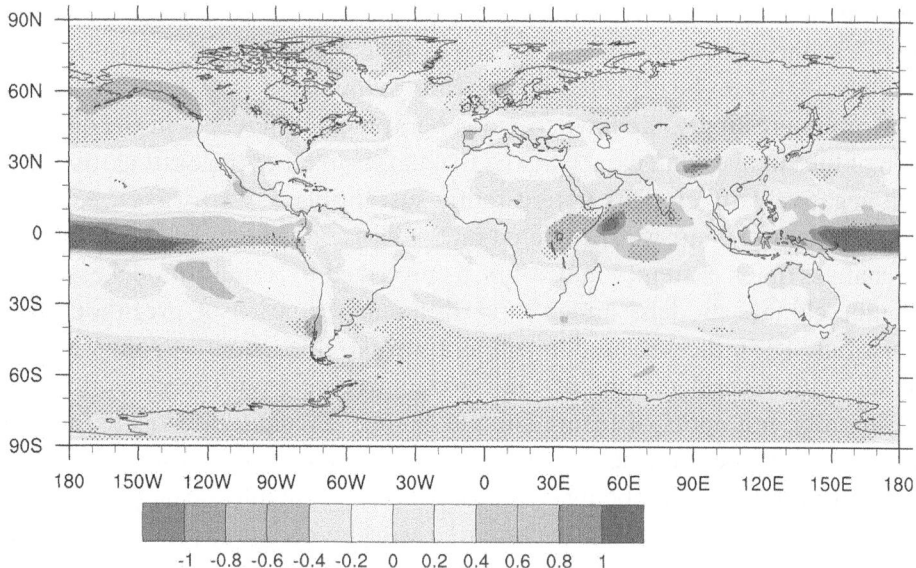

Figure 6.8. Multi-model ensemble means of spatial patterns of change in average precipitation under the A1B scenario. Units are millimeters per day. Dotted areas are statistically significant. First published in Meehl *et al.* (2004). For color version, see plate section.

It is of course straightforward to infer that the greater availability of water vapor from the warmer oceans in the tropics directly affects the positive changes in the low latitudes in both average precipitation and intensity of precipitation. For the mid- to high latitudes, it is necessary to invoke changes in the large-scale circulation patterns that have already been linked to greenhouse gas concentration increases by earlier studies (Meehl *et al.*, 2004), among these a poleward shift of the storm track as is documented in Yin (2005).

6.4 Conclusions

We have presented a range of results from global climate models that are pertinent to expected changes in extreme climate characteristics. In recent years, AOGCMs have become more accurate and reliable in the simulation of climate parameters at regional scales, and with respect to statistical properties of their distribution other than the means. We thus argue that valuable information on future trends can be derived by analyzing either single-model ensembles or multimodel ensembles, which are becoming more easily available through coordinated efforts and centralized archives (thanks to the activity related to the IPCC Fourth Assessment and PCMDI). Specifically, we have summarized a study of heat wave characteristics in simulations of future climate under a business as usual scenario; from our results, we can argue that in all likelihood heat waves will worsen in all aspects (intensity, duration, and frequency), as can be expected in a warming climate.

More interestingly, we have found geographical differences in the distribution of the changes that can be linked to changes in large-scale circulation, and will negatively affect areas not currently susceptible to heat waves. An additional series of results was presented on the basis of a survey of characteristics of changes in ten extremes indicators, as projected by a multi-model ensemble of nine climate models, under three different scenarios of greenhouse gas emissions. The indices related to temperature extremes show a strong warming signal, with several regions of the Earth emerging from the general results as, literally, hot spots of future climate change. Precipitation-related indices may not show the same degree of consistency across models when it comes to the pattern of changes, being by nature linked to smaller-scale phenomena, but they do convey a general picture of a world where precipitation intensity is bound to increase, especially in the tropics and higher latitudes of the Northern Hemisphere.

Even if we could not yet present an in-depth validation study of the model's ability to accurately reproduce the absolute values of these indices as observed, we found a degree of commonality in the simulations and a general qualitative agreement with observed trends that lets us put a fair degree of reliance on the summary results presented here.

References

Alexander, L. V., Zhang, X., Peterson, T. C., *et al.* (2006). Global observed changes in daily climate extremes of temperature and precipitation. *Journal of Geophysical Research – Atmosphere,* **111**, D5, D05109, doi:10.1029/2005JD006290.

Coles, S. (2001). *An Introduction to Statistical Modeling of Extreme Values.* London: Springer-Verlag.

Covey, C. K., AchutaRao, K. M., Cubasch, U., *et al.* (2003). An overview of the results from the Coupled Model Intercomparison Project. *Global and Planetary Change,* **37**, 103–33.

Diaz, H. F., and Bradley, R. S. (eds.) (2005). *The Hadley Circulation: Present, Past and Future (Advances in Global Change Research).* Dordrecht, The Netherlands: Kluwer Academic Publishers.

Duffy, P. B., Govindasamy, B., Iorio, J. P., *et al.* (2003). High-resolution simulations of global climate, part 1: present climate. *Climate Dynamics,* **21**, 371–90.

Easterling, D. R., Meehl, G. A., Parmesan, C., *et al.* (2000). Climate extremes: observations, modeling, and impacts. *Science,* **289**, 2068–74.

Frich, P., Alexander, L. V., Della-Marta, P., *et al.* (2002). Observed coherent changes in climatic extremes during the second half of the twentieth century. *Climate Research,* **19**, 193–212.

Govindasamy, B., Duffy, P. B., and Coquard, J. (2003). High-resolution simulations of global climate, part 2: effects of increased greenhouse gases. *Climate Dynamics*, **21**, 391–404.

Huth, R., Kysely, J., and Pokorna, L. (2000). A GCM simulation of heat waves, dry spells and their relationships to circulation. *Climate Change*, **46**, 29–60.

Karl, T. R., and Knight, R. W. (1997). The 1995 Chicago heat wave: how likely is a recurrence? *Bulletin of the American Meteorological Society*, **78** (6), 1107–19.

Katz, R. W., and Brown, B. G. (1994). Sensitivity of extreme events to climate change: the case of autocorrelated time-series. *Environmetrics*, **5**, 451–62.

Kiktev, D., Sexton, D., Alexander, L., and Folland, C. (2003). Comparison of modeled and observed trends in indices of daily climate extremes. *Journal of Climate*, **16**, 3560–71.

Kunkel, K. E., Pielke, R., Jr., and Changnon, S. A. (1999). Temporal fluctuations in weather and climate extremes that cause economic and human health impacts: a review. *Bulletin of the American Meteorological Society*, **80**, 1077–98.

Meehl, G. A., and Tebaldi, C. (2004). More intense, more frequent, and longer lasting heat waves in the twenty-first century. *Science*, **305**, 994–7.

Meehl, G. A., Arblaster, J. M., and Tebaldi, C. (2005). Understanding future patterns of increased precipitation intensity in climate model simulations. *Geophysical Research Letters*, **32**, Art. No. L18719.

Meehl, G. A., Karl, T., Easterling, D., *et al.* (2000). An introduction to trends in extreme weather and climate events: observations, socioeconomic impacts, terrestrial ecological impacts, and model projections. *Bulletin of the American Meteorological Society*, **81**, 413–16.

Meehl, G. A., Tebaldi, C., and Nychka, D. (2004). Changes in frost days in simulations of twenty-first-century climate. *Climate Dynamics*, **23**, 495–511.

Nakicenovic, N., and Swart, R. (eds.) (2000). *Emissions Scenarios: A Special Report of Working Group III of the Intergovernmental Panel on Climate Change (IPCC). Special Report on Emissions Scenarios.* Cambridge, UK, and New York: Cambridge University Press.

Schär, C., Vidale, P. L., Lüthi, D., *et al.* (2004). The role of increasing temperature variability in European summer heat waves. *Nature*, **427**, 332–6.

Scherrer, S. C., Appenzeller, C., Liniger, M. A., and Schär, C. (2005). European temperature distribution changes in observations and climate change scenarios. *Geophysical Research Letters*, **32**, Art. No. L19705.

Semenov, V., and Bengtsson, L. (2002). Secular trends in daily precipitation characteristics: greenhouse gas simulation with a coupled AOGCM. *Climate Dynamics*, **19**(2), 123–40.

Tebaldi, C., Hayhoe, K., Arblaster, J. M., and Meehl, G. A. (2007). Going to the extremes: an intercomparison of model-simulated historical and future changes in extreme events. *Climatic Change*, **79**, 185–211.

Washington, W. M., Weatherly, J. W., Meehl, G. A., *et al.* (2000). Parallel climate model (PCM) control and transient simulations. *Climate Dynamics*, **16**, 755–74.

Yin, J. (2005). A consistent poleward shift of the storm tracks in simulations of twenty-first-century climate. *Geophysical Research Letters*, **32**, Art. No. L18701.

Zhang, X., Hegerl, G., Zwiers, F. W., and Kenyon, J. (2005). Avoiding inhomogeneity in percentile-based indices of temperature extremes. *Journal of Climate*, **18** (11), 1641–51.

7

Tropical cyclones and climate change: revisiting recent studies at GFDL

THOMAS R. KNUTSON AND ROBERT E. TULEYA

Condensed summary

In this chapter, we revisit two recent studies performed at the Geophysical Fluid Dynamics Laboratory (GFDL), with a focus on issues relevant to tropical cyclones and climate change. The first study was a model-based assessment of twentieth-century regional surface temperature trends. The tropical Atlantic Main Development Region (MDR) for hurricane activity was found to have warmed by several tenths of a degree Celsius over the twentieth century. Coupled model historical simulations using current best estimates of radiative forcing suggest that the century-scale warming trend in the MDR may contain a significant contribution from anthropogenic forcing, including increases in atmospheric greenhouse gas concentrations. The results further suggest that the low-frequency variability in the MDR, apart from the trend, may contain substantial contributions from both radiative forcing (natural and anthropogenic) and internally generated climate variability. The second study used the GFDL hurricane model, in an idealized setting, to simulate the impact of a pronounced CO_2-induced warming on hurricane intensities and precipitation. A 1.75 °C warming increases the intensities of hurricanes in the model by 5.8% in terms of surface wind speeds, 14% in terms of central pressure fall, or about one half category on the Saffir–Simpson Hurricane Scale. A revised storm-core accumulated (six-hour) rainfall measure shows a 21.6% increase under high-CO_2 conditions. Our simulated storm intensities are substantially less sensitive to sea surface temperature (SST) changes than recently reported historical observational trends are – a difference we are not able to completely reconcile at this time.

7.1 Introduction

In a recent study (Knutson and Tuleya, 2004; hereafter referred to as KT04), we reported results from an extensive set of idealized hurricane model

Climate Extremes and Society, ed. H. F. Diaz and R. J. Murnane. Published by Cambridge University Press. © Cambridge University Press 2008.

simulations that explored the impact of a substantial CO_2-induced climate warming on hurricane intensities and precipitation. We found increased hurricane intensities under warmer climate conditions, with an average maximum surface winds sensitivity of 5.8%, for an average sea surface temperature (SST) increase of 1.75 °C. Since that paper was published, several provocative observational studies have appeared (e.g., Emanuel, 2005a; Webster *et al.*, 2005; Mann and Emanuel, 2006) that presented some evidence for increasing trends in intensity-related tropical cyclone measures in historical observations.

In this study, we review our earlier work (Section 7.3) and discuss the prospects for reconciling our earlier results with the recently reported observed trends of hurricane intensity measures (Section 7.4). We also present some further analysis of precipitation from the KT04 experiments (Section 7.3), through which we find that a recently identified analysis problem with our original inner-core precipitation measure – when properly addressed – has little impact on the percent changes in inner-core rainfall rates under high-CO_2 conditions as reported in KT04 (although it does affect the absolute values of the precipitation rates).

In a separate recent study, Knutson *et al.* (2006; hereafter referred to as K06) used a new, coupled ocean–atmosphere climate model to assess potential causes of twentieth-century surface temperature trends in a number of regions. With regard to tropical cyclone behavior, a potentially important result emerging from this study was the identification of a century-scale surface warming trend in the tropical Atlantic Main Development Region (MDR). The model assessment results suggest that this warming may contain a significant contribution from anthropogenic forcing, including increasing greenhouse gas concentrations. In this chapter, the K06 analysis related to the MDR warming trends is reviewed and expanded upon (Section 7.2).

7.2 Tropical Atlantic (Main Development Region) temperature trends

The region of the tropical North Atlantic and Caribbean extending from 10° N to 20° N is often referred to as the Main Development Region (MDR) for Atlantic hurricane activity, owing to the large portion of major hurricanes that can be traced to disturbances originating there (e.g., Goldenberg *et al.*, 2001; Bell and Chelliah, 2006). Here we explore possible causes for a warming of this region during the twentieth century. Shown in Figure 7.1 are 10-year running mean surface temperature indices for the MDR and for the globe obtained from the HadCRUT3 combined SST/land surface air temperature dataset (www.cru.uea.ac.uk/cru/data/temperature/; Parker *et al.*, 1995; Jones and Moberg, 2003; Rayner *et al.*, 2003; Brohan *et al.*, 2006).

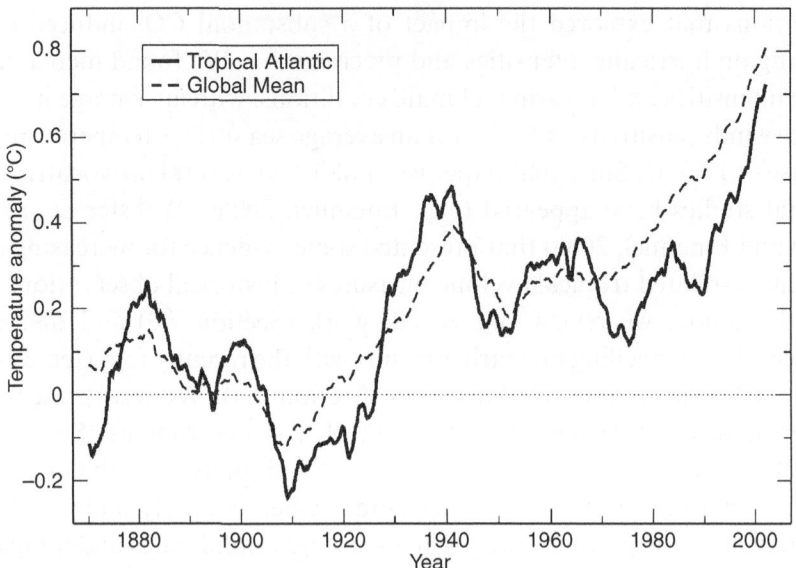

Figure 7.1. Ten-year running mean surface temperature anomalies for the Tropical Atlantic Main Development Region and global average. The MDR is defined here as the region 10° N–20° N, 80° W–20° W. The dataset used is the combined land–ocean HadCRUT3. Anomalies (in degrees Celsius) have been adjusted to have zero mean for the period 1881–1920.

Low-frequency surface temperature variability in the MDR is dominated by sea surface temperature variability in this predominantly oceanic region. The surface temperature data indicate that the MDR warmed by several tenths of a degree Celsius over the twentieth century (Figure 7.1). This warming has been noted previously (e.g., Emanuel, 2005a; Trenberth, 2005; K06; Santer *et al.*, 2006). The MDR warming roughly tracks the increase in global mean surface temperature, but with larger multidecadal swings in temperature compared with those in the global mean record. Concerning the global mean temperature increase, a large body of research has assessed the possible role of increasing greenhouse gases on global mean temperature (e.g., Meehl *et al.*, 2004; International Ad Hoc Detection and Attribution Group [IADAG], 2005; K06) and concludes that most of the global warming over the past 50 years is likely due to the increase in greenhouse gases. Fewer studies have addressed this question on regional scales such as the MDR. Santer *et al.* (2006) recently examined warming trends in both the Atlantic and Pacific tropical cyclogenesis regions using 22 different climate models and concluded that there is an 84% chance that external forcing led to at least two-thirds of the observed SST increases in these regions.

In another recent model-based assessment (K06) observed twentieth-century surface temperature trends in the MDR and various other regions were compared with trends obtained by using two new Geophysical Fluid Dynamics Laboratory (GFDL) global climate models. Three different historical climate-forcing scenarios were examined (see below). Current best estimates of a number of historical climate forcings over the period 1860–2000 were specified for these scenarios. The forcings included representations of greenhouse gases, volcanic eruptions, solar variability, land cover changes, and aerosols. These forcings were incorporated more realistically than those used in previous GFDL coupled climate model experiments (e.g., Broccoli *et al.*, 2003). The aerosol forcing included the "direct effect" only and did not include effects of interactions of aerosols with clouds or precipitation processes.

Figure 7.2 shows late summer sea surface temperatures in the MDR as simulated in the K06 historical runs compared with observations from the HadISST (Rayner *et al.*, 2003) and ERSST.v2 (Smith and Reynolds, 2003, 2004) datasets. The top panel (Figure 7.2a) shows the observed MDR series in comparison to an eight-member ensemble of experiments, which used both anthropogenic and natural historical forcings (i.e., all available forcings). Natural forcings (Figure 7.2b) include only volcanic aerosols and long-term variability of solar radiation. Anthropogenic forcings (Figure 7.2c) include only changes in well-mixed greenhouse gases, ozone, aerosols, and land cover. Further details on these forcings are provided in K06. The differences among the ensemble members for each forcing scenario reflect internal climate variability as simulated by the model. Each ensemble member is initialized with different ocean initial conditions taken from a multi-century, 1860-condition control run, and thus begins in a different phase in terms of internally generated modes of the model, such as the model's El Niño, North Atlantic Oscillation (NAO), or internal Atlantic Ocean variability.

In the MDR, the observed long-term warming during the twentieth century is much more realistically simulated in the model runs, which include anthropogenic forcing (i.e., the all-forcing or anthropogenic-only-forcing scenarios) than in the natural-forcing runs. This is particularly true for the late twentieth-century warming. There is some resemblance of the temporal structure of the all-forcing ensemble mean compared to the observations beyond just the century-scale linear trend. However, the observations still exhibit pronounced multidecadal departures from the ensemble mean of the all-forcing runs. The anthropogenic-forcing ensemble mean response (c; $n = 4$) in the MDR appears fairly linear. A more nonlinear response (increasing slope over time) is evident for the global mean results (see K06, Fig. 1e). The anthropogenic- and all-forcing runs include only the direct effect of aerosols, and so the total forcing from aerosol changes (including indirect effects) is

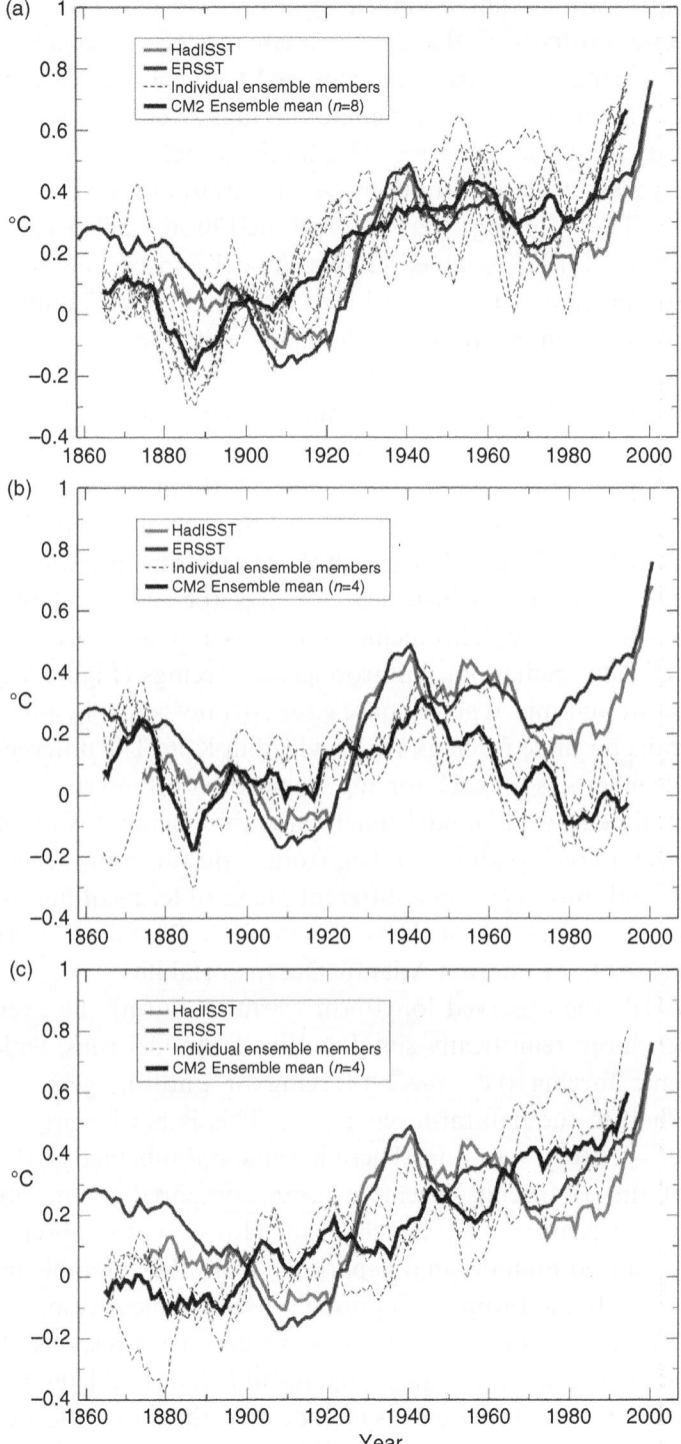

likely underestimated. The natural-forcing ensemble (b; $n = 4$) indicates a significant role for volcanic activity in producing some of the low-frequency structure in the all- and natural-forcing simulations. Similar results to these, although with less internal variability, were obtained for other regions, such as annual-mean global mean temperature, in K06. These results are generally consistent with a multi-model analysis of a region similar to our MDR by Santer *et al.* (2006), who also show reasonable agreement of twentieth-century all-forcing runs with observations, and a notable secondary impact of volcanic eruptions.

Further analysis of the MDR warming was presented in K06, where maps of observed trends over the periods 1901–2000 and 1949–2000 were quantitatively compared to model forcing scenarios and internal variability, using local t-tests. In the MDR and vicinity, the 1901–2000 warming trends (based on annual means) were generally significantly larger than the model's internally generated variability, and significantly different from trends in the natural-forcing runs. In contrast, these trends were not significantly different from model trends in forcing runs that included anthropogenic forcing. For the more recent period (1949–2000), the relatively smaller observed trends in the MDR and vicinity were not distinguishable from the model's internal variability, highlighting the fact that a longer (century-scale) record is useful for distinguishing a long-timescale warming trend from other climate variations in the MDR. The relatively cool SSTs in the MDR in the first half of the twentieth century, combined with generally warmer SSTs in the second half of the century, produce a pronounced century-scale warming trend that is larger than expected from internal climate variability or natural forcing alone, according to the model-based assessment. The model-observation statistical comparisons in K06 are limited to periods ending in the year 2000, since the CM2.1 historical model runs performed for K06 ended with that year. The observed low-pass filtered results in Figure 7.2 show a continued strong warming past 2000 (here using data through 2005), which serves to strengthen

---

Caption for Figure 7.2. Observed sea surface temperature variations in the MDR from HadISST (red) and ERSST (blue) datasets vs. CM2 historical climate simulations using (a) all forcings, (b) natural forcings only, or (c) anthropogenic forcings only. Ten-year running mean anomalies for the August–October season, referenced to 1881–1920 means in degrees Celsius, are shown. Black dashed curves are individual CM2.0 or CM2.1 ensemble members; thick black curves are the CM2.0/CM2.1 ensemble means ($n = 8$ experiments with all forcings; $n = 4$ experiments with natural- or anthropogenic-only forcings). For color version, see plate section.

the conclusions of K06 concerning the unusual nature of this warming compared with internal or natural climate variability.

In total, the results in K06 and in Figures 7.1 and 7.2 here suggest that the century-scale warming trend in the MDR may contain a significant contribution from anthropogenic forcing, including increases in atmospheric greenhouse gas concentrations. Furthermore, the model results suggest that the low-frequency variability in the MDR, apart from the trend, may contain substantial contributions from both radiative forcing (both natural and anthropogenic) and internally generated climate variability. Further work is under way to explore the relative roles of these factors – a topic of recent debate (e.g., Goldenberg *et al.*, 2001; Mann and Emanuel, 2006; Trenberth and Shea, 2006).

The observed and simulated (all-forcing) twentieth-century warming of the MDR is actually part of a much broader-scale warming pattern (e.g., Fig. 7 of K06), spanning both the Northern and Southern Hemispheres. This argues against the notion that the observed global-scale twentieth-century warming is primarily due to fluctuations in the Atlantic thermohaline circulation (THC). For example, Zhang and Delworth (2005) simulated the surface temperature response associated with a pronounced weakening of the Atlantic THC, and showed generally opposite-signed anomalies between the two hemispheres. This contrasts with the general twentieth-century warming trend evident in both hemispheres for both the observations and model simulations.

Another related finding in K06 was the very pronounced observed warming trend in the Indian Ocean – western Pacific warm pool region: an even stronger warming signal than that in the Atlantic MDR. As with the MDR, the model assessment results suggested that anthropogenic forcing has played a significant role in the Indian Ocean – western Pacific warming.

The long-term warming that has occurred in various tropical ocean regions is relevant to our discussion of tropical cyclone intensities in light of the strong correlation between tropical SSTs and several tropical cyclone measures, as has been reported by Emanuel (2005a), Webster *et al.* (2005), Hoyos *et al.* (2006), and Mann and Emanuel (2006) for several tropical ocean basins. The reliability of the tropical cyclone databases used to infer these relationships has been questioned (e.g., Landsea *et al.*, 2006). According to Landsea (2005; 2007), the reliability of basin-wide tropical cyclone statistics for the Atlantic basin decreases as one goes back in time, particularly in the pre-satellite era. Longer records, extending back to 1900, of tropical cyclone intensity measures for US landfalling tropical and subtropical systems have also been examined (e.g., Landsea 2005) but show no apparent trend over the period, in contrast to the MDR SSTs shown in Figures 7.1 and 7.2. However, Emanuel (2005b) noted that US landfalling storm statistics are composed of only a very small fraction of tropical cyclone observations over the whole basin, and thus any trends present could well be masked by noise effects due to the small sample size.

7.3 Review of KT04 results

7.3.1 Methodology for idealized hurricane simulations

In KT04, we used an idealized simulation approach to investigate the impact of CO_2-induced warming on hurricane precipitation and intensity. The model used was an idealized version of the GFDL Hurricane Prediction System, which has been run operationally for hurricane prediction at the US National Centers for Environmental Prediction (NCEP) for several years. The model has variable resolution, with grid spacing of about 9 km in a $5° \times 5°$ fine-mesh grid that moves with the hurricane through a larger coarse-grid domain. No ocean coupling was used in the experiments. Knutson *et al.* (2001) found that the inclusion of ocean coupling has a minimal impact on the percentage increases in hurricane intensity simulated for warm climate conditions in such experiments. In the idealized experiments, the hurricanes were simulated for 5 days as they traveled over a uniform sea surface (no land) in a large-scale atmospheric environment consisting of a uniform $5 \, \text{m s}^{-1}$ easterly flow. Therefore, no effects of the storms' interactions with land, topography, vertical wind shear,[1] or large weather systems were included in our experimental design. KT04 (its Fig. 15) found that climate model-projected changes in Atlantic basin vertical wind shear in response to increasing CO_2 were quite model-dependent in the nine climate models they examined, although the changes were not very dramatic, even for the most sensitive models. Vecchi and Soden (2007) recently reported that increased vertical wind shear in the Caribbean was a response appearing consistently in most twenty-first-century projections using a newer set of climate models.

To specify the large-scale SSTs, atmospheric temperatures, and atmospheric moisture conditions for our hurricane model experiments, we obtained present-day and high-CO_2 climatologies from nine global climate models, which took part in the international Coupled Model Intercomparison Project (CMIP2+). The high-CO_2 environments were based on the control run climatologies plus 80-year net linear trends from +1% per year compounded CO_2 experiments run for each model (KT04). The present-day conditions were based on 80-year averages from the models' control runs

---

[1] Vertical wind shear refers to a change in wind speed and/or direction with height. Larger values of vertical wind shear are believed to be detrimental for tropical cyclone development, as they disrupt the vertical organization of the storm.

with constant (present-day) CO_2 levels. A 1% per year compounded increase of CO_2 is a strong, though not extreme, idealized future global radiative forcing scenario (Knutson and Tuleya, 2005). For each climate model, we obtained area-averaged climatologies for the Atlantic, northeastern Pacific, and northwestern Pacific tropical storm basins, in each case time-averaged over the months of July through November.

In addition to sampling high-CO_2 climate states from nine global models, we also tested four different moist physics options in the GFDL hurricane model to assess the potential impact of changes in the treatment of precipitation processes – a crucial process for tropical cyclones – on our simulation results. The four treatments included two mass flux schemes, one convective adjustment scheme, and resolved convection (i.e., using no convective parameterization in the fine-scale inner grid). For each combination of climate model and hurricane model physics, we conducted experiments for two climate states (present-day and high-CO_2), and three basins (Atlantic, northwestern Pacific, and northeastern Pacific). The three basins, nine models, and four moist convection physics treatments yielded $3 \times 9 \times 4 = 108$ configurations to test for each climate state. For each of the 108 configurations, we ran a six-member ensemble of experiments, differing only in terms of small random perturbations to the atmospheric initial conditions for the five-day runs. Thus, in all we ran 1,296 simulations (108 configurations $\times$ 2 climates $\times$ 6 ensemble members). The initial disturbance used was based on a fairly robust initial hurricane condition (maximum wind speeds of approximately $35 \, \mathrm{m \, s^{-1}}$) at a radius of 55 km.

The SST changes for the three tropical cyclone basins obtained from the nine CMIP2+ global models averaged 1.75 °C, with a range of 0.8 °–2.4 °C across the 27 (3 basins $\times$ 9 models) samples. A prominent feature of the tropical climate changes in these global models was an enhancement of the warming in the upper troposphere, relative to the warming at the surface (Figure 7.3). This "tropospheric stabilization" effect has been shown by Shen *et al.* (2000) to reduce the intensity of simulated hurricanes in the GFDL model. On the other hand, warmer SSTs led to more intense storms in their study. A key purpose of our experiments was to use the hurricane model to simulate which of these two effects dominates in a CO_2-induced climate-warming scenario. Our results show that the SST warming effect (increased intensity) exceeds the tropospheric stabilization effect (decreased intensity), yielding a net increase of hurricane intensity in almost all (107 out of 108) ensemble mean combinations examined, as is discussed in detail below.

Further details on all aspects of our models, the CO_2-induced changes in the tropical mean environments from the CMIP2+ models, and other experimental design considerations are contained in KT04.

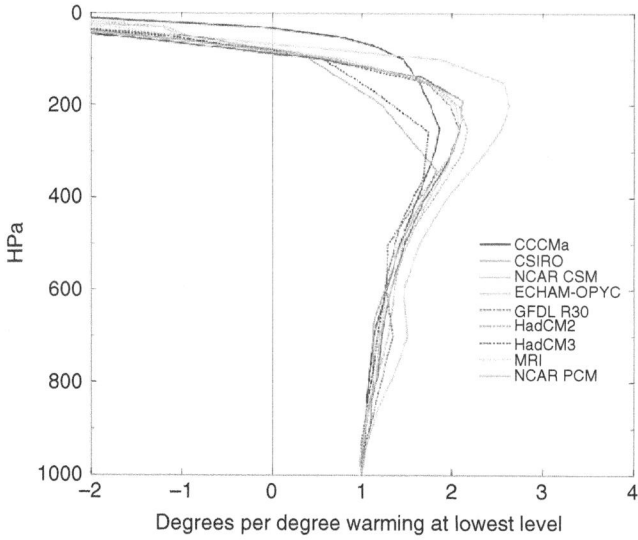

Figure 7.3. Vertical profiles of normalized atmospheric temperature change (high-CO_2 minus control) zonally averaged over $20°$ N–$20°$ S from a set of transient $+1\%$ per year CO_2 increase experiments using nine global climate models. The difference is based on years 61–80 of the high-CO_2 runs minus years 61–80 of the control runs for each model (see legend). The difference at each model level is normalized by dividing by the difference at the lowest available level for that model. For color version, see plate section.

7.3.2 *Intensity simulation results*

The maximum hurricane intensity (minimum central pressure) results from the 1,296 five-day experiments are summarized in Figure 7.4. The light curve shows the distribution of minimum central pressures (one per simulated storm) obtained for the control (present-day) conditions. The dark curve shows the distribution for the high-CO_2 conditions. The high-CO_2 distribution is shifted systematically to the left, toward lower pressures and higher intensities, compared with the control. The size of the shift is 10.4 millibars (mb) for the mean, and represents about a one half category shift on the Saffir–Simpson Hurricane Scale. Substantially more storms in our idealized experiments reached category 5 for the high-CO_2 conditions than for the control conditions. Although our experiments did not address the question of possible future changes in storm frequency, they did suggest an increasing relative risk in the occurrence of category 5 hurricanes under high-CO_2 conditions. Analysis of subsets of experiments indicated that the increased simulated intensity occurred for nearly all combinations (107 out of 108) of climate model boundary condition, tropical storm basin, and hurricane model moist convection treatment tested. The simulated hurricane intensity increase was

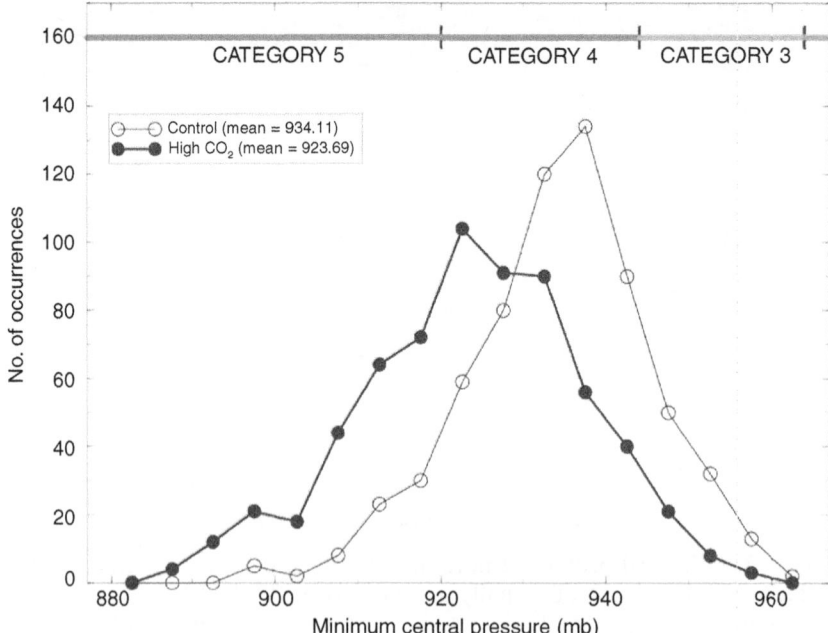

Figure 7.4. Frequency histogram showing hurricane intensity results (millibars) aggregated across all 1,296 experiments of KT04 (9 GCMs, 3 basins, 4 parameterizations, 6-member ensembles). The histograms are formed from the minimum central pressures, averaged over the final 24 h from each 5-day experiment. The light (dark) curve with open (solid) circles depicts the histogram from the control (high-CO_2) cases. See text for further details. For color version, see plate section.

relatively insensitive to the moist convection treatment (e.g., Fig. 7 and Table 2 of KT04).

Although it is not shown here, a similar intensity analysis based on maximum surface wind speeds also indicates higher-intensity storms under high-CO_2 conditions. In these experiments, the mean maximum surface wind intensity increased by 5.8% for high-CO_2 conditions. Based on the average change in SST of 1.75 °C between our control and high-CO_2 runs, our experiments indicated a normalized tropical cyclone wind intensity sensitivity of +3.3% per degree Celsius. Using central pressures from the model and inferring maximum sustained surface wind speeds following Landsea (1993), based on Kraft's (1961) analysis of Atlantic tropical cyclones, we obtained a slightly higher sensitivity of 3.7% per degree Celsius. These sensitivities are slightly less than the 5% per degree Celsius sensitivity reported by Emanuel (2005a) for his hurricane potential intensity theory, and are significantly less than the sensitivities inferred from recent observational studies, as will be discussed later. In

KT04, we also presented calculations of potential intensity changes for our environmental fields using both the Emanuel (1987, 1988) and the Holland (1997) methodologies. Those calculations indicated broadly similar sensitivity results from potential intensity theory, as compared to our model simulations (see KT04 for details).

7.3.3 *Revised precipitation results*

In KT04, three different types of precipitation measures for the simulated hurricanes were presented, each indicating enhanced rainfall rates for warmer, high-CO_2 conditions. The specific results reported in KT04 included statistics for an area-averaged near-storm rainfall rate (within 100 km of the storm center), the maximum precipitation rate anywhere in the model domain, and the average precipitation rate in a 5° latitude × 5° longitude inner mesh centered on the storm.

We have since discovered some problems with these precipitation statistics as reported in KT04. The first problem stems from the application of a spatial smoothing procedure once per simulation hour during the time integration of the hurricane model. The precipitation statistics were affected by the unrepresentative high values occurring once per hour, which were artifacts of the spatial smoothing procedure. The second problem was an incorrect description of the large-scale precipitation measure. Our large-scale rainfall measure from KT04 is now correctly identified (Table 7.1) as the average storm total precipitation averaged over the entire (70° × 70°) model domain, as opposed to just the 5° × 5° inner mesh.

Table 7.1 presents our revised precipitation statistics, which supersede the precipitation entries in Table 2 of KT04 (see the Appendix to this chapter for details). The problem related to spatial smoothing mentioned above had only a relatively minor effect on the percentage increase in storm-core rainfall induced by CO_2-related global warming, which was the main precipitation conclusion we emphasized in KT04. With the revised, more representative statistics discussed in the Appendix, we obtain an increase in storm-core rainfall of 21.6% for high-CO_2 conditions, compared with 18.3% for our original analysis in KT04. The percentage change is still fairly sensitive to the moist convection scheme used, varying from 18% to 27% for the three treatments analyzed (Table 7.1). The smoothing-induced problem significantly affected the absolute values of the precipitation rates as shown in Table 2 of KT04 and in some of the figures of that report. For example, the magnitudes of the precipitation rates shown in Figures 1, 8, and 14, and Table 1 (100 km radius) of KT04 overestimate the true time averages from the model by roughly 80%.

Table 7.1. *Revised precipitation statistics for KT04 simulations*[a]

	All convection schemes		HPAN		EMAN		KURI		Resolved	
	Control	Percent change	Control	Percent change	Control	Percent change	Control	Percent change	Control	Percent change
6-hour accumulated (32,700 km^2 region; cm)	12.4	21.6%	12.8	26.5%	n/a	n/a	12.6	19.9%	11.8	18.0%
Maximum rate in domain	n/a		n/a		n/a		n/a		n/a	
5-day accumulated (averaged over 70° × 70° domain; cm)	1.62	6.7%	1.74	8.8%	1.27	8.7%	2.18	2.7%	1.28	8.6%

[a] HPAN, EMAN, and KURI refer to different moist convection physics schemes in the model, which were varied as a sensitivity test in KT04. "Resolved" refers to model runs with no convection parameterization for the model's inner grid so that only condensation processes resolved by the model are included. See KT04 for further details on the moist physics schemes.

The question arises whether the GFDL hurricane model has skill at predicting hurricane-related rainfall. Recently, Tuleya *et al.* (2007) and Marchok *et al.* (2007) have evaluated this aspect of the operational performance of the model using several rainfall measures. They found that the model has a high bias in terms of its inner-core (100 km radius) 6-hourly accumulated rainfall. The high bias they identified is over and above the (artificial) high bias we introduced in KT04, as discussed in Appendix 7.1. Despite the high bias identified by Marchok *et al.* (2007), they found the overall predictive skill of rainfall forecasts using the GFDL model to be competitive with that of other operational models. The operational models evaluated, including the Global Forecast System (GFS) and North American Mesoscale (NAM) models of the National Centers for Environmental Prediction and the GFDL hurricane model, each appear to have significant skill, relative to a "climatology/persistence" benchmark. The comparison benchmark consisted of a climatological hurricane rainfall rate pattern, decreasing radially outward from the storm center, which is projected along the storm's path. Their analysis also showed that the heavier rainfall totals (e.g., exceeding 9 inches' accumulation) were often not predicted at the correct locations by the models.

The heavy rainfall accumulations from the storms in the KT04 idealized experiments were generally confined to a swath within about 100 km of the storm track. This feature is well captured by our storm core precipitation measure described in Appendix 7.1. In the real world, areas of significant rainfall accumulation from hurricanes may not be as confined to the storm track region as they were in our idealized experiments. Under real-world conditions, interactions with extratropical weather systems, landfall effects, interactions with topography, stalling storm motion, and other phenomena (all of which are absent from our idealized experimental design), can lead to more spatially extensive regions with high rainfall accumulations. Future studies with less idealized design will likely produce examples of more areally extensive high-rainfall regions, in which case other aspects of the rainfall response can be examined in more detail.

A key physical mechanism that produces enhanced precipitation rates under warmer, high-CO_2 conditions in the GFDL hurricane model is enhanced boundary layer and lower tropospheric moisture associated with the warmer atmosphere. Since moisture convergence is an important component of the moisture budget for the model, the enhanced moisture leads to enhanced moisture convergence, and thus enhanced core precipitation rates, independent of the effect of intensified circulation. Storm intensification appears to be a secondary influence and acts to further enhance the moisture convergence.

To date, no observational evidence has been reported for an increase in precipitation from hurricane activity (e.g., Groisman *et al.*, 2004), although per-storm precipitation rates have not been evaluated. This is likely to be a challenging measurement and trend detection problem. Trenberth *et al.* (2005) reported a substantial increase during 1988–2003 in column-integrated atmospheric water vapor over the global oceans as derived from the special sensor microwave imager (SSM/I) satellite dataset. Thus it appears that tropical precipitable water vapor (an important ingredient of our simulated precipitation increase) is in fact increasing along with tropical SSTs in a manner consistent with the notion of approximately constant relative humidity, and in accord with model simulations of tropical relative humidity under CO_2 warming conditions (e.g., KT04).

7.3.4 Comparison of KT04 with observed intensity trends

In this section, we consider how our modeled hurricane intensity sensitivity compares with some recently reported observed trends in tropical cyclone (TC) measures.

Emanuel (2005a) introduced a power dissipation index (PDI) of tropical cyclones, based on the time-integrated cube of the maximum surface wind speeds (reported or inferred from central pressure reports and/or satellite data) for the Atlantic and Northwest Pacific tropical cyclone basins from the late 1940s to 2003. After adjusting for time-dependent biases due to changes in measurement and reporting practices, Emanuel reported an approximately 50% increase over the period of record in the annual mean maximum intensity (specifically, the velocity cubed) of the storms. This increase in intensity implies about a 15% increase in terms of wind speed ($1.15^3 = 1.52$). The PDI had a near doubling over the period, with contributions from increases in frequency, intensity, and mean storm duration. The low-pass filtered PDI series were significantly correlated with large-scale tropical SST indices for both basins.

A subsequent comment by Landsea (2005) resulted in adjustments, removing much of the large post-2000 upswing in the Atlantic PDI series through 2003. Emanuel (2005b) reported that these adjustments had a minimal impact on the Northwest Pacific results, or on the multi-basin series or on the correlations with SSTs. A revised and updated analysis (Emanuel, 2007) still shows a clear long-term rise in Atlantic PDI between ~1950 and 2005, correlated with increasing tropical Atlantic sea surface temperature.

Emanuel (2006) restricting the analysis to the Atlantic since about 1980, found a potential intensity increase of about 10% accompanying an SST increase of about 0.5 °C, yielding a potential intensity sensitivity of about 20% per degree Celsius, using NCEP reanalyses of the large-scale climate conditions. Emanuel noted that the change in the actual intensities (average storm lifetime maximum wind speeds) was consistent with that of the potential intensity during this latter period in the Atlantic.

Mann and Emanuel (2006) presented a low-pass filtered time series of annual TC numbers for the Atlantic basin extending back into the late 1800s. This measure tracks the long-term variation in Atlantic MDR SSTs – an independently observed, but physically related, environmental variable – fairly closely, particularly for the century-scale warming trend. This correlation lends support to the notion that the trends in both series are real. On the other hand, one can question what impact changes in observing capabilities have had on the annual TC counts, particularly extending into the late 1800s. Landsea *et al.* (2004) had earlier estimated the number of "missed" Atlantic basin tropical storms and hurricanes per year to be on the order of zero to six for the period 1851–85 and zero to four for the period 1886–1910.

Webster *et al.* (2005) reported that the number of category 4 and 5 hurricanes has almost doubled globally over the past three decades. Although their analysis spans a shorter time period than Emanuel's, their results indicate that a substantial increase has occurred in all six tropical storm basins. In a follow-on study, Hoyos *et al.* (2006) found that the increasing trends in category 4 and 5 tropical cyclones are principally correlated with SST as opposed to other environmental factors. Chan (2006) extended the analysis of Webster *et al.* for the Northwest Pacific basin back to earlier years and argued that the "trend" in that basin is part of a large interdecadal variation. Chan used unadjusted data from the earlier part of the record, in contrast to the adjustments for this period proposed by Emanuel (2005a) for this basin.

A precise comparison of the TC statistics in the above studies with those in the KT04 study is beyond the scope of the present work. The experimental design in KT04 does not consider frequency changes, for example, and so is not directly comparable to the Mann and Emanuel (2006) finding. The PDI measures reported in Emanuel's studies depend on frequency and duration as well as on the intensities of storms. The findings of Webster *et al.* show a redistribution of hurricane intensities preferentially toward more frequent occurrences of category 4 and 5 storms, but no discernible trend in maximum intensities was found. Therefore, for comparison to KT04, we focus on the observed intensity or potential intensity changes as reported in Emanuel (2005a) and Emanuel (2006).

Based on Emanuel (2005a), we assume that a 15% increase in maximum surface wind speeds, as inferred from that study, is representative of the multi-basin change of intensity over the second half of the twentieth century and corresponds to a tropical (30° S–30° N) SST increase of approximately 0.5 °C. This figure yields an approximate sensitivity of 30% per degree Celsius. Alternatively Emanuel's (2006) Atlantic observed potential intensity trend since about 1980 corresponds to a sensitivity of about 20% per degree Celsius. In comparison, the mean intensity result from KT04 is a 5.8% increase in maximum surface wind speeds for a CO_2-induced warming of 1.75 °C, yielding a sensitivity of 3.3% per degree Celsius (or 3.7% if winds are inferred from surface pressure). Emanuel (2005a) reported a theoretically derived potential intensity sensitivity of 5% per degree Celsius, similar to our model results. Although these are admittedly crude comparisons, large differences in sensitivity (by a factor of roughly 5–8 in these examples) remain to be reconciled between our model results and Emanuel's observational findings. Emanuel (2005a, 2006) presented a similar discussion on this topic, proposing that surface wind speed changes could help reconcile the observed Atlantic intensity data with existing theory.

We speculate that the large difference between our model and Emanuel's observed trends in the apparent sensitivity of TC intensity to SST changes may arise from three potential causes: (1) an overestimation of the observed trend due to potential data problems, (2) an underestimation by our model of the sensitivity of hurricane intensities to CO_2-induced SST changes, or (3) impacts of changes in other related environmental factors besides SST on hurricane intensity trends. Each of these possibilities is elaborated on below.

As a first possibility, we speculate that the reported observed intensity trends are overestimated owing to data problems. A recurrent problem in climate studies examining past records for evidence of trends is the impact of changes in instrumentation and reporting practices, which can produce artificial trends. Emanuel (2005a) provided considerable discussion of this issue in the supplemental notes to his study, including a description of adjustments that he made to the data to obtain a more homogeneous record. Webster *et al.* (2005) and Hoyos *et al.* (2006) restricted their analysis to the satellite era in an effort to reduce data homogeneity problems. However, the possibility that other data-related problems are significantly biasing the reported trends continues to be a major issue of concern in the tropical cyclone historical database community (e.g., Landsea *et al.*, 2004; Landsea, 2005; Knaff and Sampson, 2006; Landsea *et al.*, 2006; Kossin *et al.*, 2007). These concerns have focused particularly on basins outside of the Atlantic and on Atlantic intensities prior to the satellite era. Thus Emanuel's (2006) intensity results for the Atlantic basin for the period since about 1980 are likely more reliable than multi-basin measures of intensity changes.

Model deficiencies are a second possible explanation of the differences between observed trends and the modeling studies. Could these lead to an unrealistically low sensitivity to SST changes? In KT04, we presented alternative calculations of CO_2 warming-induced intensity changes based on the potential intensity theories of Emanuel (1987, 2000) and Holland (1997). Despite differences between these theories in the sensitivity of potential intensity to different environmental factors (e.g., Camp and Montgomery, 2001), both formulations gave results that were similar to those of our model calculations, again supporting the general magnitude of our model estimates. The performance of these hurricane intensity frameworks has been assessed to varying degrees based on geographical or seasonal variations in real-world tropical cyclone intensities (e.g., Emanuel, 1987; Holland, 1997; Knutson *et al.*, 1998; Emanuel, 2000; Tonkin *et al.*, 2000). The skill of the GFDL hurricane model for operational intensity forecasts and its relevance for the research issue assessed here has been a subject of debate (Knutson and Tuleya, 2005; Michaels *et al.*, 2005). The 9 km horizontal grid spacing of our model could conceivably be an important limitation of our sensitivity results: a topic that deserves further study. However, we see little evidence at this time that the model used in KT04 is under-sensitive to environmental changes to a degree large enough to explain the factor of 5–8 discrepancy with Emanuel's (2005a, 2006) reported intensity trends.

A third possible explanation of the differences between the results of these studies is that our previous discussion, by focusing on SST sensitivity alone, is too simplistic. In particular, other factors besides SST (e.g., environmental lapse rate,[2] convective available potential energy [CAPE[3]], or tropical cyclone potential intensity) may have changed over the past 25–50 years in a manner different from that simulated by climate models in response to CO_2-only forcing. For example, models and theory indicate that an enhanced warming of the troposphere relative to the sea surface under climate change should act to limit any increase of hurricane intensity for a given SST increase. Therefore, if the tropical upper troposphere, on the whole or in part, has not warmed more than near the surface for some reason during recent decades, the estimated TC intensity change obtained by scaling the KT04 results to recent observed tropical SST changes would be too small. Aside from potential intensity, other investigators emphasize the role of vertical wind shear or other dynamical influences on TC activity (e.g., Gray, 1990; Goldenberg *et al.*, 2001; Bell and Chelliah, 2006). These latter studies suggest that a variety of thermodynamic and dynamic influences, in addition to SST and potential intensity, could be

---

[2] Lapse rate is the rate of change of temperature with height in the atmosphere.

[3] Convective available potential energy is the maximum energy available to an ascending idealized parcel according to parcel theory, and is an integrated measure of moist atmospheric stability in the vertical direction.

influencing any observed trends in TC activity. Holland and Webster (2007) interpret the strong increase in Atlantic hurricane activity in recent years as arising from a combination of a long-term upward trend in total numbers of Atlantic TCs forced by greenhouse warming and an internal oscillation currently favoring more low-latitude developments, which results in a greater proportion of TCs becoming major hurricanes. From these perspectives, explaining past variations of tropical cyclone indices would require more sophisticated statistical or modeling approaches than KT04 to account for the relative roles of these various factors.

We now briefly return to the question of whether an increasingly unstable tropical atmosphere, as is implied by trend profiles computed by using radiosonde data (e.g., Santer *et al.*, 2005) could help explain the large increase in tropical cyclone intensities. Unfortunately, the picture emerging to date from observational studies of trends in tropical tropospheric lapse rates, CAPE, and potential intensity is inconclusive. For example, Gettelman *et al.* (2002) found a preponderance of upward trends in tropical CAPE since roughly the early 1960s. DeMott and Randall (2004) examined a larger number of tropical stations over a shorter period (1973–99) and reported a more evenly divided mixture of increasing and decreasing CAPE trends. Trenberth (2005) questioned the reliability of the radiosonde data in DeMott and Randall's larger sample. Free *et al.* (2004), using a selected set of 14 tropical island radiosonde stations, found only small, statistically insignificant trends in potential intensity over the periods 1975–95 and 1980–95. Emanuel's (2006) reported 10% increase in Atlantic Main Development Region potential intensity since about 1980 was based on HadISST and NCEP reanalysis data. The relation of this multidecadal increase in potential intensity to century-scale tropical SST warming (e.g., Mann and Emanuel, 2006) remains unclear. Also, Santer *et al.* (2005) suggest that there are further problems with radiosonde-derived and satellite-derived temperature trends for the period since 1979 – a conclusion also receiving some support from two additional recent studies, which examined issues with radiosonde-based observations (Sherwood *et al.*, 2005) and satellite-based analyses (Mears and Wentz, 2005). Such problems could conceivably affect trends in atmospheric stability or potential intensity derived from various reanalysis products.

In summary, we have examined three general possibilities for reconciling our model results for intensity with recently reported observational work, but we are unable to reconcile those differences at this time. Both our model and our experimental design may be questioned. On the other hand, they may also have contributed to the discrepancies, particularly for data from outside of the Atlantic basin, or from the presatellite era (pre-1966).

7.4 Conclusions

The main conclusions of this chapter are as follows.

- The tropical Atlantic MDR has warmed by several tenths of a degree Celsius over the twentieth century, including a more rapid rise than for global temperature since about 1970. Coupled model historical simulations using current best estimates of radiative forcings suggest that the warming trend in the MDR may contain a significant contribution from anthropogenic forcing, including increases in atmospheric greenhouse gas concentrations.
- The GFDL hurricane model, in an idealized setting, simulates that a CO_2-induced warming of 1.75 °C causes hurricanes to have increased intensities by 5.8% in terms of surface wind speeds, 14% in terms of central pressure fall, or about one half category on the Saffir–Simpson scale. The 1.75 °C warming found in KT04 is the three-basin average warming simulated by nine global climate models in response to a 1% per year compounded buildup of CO_2 over 80 years: a strong though not extreme scenario for future global mean radiative forcing. Normalizing by the SST change and using wind speeds inferred from central pressure, the wind speed sensitivity of the model is 3.7% per degree Celsius. The storm intensification due to increased SSTs exceeds the moderating effects of more stable tropospheric lapse rates in these hurricane model experiments.
- A measure of the primary core of accumulated precipitation from the idealized hurricanes (the accumulated precipitation during the final 6 hours over a 32,700 km^2 region) shows a 21.6% increase under high-CO_2 conditions.
- Our simulated storm intensities are substantially less sensitive to SST changes compared with historical trends of TC intensity or potential intensity as reported by Emanuel (2005a, 2006). We speculate that the large (factor of 5–8) difference in apparent sensitivity may arise from three potential causes: (i) an overestimation of the observed trends due to potential data problems, (ii) an underestimation by our model of the sensitivity of hurricane intensities to CO_2 warming-induced SST changes, or (iii) impacts of changes in other environmental factors besides SST on hurricane intensity or potential intensity trends. We are unable to reconcile these differences at this stage.

Future work on this topic should include both more extensive evaluations of historical tropical cyclone databases, and simulation efforts aimed at increasing the realism of tropical cyclone climatological behavior. For example, we have not addressed issues of tropical cyclogenesis (frequency), duration, or tracks under modified climate conditions. In the KT04 study, the effects on intensity of wind shear, landfall, topography, and interactions with other weather systems were not addressed.

As a final reminder of the complexities that can arise in the case of Atlantic basin hurricanes, recent climate model simulations point to a significant role for anthropogenic forcing in producing both past and future drought conditions in the Sahel (Held *et al.*, 2005). West African monsoonal activity is believed to be related to Atlantic hurricane activity (e.g., Gray, 1990; Bell and Chelliah, 2006), and indices of atmospheric dust cover emanating from the Sahara region also correlate significantly with the number of TC days (Evans *et al.*, 2006). Furthermore, our results suggest that the low-frequency variability in the MDR, apart from the trend, may contain substantial contributions from both radiative forcing (natural and anthropogenic) and internally generated climate variability. Efforts are now ongoing to attempt to understand and evaluate this complex set of physical mechanisms, which should lead to increased understanding of these important aspects of the tropical climate.

References

Bell, G. D., and Chelliah, M. (2006). Leading tropical modes associated with interannual and multidecadal fluctuations in North Atlantic hurricane activity. *Journal of Climate*, **19**, 590–612.

Broccoli, A. J., Dixon, K. W., Delworth, T. L., Knutson, T. R., and Stouffer, R. J. (2003). Twentieth-century temperature and precipitation trends in ensemble climate simulations including natural and anthropogenic forcing. *Journal of Geophysical Research*, **108**, D24, 4798, doi:10.1029/2003JD003812.

Brohan, P., Kennedy, J. J., Haris, I., Tett, S. F. B., and Jones, P. D. (2006). Uncertainty estimates in regional and global observed temperature changes: a new dataset from 1850. *Journal of Geophysical Research*, **111**, D12106, doi:10.1029/2005JD006548.

Camp, J. P., and Montgomery, M. T. (2001). Hurricane maximum intensity: past and present. *Monthly Weather Review*, **129**, 1704–17.

Chan, J. C. L. (2006). Comment on "Changes in tropical cyclone number, duration, and intensity in a warming environment". *Science*, **311**, 1713.

DeMott, C. A., and Randall, D. A. (2004). Observed variations of tropical convective available potential energy. *Journal of Geophysical Research*, **109**, D02102, doi:10.1029/2003JD003784.

Emanuel, K. A. (1987). The dependence of hurricane intensity on climate. *Nature*, **326**, 483–5.

Emanuel, K. A. (1988). The maximum intensity of hurricanes. *Journal of Atmospheric Science*, **45**, 1143–55.

Emanuel, K. (2000). A statistical analysis of tropical cyclone intensity. *Monthly Weather Review*, **128**, 1139–52.

Emanuel, K. A. (2005a). Increasing destructiveness of tropical cyclones over the past 30 years. *Nature*, **436**, 686–8.

Emanuel, K. A. (2005b). Emanuel replies. *Nature*, **438**, doi:10.1038/nature04427.

Emanuel, K. (2006). Environmental influences on tropical cyclone variability and trends. *Proceedings of 27th AMS Conference on Hurricanes and Tropical Meteorology*, No. 4.2. Available online at http://ams.confex.com/ams/pdfpapers/107575.pdf.

Emanuel, K. (2007). Environmental factors affecting tropical cyclone power dissipation. *Journal of Climate* (in press).

Evans, A., Dunion, J., Foley, J. A., Heidinger, A. K., and Velden, C. S. (2006). New evidence for a relationship between Atlantic tropical cyclone activity and African dust outbreaks. *Geophysical Research Letters*, **33**, L19813, doi:10.1029/2006GL026408.

Free, M., Bister, M., and Emanuel, K. (2004). Potential intensity of tropical cyclones: comparison of results from radiosonde and reanalysis data. *Journal of Climate*, **17**, 1722–7.

Gettelman, A., Seidel, D. J., Wheeler, M. C., and Ross, R. J. (2002). Multidecadal trends in tropical convective available potential energy. *Journal of Geophysical Research*, **107**, 4606, doi:10.1029/2001JD001082.

Goldenberg, S. B., Landsea, C. W., Mestas-Nuñez, A.M., and Gray, W. M. (2001). The recent increase in Atlantic hurricane activity: causes and implications. *Science*, **293**, 474–9.

Gray, W. M. (1990). Strong association between West African rainfall and U.S. landfall of intense hurricanes. *Science*, **249**, 1251–6.

Groisman, P. Y., Knight, R. W., Karl, T. R., *et al.* (2004). Contemporary changes of the hydrological cycle over the contiguous United States: trends derived from *in situ* observations. *Journal of Hydrometeorology*, **5**, 64–85.

Held, I. M., Delworth, T. L., Lu, J., Findell, K. L., and Knutson, T. R. (2005). Simulation of Sahel drought in the twentieth and twenty-first centuries. *Proceedings of the National Academy of Sciences, USA*, **102**(50), 17 891–6.

Holland, G. J. (1997). The maximum potential intensity of tropical cyclones. *Journal of Atmospheric Science*, **54**, 2519–41.

Holland, G. J., and Webster, P. J. (2007). Heightened tropical cyclone activity in the North Atlantic: natural variability or climate trend? *Philosophical Transactions of the Royal Society A*, doi:10.1098/rsta.2007.2083.

Hoyos, C. D., Agudelo, P. A., Webster, P. J., and Curry, J. A. (2006). Deconvolution of the factors contributing to the increase in global hurricane intensity. *Science*, **312**, 94–7.

International Ad Hoc Detection and Attribution Group (IADAG) (2005). Detecting and attributing external influences on the climate system: a review of recent advances. *Journal of Climate*, **18**, 1291–314.

Jones, P. D., and Moberg, A. (2003). Hemispheric and large-scale surface air temperature variations: an extensive revision and an update to 2001. *Journal of Climate*, **16**, 206–23.

Knaff, J. A., and Sampson, C. R. (2006). Reanalysis of West Pacific tropical cyclone maximum intensity 1966–1987. *Proceedings of 27th AMS Conference on Hurricanes and Tropical Meteorology*, No. 5B.5. Available online at http://ams.confex.com/ams/pdfpapers/108298.pdf.

Knutson, T. R., and Tuleya, R. E. (2004) (KT04). Impact of CO_2-induced warming on simulated hurricane intensity and precipitation: sensitivity to the choice of climate model and convective parameterization. *Journal of Climate*, **17**, 3477–95.

Knutson, T. R., and Tuleya, R. E. (2005). Reply. *Journal of Climate*, **18**(23), 5183–7.

Knutson, T. R., Delworth, T. L., Dixon, K. W., *et al.* (2006) (K06). Assessment of twentieth-century regional surface temperature trends using the GFDL CM2 coupled models. *Journal of Climate*, **19**(9), 1624–51.

Knutson, T. R., Tuleya, R. E., and Kurihara, Y. (1998). Simulated increase of hurricane intensities in a CO_2-warmed climate. *Science*, **279**, 1018–21.

Knutson, T. R., Tuleya, R. E., Shen, W., and Ginis, I. (2001). Impact of CO_2-induced warming on hurricane intensities as simulated in a hurricane model with ocean coupling. *Journal of Climate*, **14**, 2458–68.

Kossin, J. P., Knapp, K. R., Vimont, D. J., Murnane, R. J., and Harper, B. A. (2007). A globally consistent reanalysis of hurricane variability and trends, *Geophysical Research Letters*, **34**, LO4815, doi:10.1029/2006GL028836.

Kraft, R. H. (1961). The hurricane's central pressure and highest wind. *Marine Weather Log*, 5, 155.

Landsea, C. W. (1993). A climatology of intense (or major) Atlantic hurricanes. *Monthly Weather Review*, 121, 1703–13.

Landsea, C. W. (2005). Hurricanes and global warming. *Nature*, 438, doi:10.1038/nature04477.

Landsea, C. W. (2007). Counting Atlantic tropical cyclones back to 1900. *EOS*, 88, 197, 202.

Landsea, C. W., Anderson, C., Charles, N., *et al.* (2004). The Atlantic hurricane database re-analysis project: documentation for the 1851–1910 alterations and additions to the HURDAT database. In *Hurricanes and Typhoons: Past, Present, and Future*, ed. R. J. Murnane and K.-B. Liu. New York: Columbia University Press, pp. 177–221.

Landsea, C. W., Harper, B. A., Hoarau, K., and Knaff, J. A. (2006). Can we detect trends in extreme tropical cyclones? *Science*, 313, 452–4.

Mann, M., and Emanuel, K. (2006). Atlantic hurricane trends linked to climate change. *Eos*, 87, 233–41.

Marchok, T., Rogers, R., and Tuleya, R. (2007), Validation schemes for tropical cyclone quantitative precipitation forecasts: evaluation of operational models for US landfalling cases. *Weather and Forecasting*, 22, 726–46.

Mears, C. A., and Wentz, F. J. (2005). The effect of diurnal correction on satellite-derived lower tropospheric temperature. *Science*, 309, 1548–51.

Meehl, G. M., Washington, W. M., Amman, C. M., *et al.* (2004). Combinations of natural and anthropogenic forcings in twentieth-century climate. *Journal of Climate*, 17, 3721–7.

Michaels, P. J., Knappenberger, P. C., and Landsea, C. (2005). Comments on "Impacts of CO_2-induced warming on simulated hurricane intensity and precipitation: sensitivity to the choice of climate model and convective scheme." *Journal of Climate*, 18, 5179–82.

Parker, D. E., Folland, C. K., and Jackson, M. (1995). Marine surface temperature: observed variations and data requirements. *Climate Change*, 31, 559–600.

Rayner, N. A., Parker, D. E., Horton, E. B., *et al.* (2003). Global analyses of sea surface temperature, sea ice, and night marine air temperature since the late nineteenth century. *Journal of Geophysical Research*, 108 (D14), 4407, doi:10.1029/2002JD002670.

Santer, B. D., *et al.* (2005). Amplification of surface temperature trends and variability in the tropical atmosphere. *Science*, 309, 1551–6.

Santer, B. D., *et al.* (2006). Forced and unforced ocean temperature changes in Atlantic and Pacific tropical cyclogenesis regions. *Proceedings of the National Academy of Sciences, USA*, 103, 13905–10, 10.1073/pnas.0602861103.

Shen, W., Tuleya, R. E., and Ginis, I. (2000). A sensitivity study of the thermodynamic environment on GFDL model hurricane intensity: implications for global warming. *Journal of Climate*, 13, 109–21.

Sherwood, S. C., Lanzante, J. R., and Meyer, C. L. (2005). Radiosonde daytime biases and late-twentieth-century warming. *Science*, 309, 1556–9.

Smith, T. M., and Reynolds, R. W. (2003). Extended reconstruction of global sea surface temperatures based on COADS data (1854–1997). *Journal of Climate*, 16, 1495–510.

Smith, T. M., and Reynolds, R. W. (2004). Improved extended reconstruction of SST (1854–1997). *Journal of Climate*, 17, 2466–77.

Tonkin, H., Holland, G. J., Holbrook, N., and Henderson-Sellers, A. (2000). An evaluation of thermodynamic estimates of climatological maximum potential tropical cyclone intensity. *Monthly Weather Review*, 128, 746–62.

Trenberth, K. (2005). Uncertainty in hurricanes and global warming. *Science*, **308**, 1753–4.

Trenberth, K. E., and Shea, D. J. (2006). Atlantic hurricanes and natural variability in 2005. *Geophysical Research Letters*, **33**, L12704, doi:10.1029/2006GL026894.

Trenberth, K. E., Fasullo, J., and Smith, L. (2005). Trends and variability in column-integrated atmospheric water vapor. *Climate Dynamics*, **24**, 741–58.

Tuleya, R. E., DeMaria, M., and Kuligowski, R. (2007). Evaluation of GFDL and simple statistical model rainfall forecasts for U.S. landfalling tropical storms. *Weather and Forecasting*, **22**, 56–70.

Vecchi, G. A., and Soden, B. J. (2007). Increased tropical Atlantic wind shear in model projections of global warming. *Geophysical Research Letters*, **34**, L08702, doi:10.1029/2006GL028905.

Webster, P. J., Holland, G. J., Curry, J. A., and Chang, H.-R. (2005). Changes in tropical cyclone number, duration, and intensity in a warming environment. *Science*, **309**, 1844–6.

Zhang, R., and Delworth, T. L. (2005). Simulated tropical response to a substantial weakening of the Atlantic thermohaline circulation. *Journal of Climate*, **18**(12), 1853–60.

APPENDIX 7.1 REVISED PRECIPITATION STATISTICS

Inspection of high-frequency (every model time step) precipitation time series from the hurricane model simulations of KT04 reveals periodic "spikes" in precipitation rates. The spikes occur just after application, once per integration hour, of a 1–2–1 spatial filter to the mass and pressure fields of the model during its time integration. Use of time-accumulated or time-averaged (across all time steps) precipitation fields largely eliminates the effect of this artifact of the numerical smoothing.

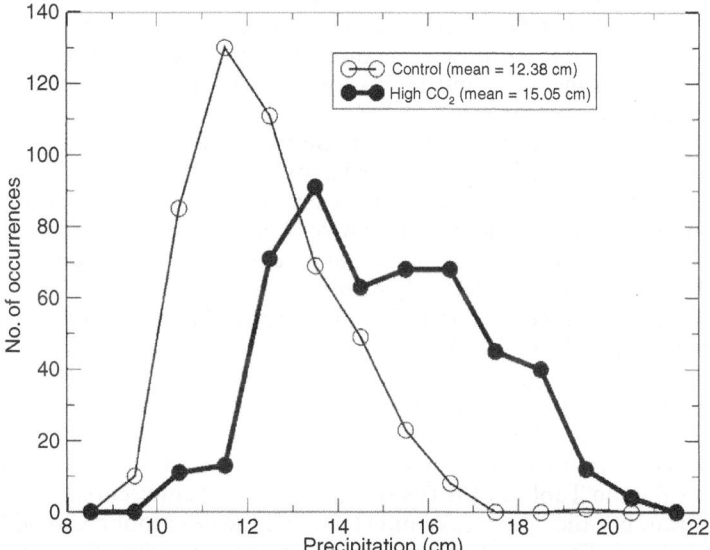

Figure 7.5. As Figure 7.4, but for precipitation (centimeters) averaged over the 102 model grid points (comprising a 32,700 km^2 area) with highest accumulated precipitation during the final 6 h of the 5-day hurricane model integrations.

However, the 100 km radially averaged precipitation rate statistic in KT04 was based on instantaneous precipitation fields from the model, and these coincidentally were computed at a time step of the model integration in which a spike in precipitation rate was occurring (i.e., at exactly hour 120). Thus the 100 km rate measure in KT04 turns out to be unrepresentative of the true average precipitation rates in the model. (Note that such spikes are not evident in storm intensity metrics for the model.)

To assess the impact of this problem on our earlier results, we reanalyzed the precipitation data and created an alternative storm-core precipitation measure from a different (time-accumulated) archived field: the 6-hour accumulated rainfall. This field was archived for each of the four cumulus convection treatments except for the Emanuel scheme. Therefore, our revised analysis and comparisons are limited to three of the four convection treatments used in KT04 (i.e, the PAN, KURI, and resolved convection schemes, see Table 7.1.)

Maps of 6-hour accumulated rainfall from the experiments (not shown) are dominated by an oblong-shaped maximum along the 6-hour storm track. Because of the oblong shape of this feature, we chose not to compute a simple circular average as in KT04 (where we were using instantaneous fields). Instead, we sorted the 6-hour accumulated precipitation field grid points (for the final 6-hour period) from largest to smallest and then averaged over the highest 102 values (out of a total of 202,500 grid points), comprising an area of 32,700 km^2, or slightly more than the area covered by a circular region of radius 100 km.

Comparisons between the revised and original measures indicate that the original control run precipitation measure (converted to units of centimeters per 6 hours for comparison) is artificially inflated by roughly 80%, which we interpret as primarily due to sampling the peaks or spurious spikes in the time series associated with the spatial smoothing procedure. However, we find that the percentage changes from control to high-CO_2 samples are quite similar for the new and original precipitation measures. The revised analysis (summarized in Figure 7.5 and Table 7.1) yields a 21.6% aggregate increase (high CO_2 vs. control) for the three available convection schemes, as compared with 17.1% for those same three schemes using the original 100 km radius instantaneous measure, and 18.3% for the original measure averaged over all four moist convection treatments.

The "maximum precipitation rate in the domain" measure was a secondary precipitation statistic presented in KT04 aimed at exploring the small-scale local maxima in precipitation. This statistic was also affected by the precipitation spikes. In this case, we could not reconstruct a suitable alternative from the available archived data. Therefore, we recommend that the maximum precipitation rate in domain statistics in KT04 not be considered quantitatively reliable.

The "inner nest average" precipitation measure in KT04 was not described accurately in the text or in Table 2 of KT04. In fact, this statistic should have been labeled as the "5-day total accumulated rainfall (in centimeters) area-averaged over the entire model domain ($70° \times 70°$)." Again, this statistic was a secondary precipitation measure designed to contrast how the larger-scale precipitation rates changed in response to CO_2-induced warming in comparison to the storm-core rainfall. The reported control run values in Table 2 of KT04 (e.g., 1.62 for "all convection schemes") are correct as given, but the units are centimeters, and the description should have been as corrected above. This statistic, being an accumulated precipitation over the storm lifetime, is only minimally affected by the spatial smoothing issue discussed above.

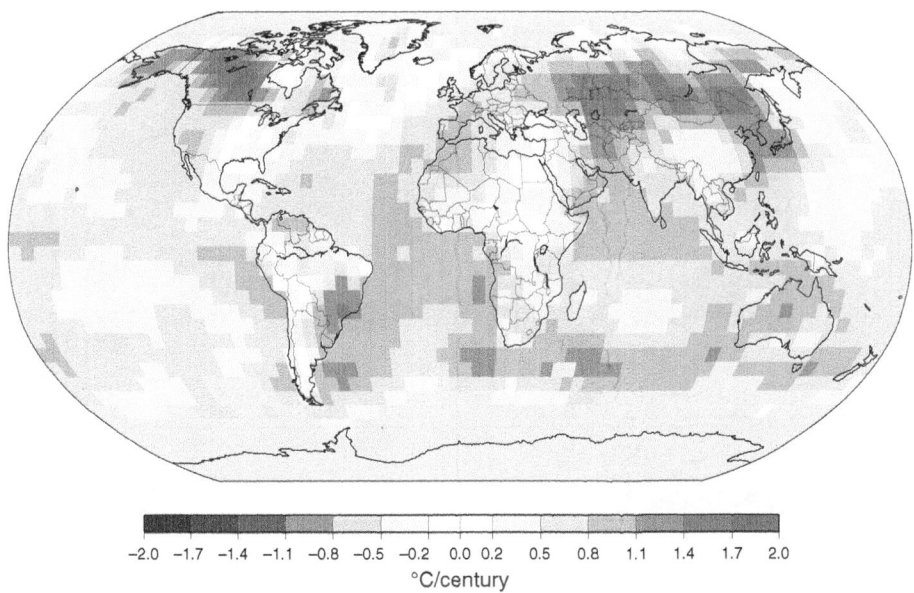

Figure 2.2. Linear trends in average annual temperature for the period 1901–2005. Areas in gray are excluded due to a lack of reliable data. (Data is from Smith and Reynolds, 2005.)

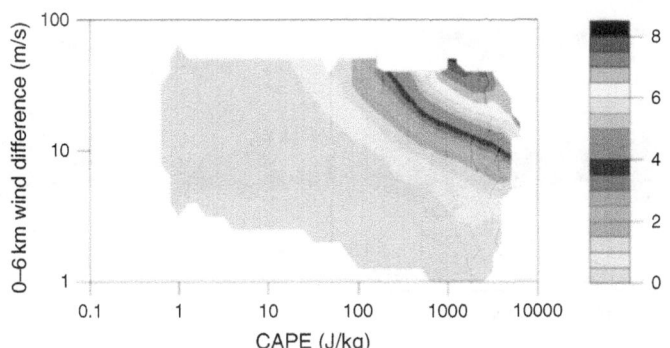

Figure 3.4. Probability in percent of environments producing severe thunderstorm with a tornado with at least F2 damage, 5 cm diameter hail, or 120 km h^{-1} wind gusts in the USA. (Based on data described by Brooks and Craven, 2002.)

This plate section is available for download in color from
www.cambridge.org/9780521870283

(a)

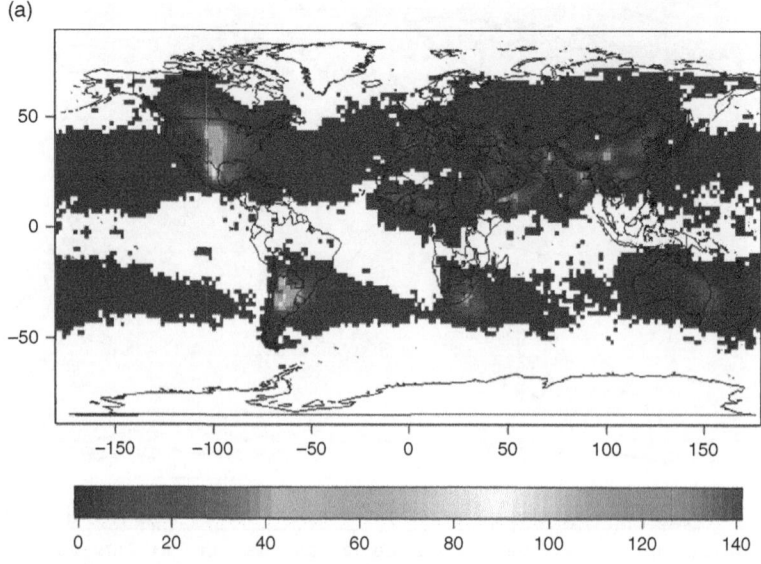

(b)

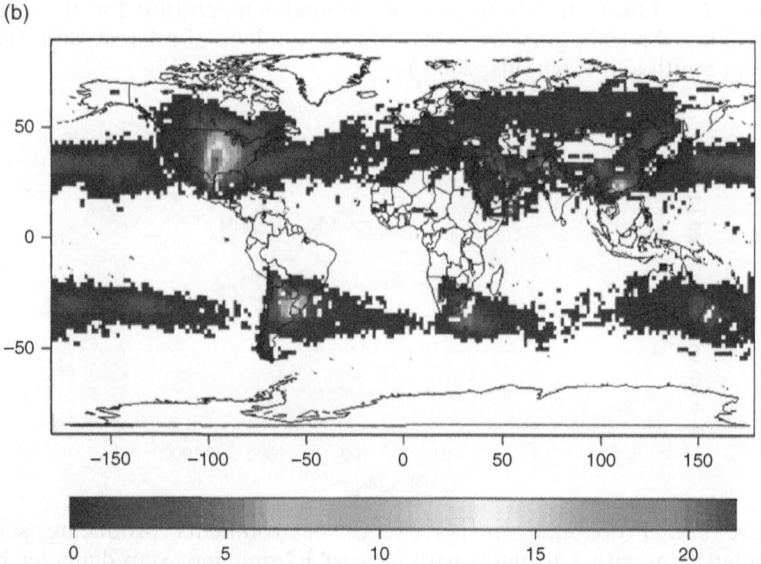

Figure 3.5. Six-hourly periods per year with environments supportive of significant severe thunderstorms (a) and significant tornadoes (b) based on NCAR/NCEP reanalysis data for 1970–1999. (Updated from Brooks *et al.* [2003b].)

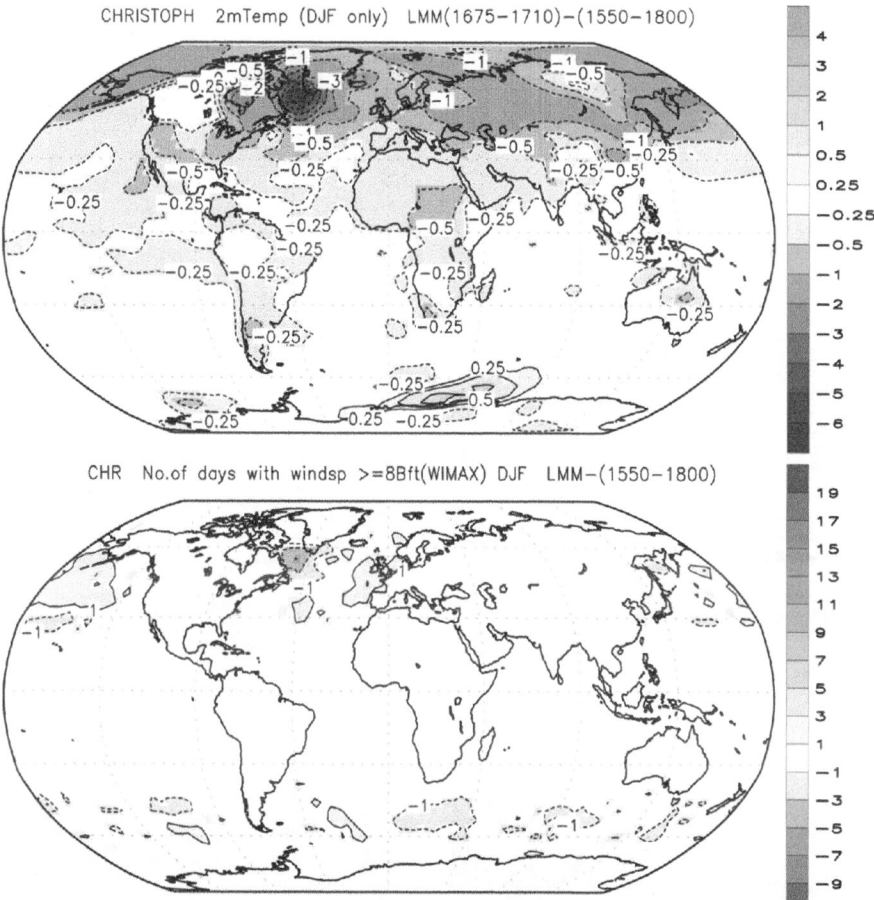

Figure 4.5. Simulated differences in winter between the Late Maunder Minimum (LMM, 1675–1710) and the pre-industrial time (1550–1850), in terms of air temperature (top, K) and of number of gale days (wind speeds of Beaufort Force 8 and more). Note that the LMM is portrayed by the model as particularly cold, but the storm activity shows little change. (Courtesy Irene Fischer-Bruns.)

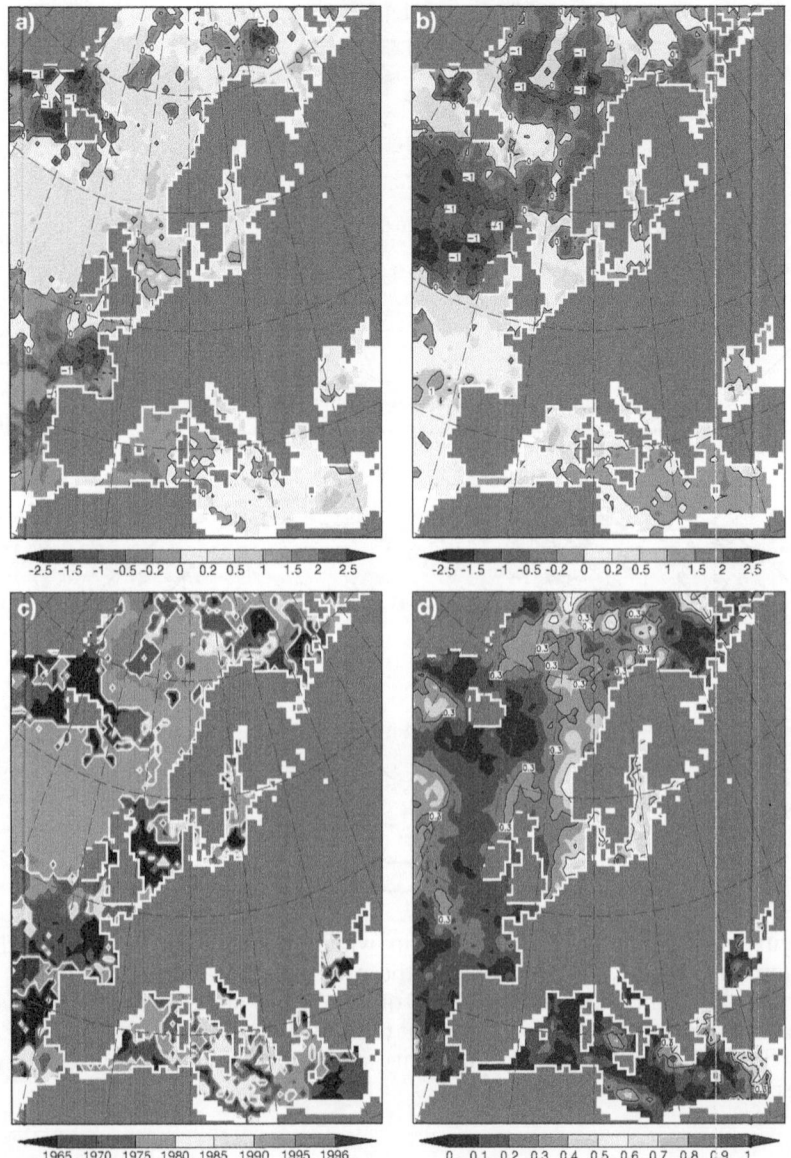

Figure 4.6. Piecewise linear trends before and after a change point T in the total number of storms per year with maximum wind speeds exceeding $17.2\,\mathrm{m\,s^{-1}}$. Both the trends and the change point are determined by a best fit to the data time series. (a) Trends for the first period 1958–T; (b) trends for the second period T–2002. Units in both cases are number of storms per year. (c) Year T at which a change in trends is indicated by the statistical model; (d) Brier skill score of the bilinear trend fitting the data as compared to using one trend for the entire period. (After Weisse *et al.*, 2005.)

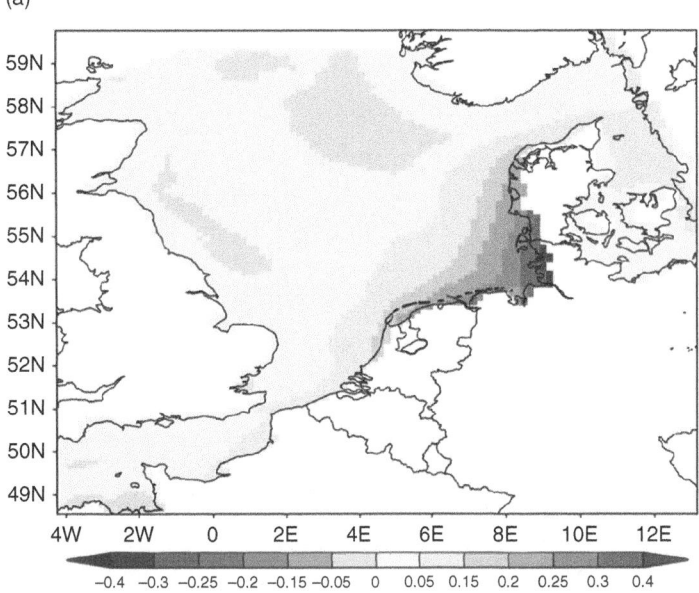

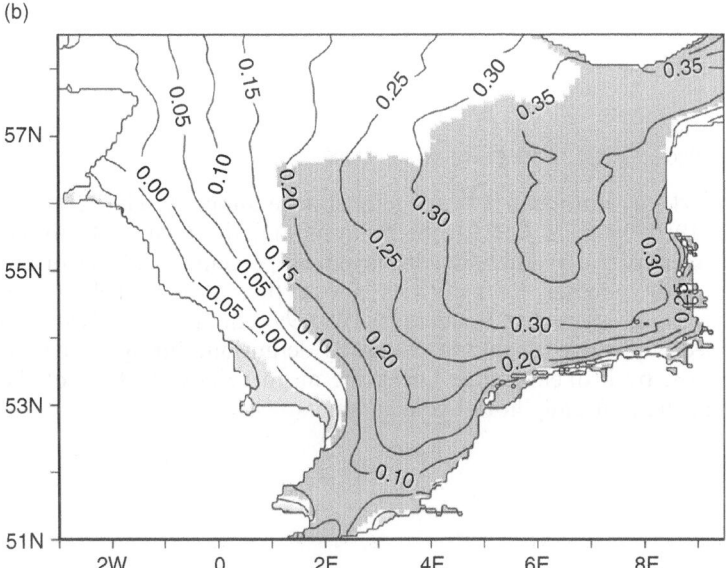

Figure 4.7. Expected changes in wind-related storm surge heights
(a) (maximum averaged across many years, Rossby Center Atmosphere
Ocean (RCAO) model, emission scenario A2) and ocean wave heights
(b) (change of 99th percentile; averaged across a series of simulations using
different models and both emission scenarios A2 and B2). Shading in (b)
indicates areas where signals from all models and scenarios have the same sign
(red, positive; blue, negative) in the North Sea at the end of the twenty-first
century. Units are meters. (Courtesy Katja Woth and Iris Grabemann.)

(a)

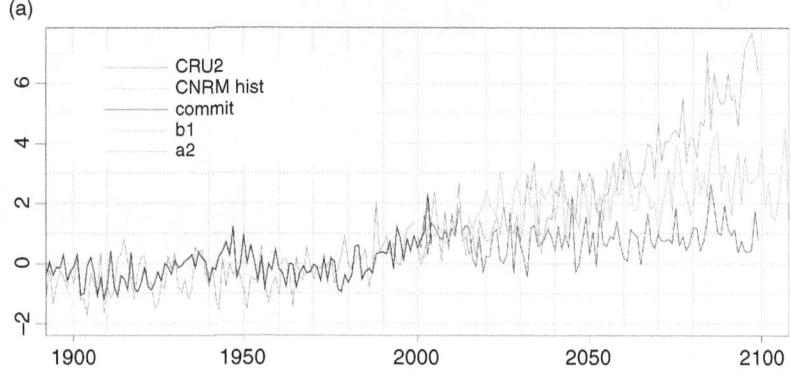

(b)

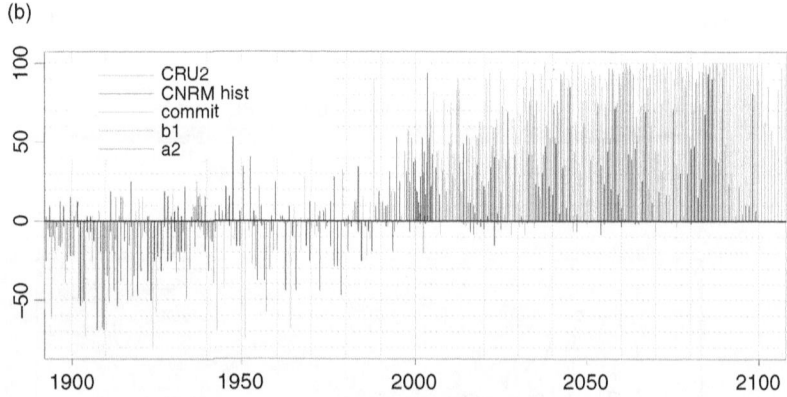

Figure 5.1. (a) European average temperature anomaly relative to the base period 1950–99 from CRU2 observations and the CGCM historical and future scenarios. (b) H&CSIs displayed as percentage of European grid points with summer temperatures above the 90th or below the 10th percentiles of their local summer 1950–99 climatology. HSI (CSI) is displayed in positive (negative) values. By definition, during the base period (delineated by broken orange lines), the mean values are 10% of the area experiencing unusually hot or cold summers.

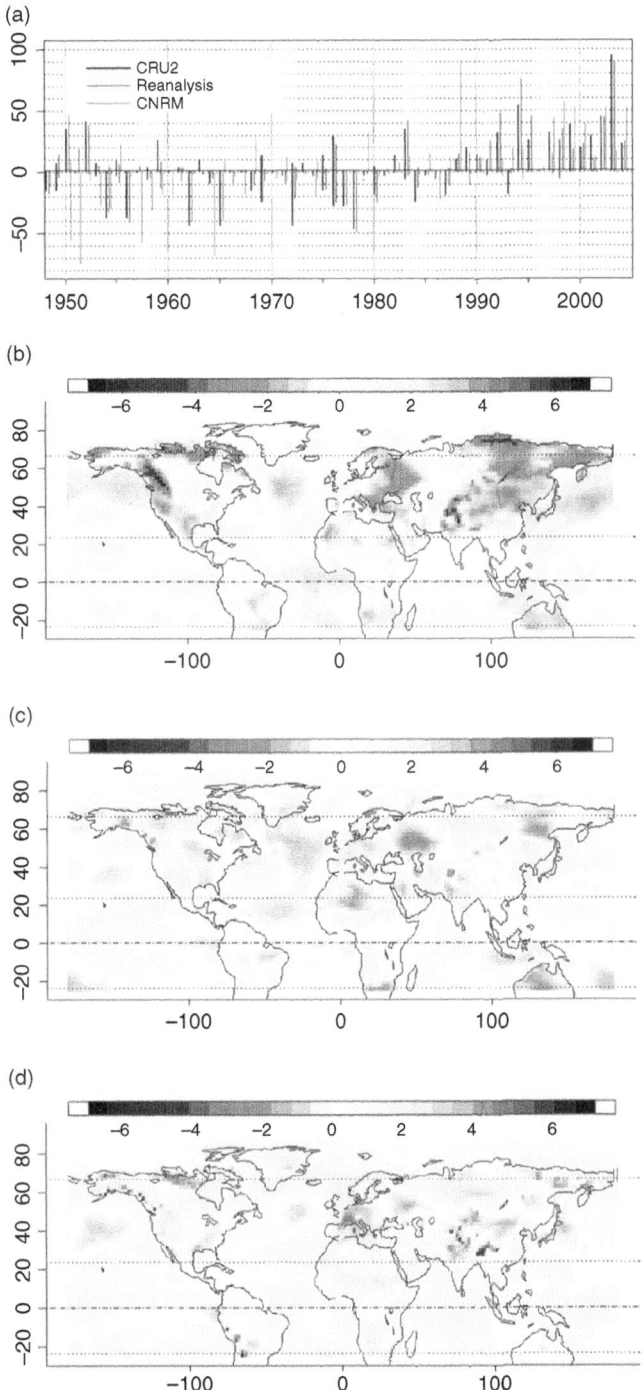

Figure 5.2. (a) H&CSI from CRU2 (black), NCEP/NCAR Reanalysis (red),
and CGCM (dashed gray) for the common period 1950–2003. (b) Reanalyzed
temperature anomalies for summer 1976, (c) 1994, and (d) 2003.

(a)

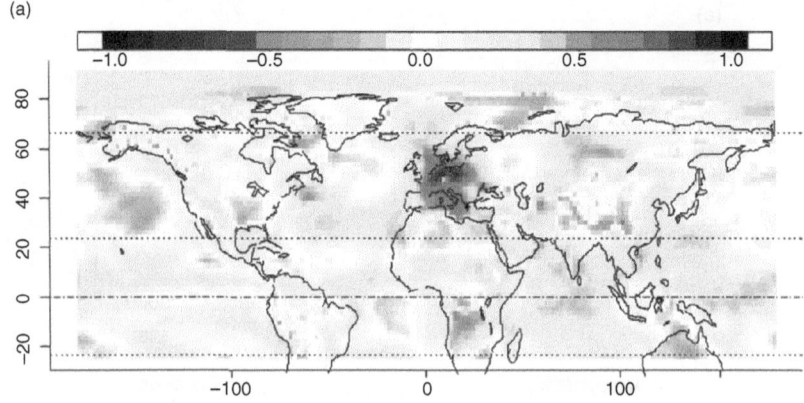

(b)

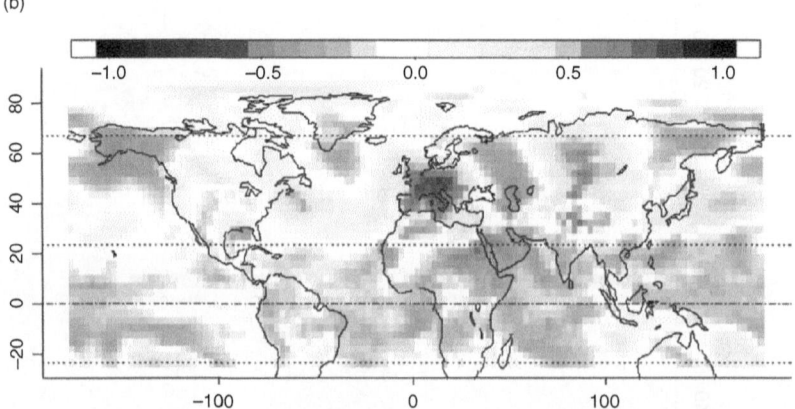

Figure 5.3. Correlation coefficient between Western European HSI and local JJA Ta2 m over the globe in (a) Reanalysis 1948–2003, and (b) the CGCM commit run 2000–99.

(a)

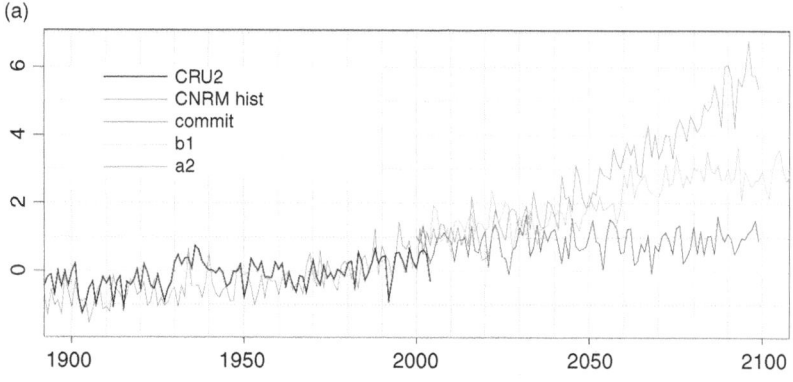

(b)

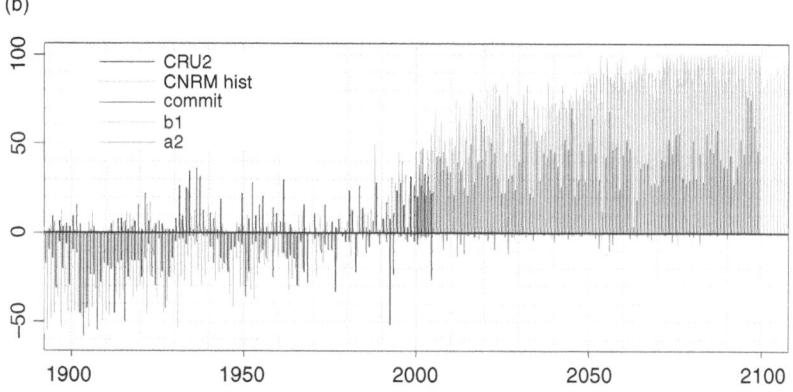

Figure 5.4. Same as Figure 5.1, but for North America; (a) North American average temperature anomaly relative to the base period 1950–99 from CRU2 observations and the CGCM historical and future scenarios. (b) H&CSI displayed as percentage of North American grid points with summer temperatures above the 90th or below the 10th percentiles of their local summer 1950–99 climatology. HSI (CSI) is displayed in positive (negative) values. By definition, during the base period (delineated by broken orange lines), the mean values are 10% of the area experiencing unusually hot or cold summers.

(a)

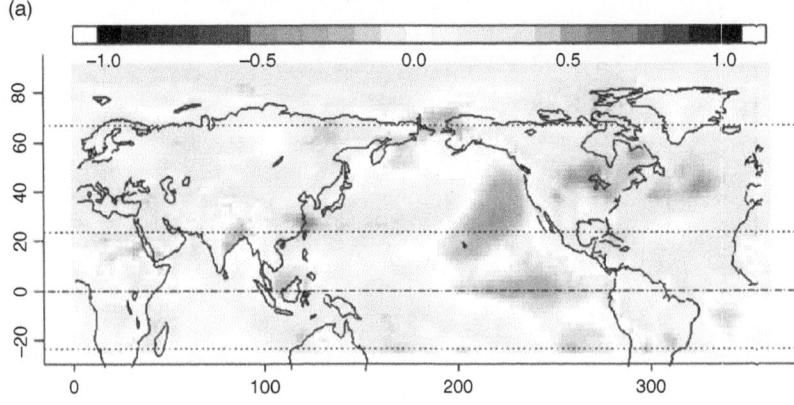

(b)

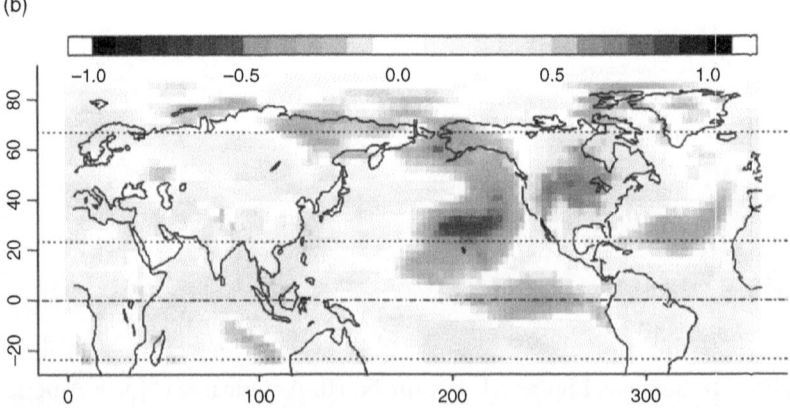

Figure 5.5. Correlation coefficient between North American HSI and local JJA Ta2 m over the globe in (a) Reanalysis, 1948–2003, and (b) the CGCM commit run, 2000–99.

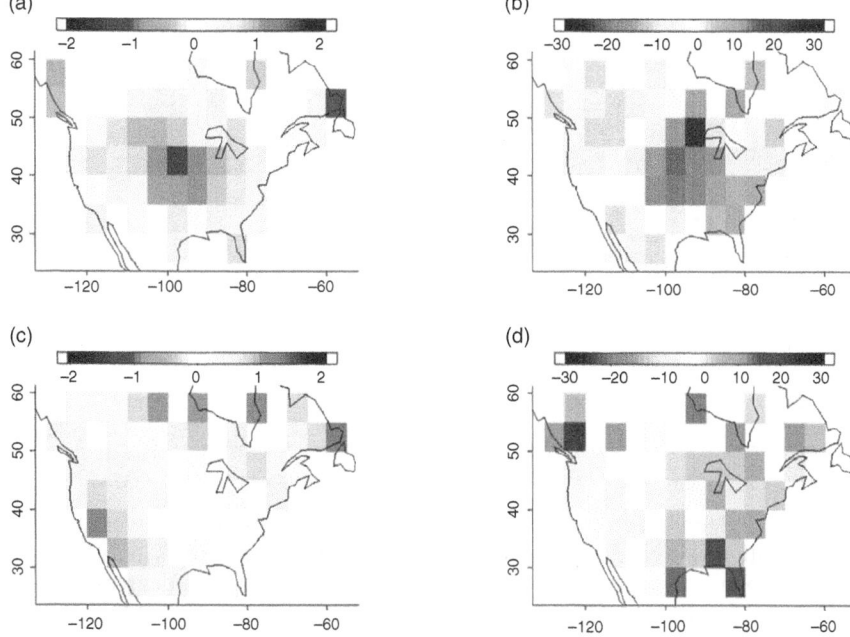

Figure 5.6. Summer temperature (a, c) and precipitation (b, d) anomalies for two 11-year periods: 1930–1940 (a, b) and the most recent period of record, 1994–2004 (c, d). Notice the reversed color scale on the precipitation plots. Canadian precipitation records suffer from missing data at the end of the record, accounting for the visible discontinuity along the US–Canada border in (d).

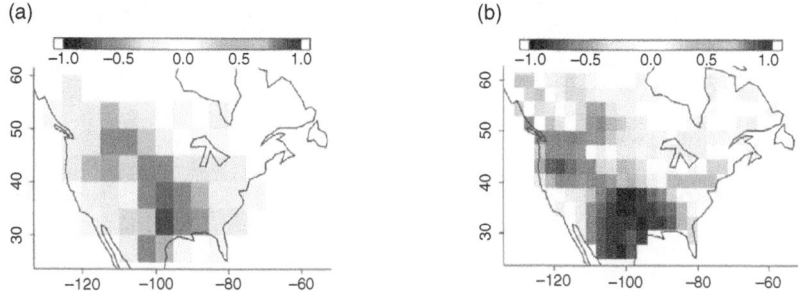

Figure 5.7. Correlation coefficient between local June–August average temperature and precipitation in (a) observations (GHCN V2 and CRU2 evaluated over the period of record, 1900–2004, or slightly shorter based on data availability for the individual grids) and (b) the model (the commit run, 100 years).

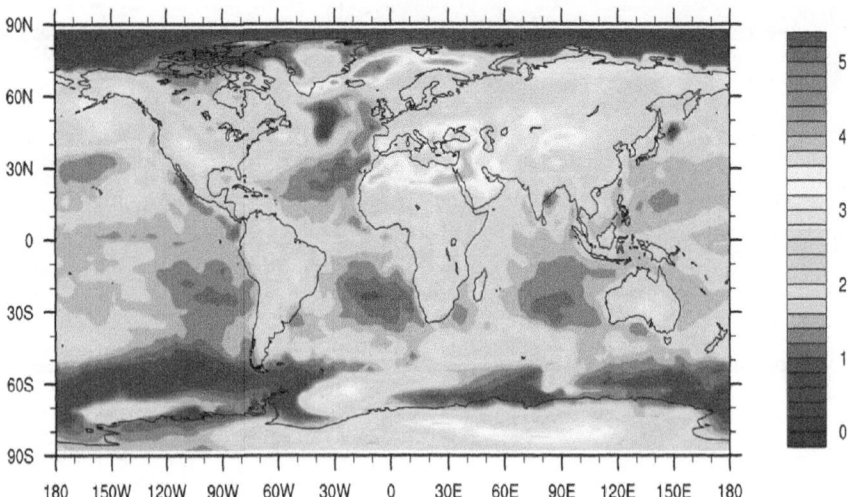

Figure 6.2. How much more intense will heat waves become, on average, in a future with no curbing of greenhouse gas emissions? The answer is in this figure. At each point on the map, the color corresponds to a value in degrees Celsius (see legend on the right of the figure). This value is the projected magnitude of the increase in temperature during the three warmest consecutive nights of the summer, in the climate of the end of the twenty-first century (the last two decades, to be precise), compared with the climate of the last three decades of the twentieth century. Summer is defined "loosely" as the 6-month period from May through October in the Northern Hemisphere and from November through April in the Southern Hemisphere.

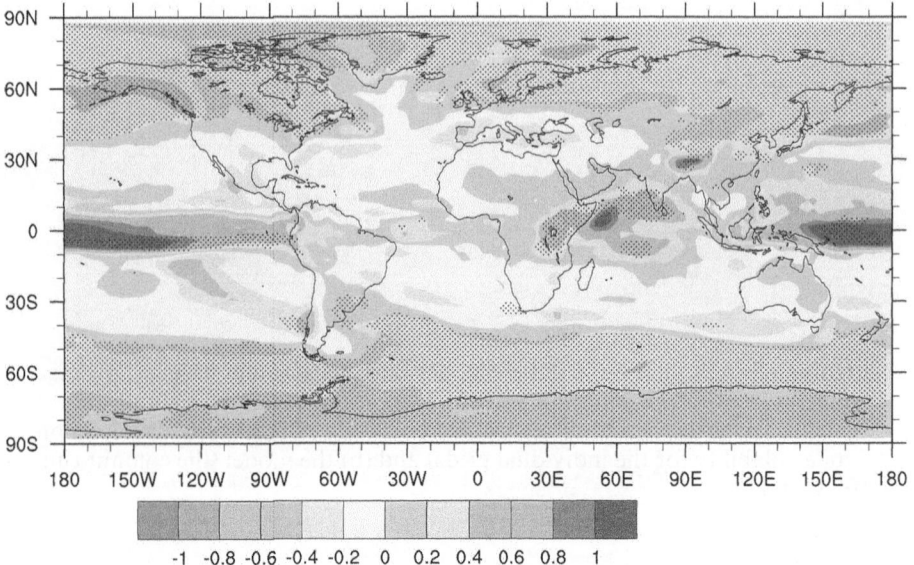

Figure 6.8. Multi-model ensemble means of spatial patterns of change in average precipitation under the A1B scenario. Units are millimeters per day. Dotted areas are statistically significant. First published in Meehl *et al.* (2004).

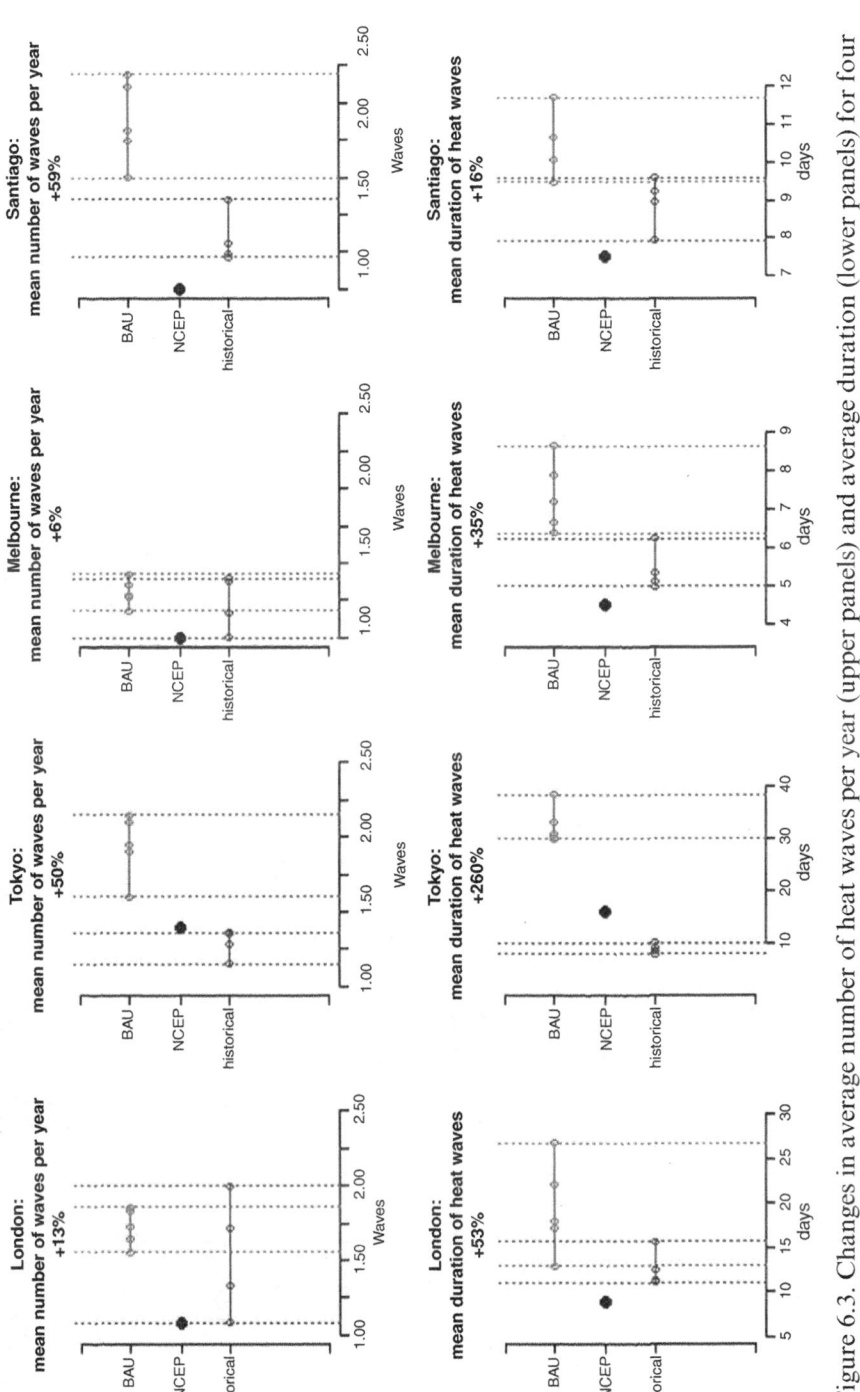

Figure 6.3. Changes in average number of heat waves per year (upper panels) and average duration (lower panels) for four grid points representative of cities at the midlatitudes of both hemispheres (London, Tokyo, Melbourne, and Santiago). The NCEP estimates are shown by a black mark labeled "NCEP" on each graph. Blue dots and segments indicate the individual members of the historical run and their range (also highlighted by blue vertical dotted lines). Similarly, future business-as-usual ensemble members are represented in red.

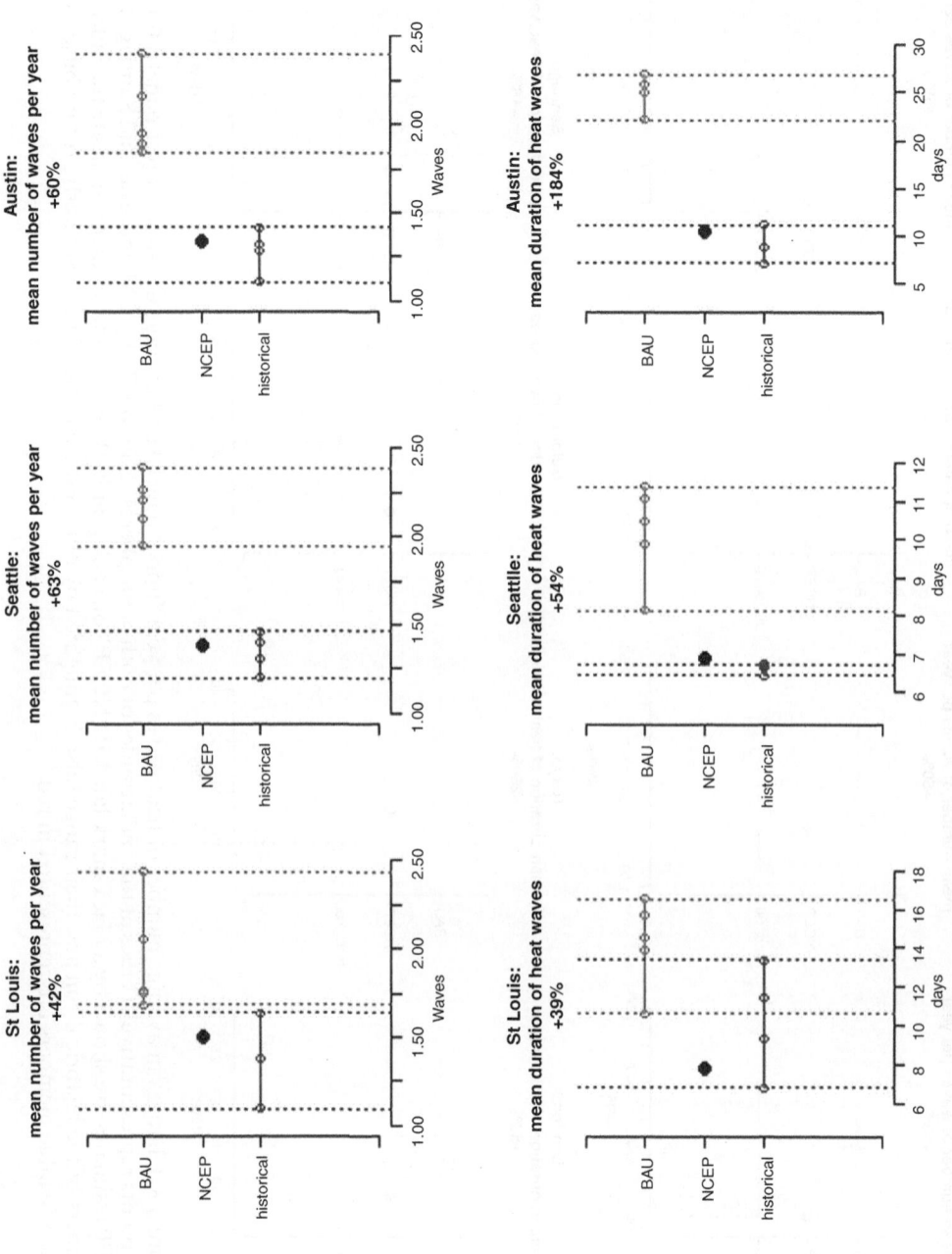

Figure 6.4. As Figure 6.3, for three grid points representative of three US cities (Seattle, Austin, and St. Louis).

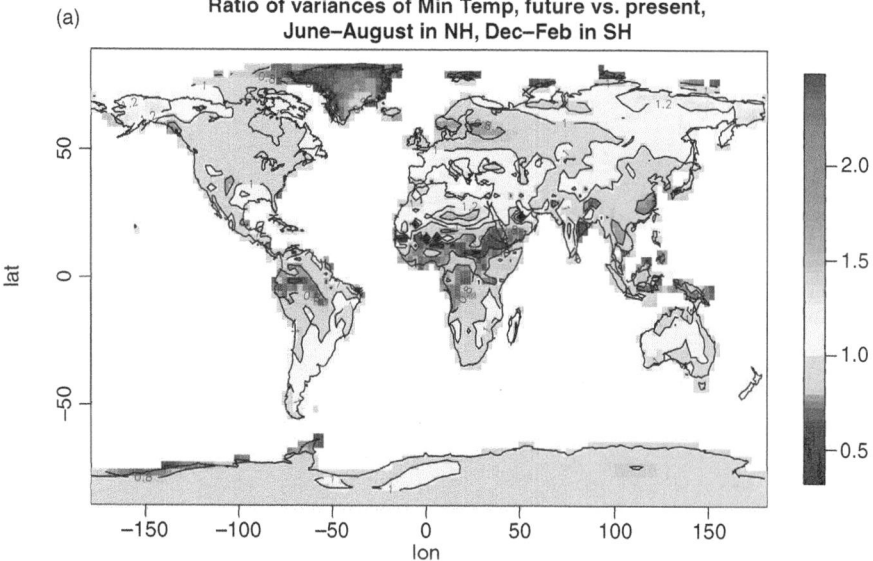

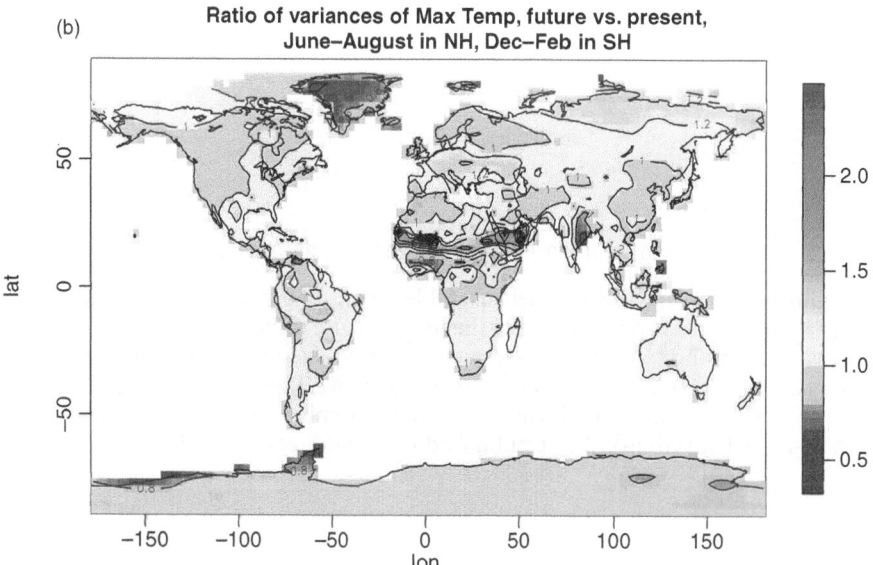

Figure 6.5. Projected changes in standard deviation of maximum (b) and minimum (a) temperature under a business-as-usual scenario, for the end of the twenty-first century compared to the baseline climatology of 1961–90.

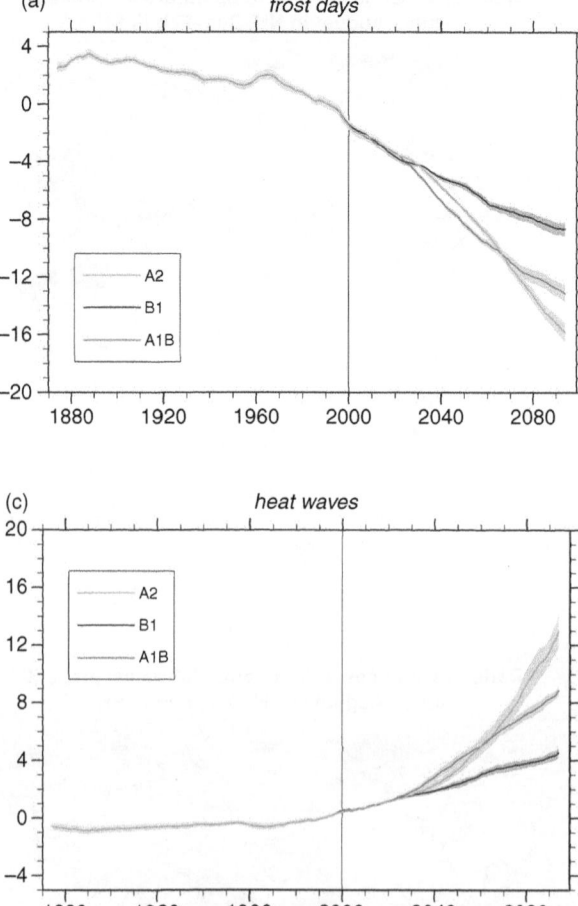

Figure 6.6a and c. Multi-model ensemble means of global average time series (after standardization) for two of the temperature indices (*frost days* and *heat waves*) (a, c) and two of the precipitation indices (*precipitation intensity* and *dry days*) (b, d). The three scenarios are shown in different colors. The shading represents one standard deviation of the ensemble mean, as a measure of inter-model variability. First published in Tebaldi *et al.* (2007).

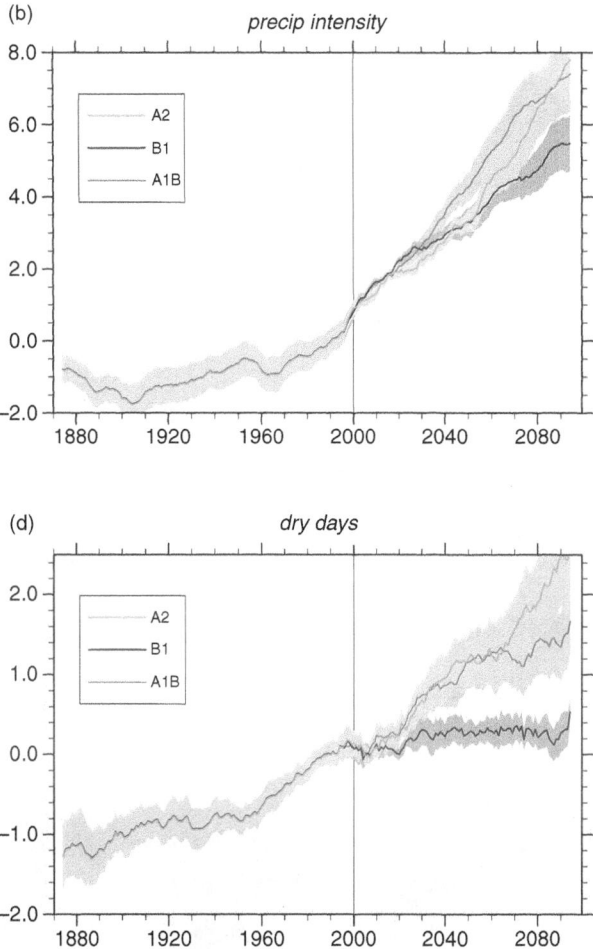

Figure 6.6b and d. Multi-model ensemble means of global average time series (after standardization) for two of the temperature indices (*frost days* and *heat waves*) (a, c) and two of the precipitation indices (*precipitation intensity* and *dry days*) (b, d). The three scenarios are shown in different colors. The shading represents one standard deviation of the ensemble mean, as a measure of inter-model variability. First published in Tebaldi *et al.* (2007).

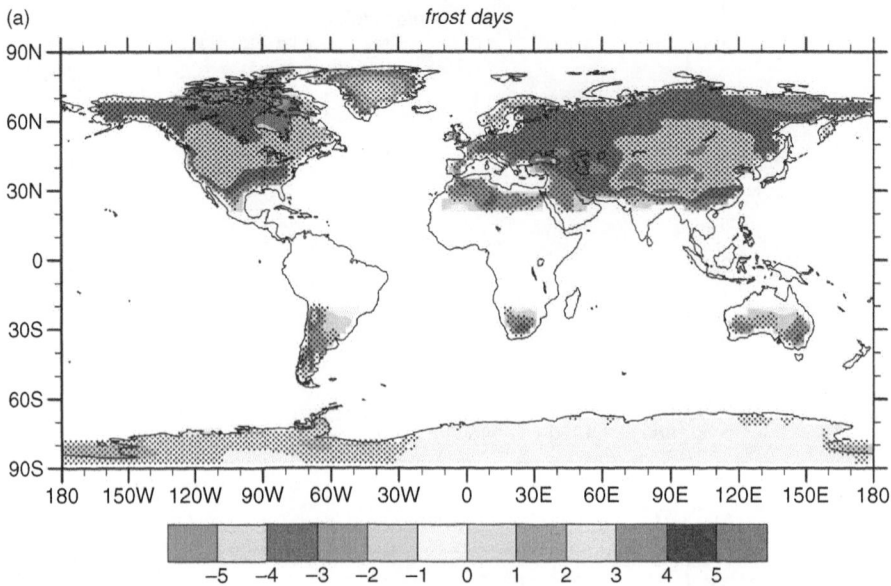

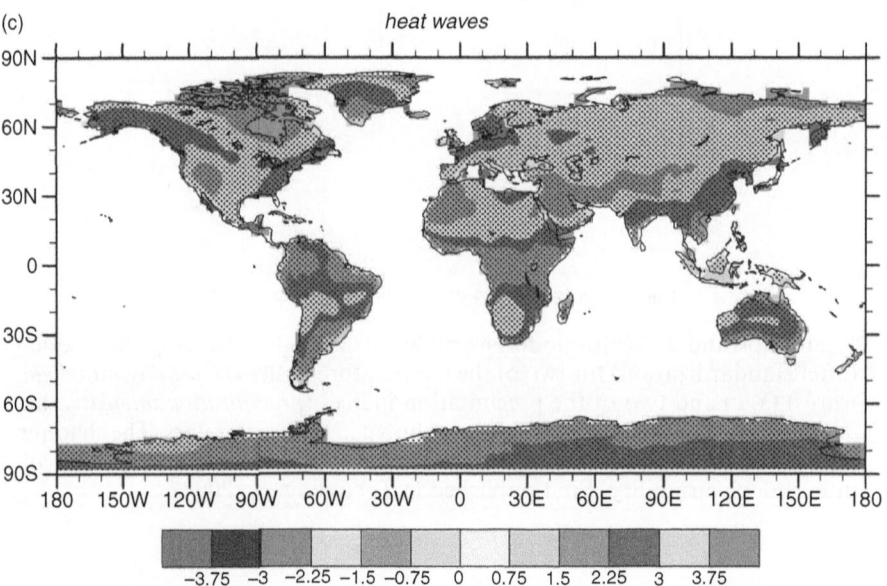

Figure 6.7a and c. Multi-model ensemble means of spatial patterns of change (after standardization) under the A1B scenario for the same four indices as in Figure 6.6. Shown are the differences between two 20-year averages (2080–99 vs. 1980–99) for the temperature indices (a, c) and the precipitation indices (b, d). Grid points over the oceans were not included in the analysis and are thus left blank. Subtropical regions are also not included in the analysis of *frost days*. Stippling indicates that a majority of the models agree on the statistical significance of the change. First published in Tebaldi *et al.* (2007).

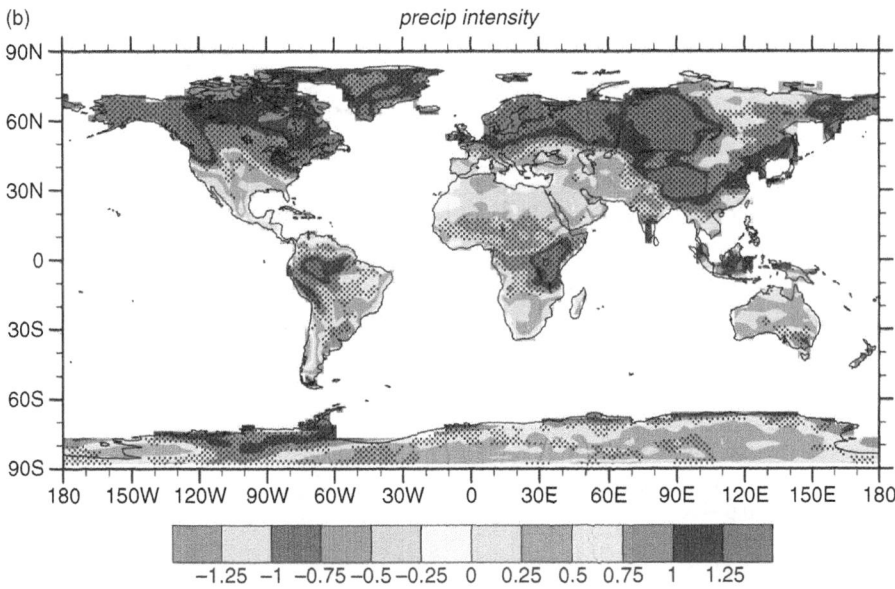

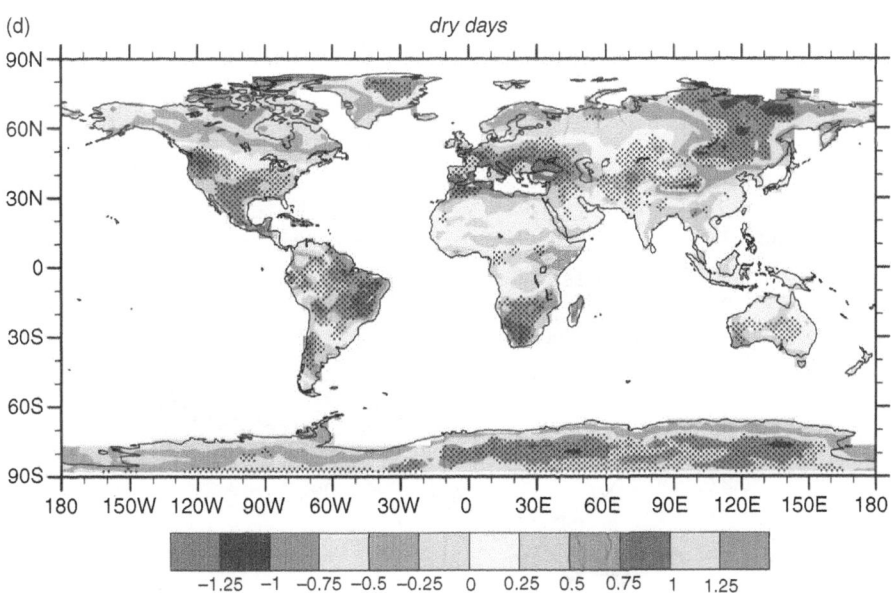

Figure 6.7b and d. Multi-model ensemble means of spatial patterns of change (after standardization) under the A1B scenario for the same four indices as in Figure 6.6. Shown are the differences between two 20-year averages (2080–99 vs. 1980–99) for the temperature indices (a, c) and the precipitation indices (b, d). Grid points over the oceans were not included in the analysis and are thus left blank. Subtropical regions are also not included in the analysis of *frost days*. Stippling indicates that a majority of the models agree on the statistical significance of the change. First published in Tebaldi *et al.* (2007).

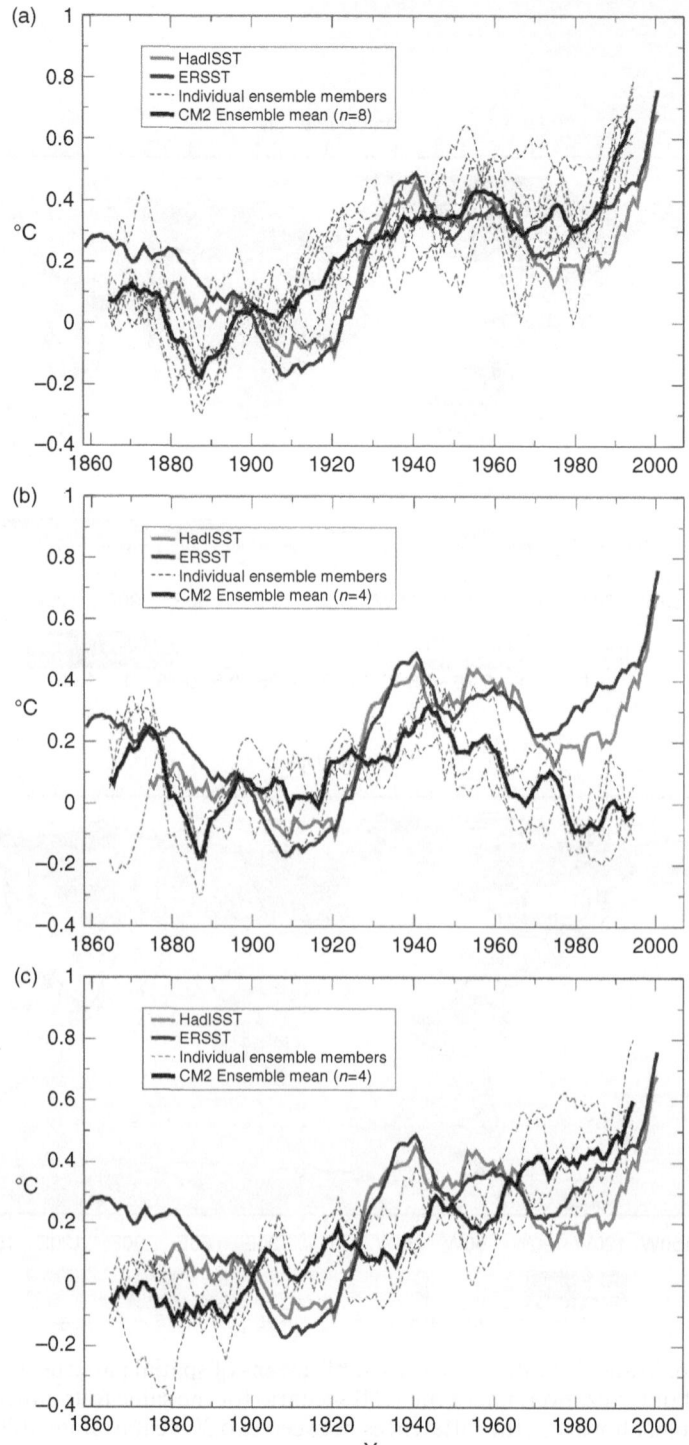

Figure 7.2. Observed sea surface temperature variations in the MDR from HadISST (red) and ERSST (blue) datasets vs. CM2 historical climate simulations using (a) all forcings, (b) natural forcings only, or (c) anthropogenic forcings only. Ten-year running mean anomalies for the August–October season, referenced to 1881–1920 means in degrees Celsius, are shown. Black dashed curves are individual CM2.0 or CM2.1 ensemble members; thick black curves are the CM2.0/CM2.1 ensemble means ($n = 8$ experiments with all forcings; $n = 4$ experiments with natural- or anthropogenic-only forcings).

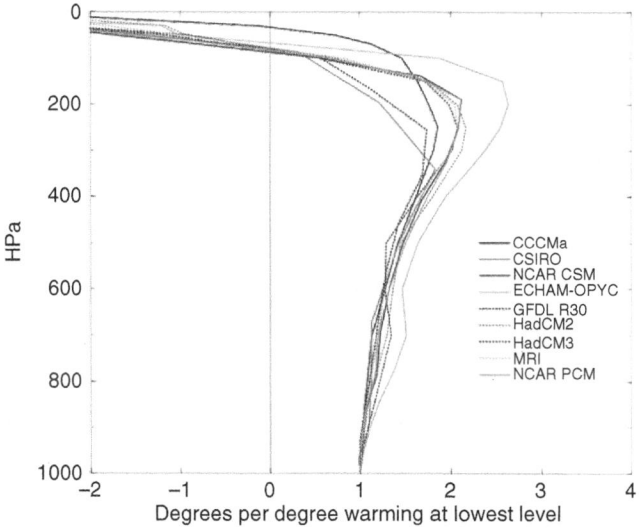

Figure 7.3. Vertical profiles of normalized atmospheric temperature change (high-CO_2 minus control) zonally averaged over $20°$ N–$20°$ S from a set of transient $+1\%$ per year CO_2 increase experiments using nine global climate models. The difference is based on years 61–80 of the high-CO_2 runs minus years 61–80 of the control runs for each model (see legend). The difference at each model level is normalized by dividing by the difference at the lowest available level for that model.

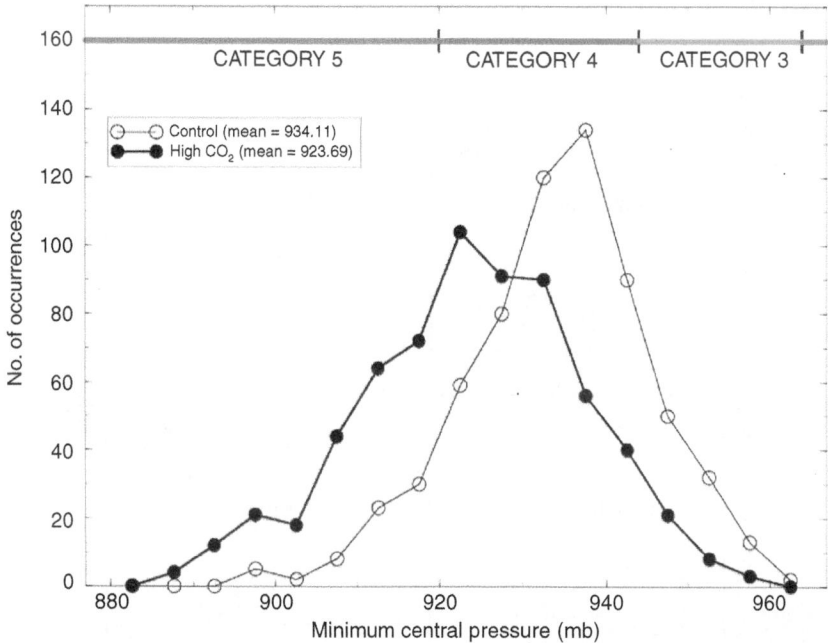

Figure 7.4. Frequency histogram showing hurricane intensity results (millibars) aggregated across all 1,296 experiments of KT04 (9 GCMs, 3 basins, 4 parameterizations, 6-member ensembles). The histograms are formed from the minimum central pressures, averaged over the final 24 h from each 5-day experiment. The light (dark) curve with open (solid) circles depicts the histogram from the control (high-CO_2) cases. See text for further details.

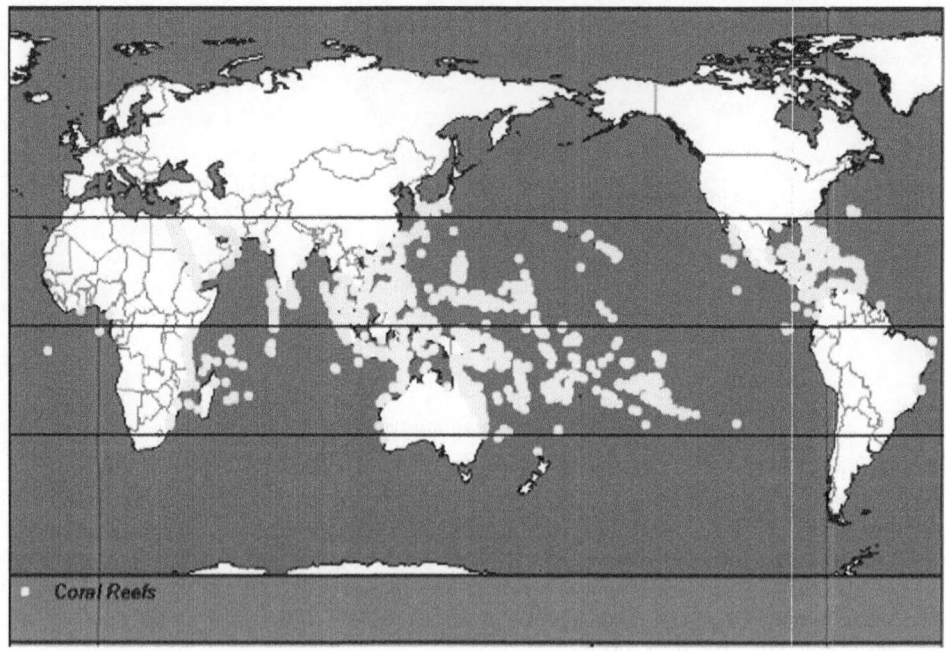

Figure 9.1. Worldwide distribution of corals and coral reefs. The dots indicate the positions of coral reefs.

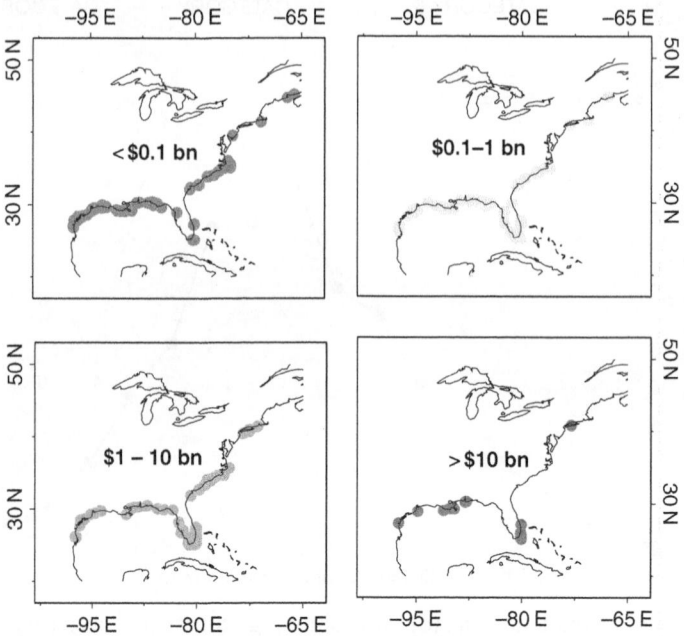

Figure 10.2. Geographic distribution of normalized insured losses from hurricanes striking the United States between 1900 and 2005.

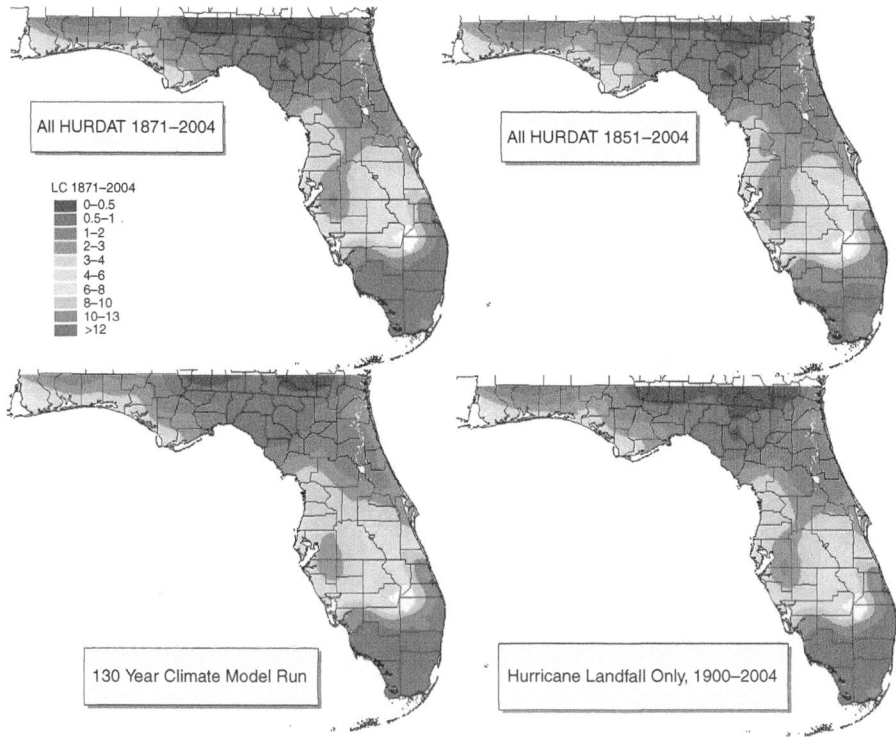

Figure 11.5. Loss costs from four different historical storm sets, including the storms generated from the climate model. The "All HURDAT" runs include all storms during the specified time period. The "Hurricane Landfall Only" run includes only storms that produced 64-knot winds over land. The legend shows the dollar loss per 1,000 dollars of exposure.

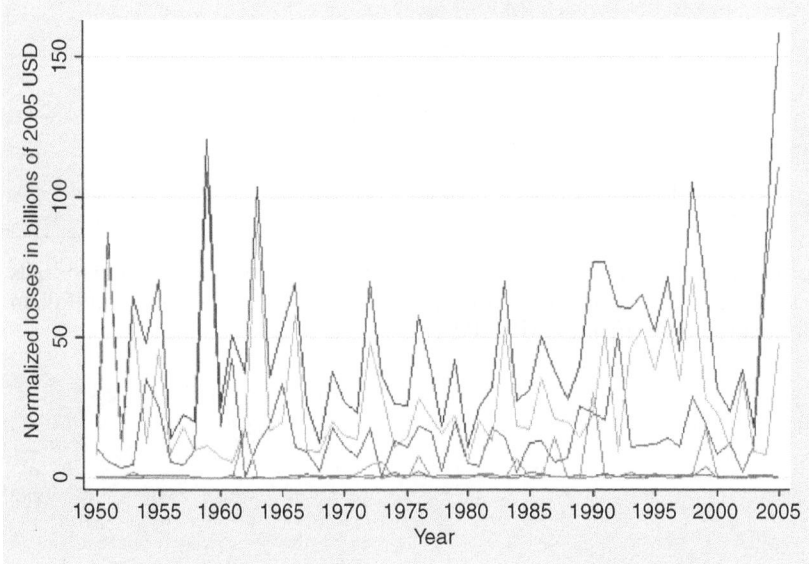

Figure 12.4. Trends in normalized losses for global totals (blue), hurricane (brown), hail (green), flood (yellow), wind (gray), and wildfire (red).

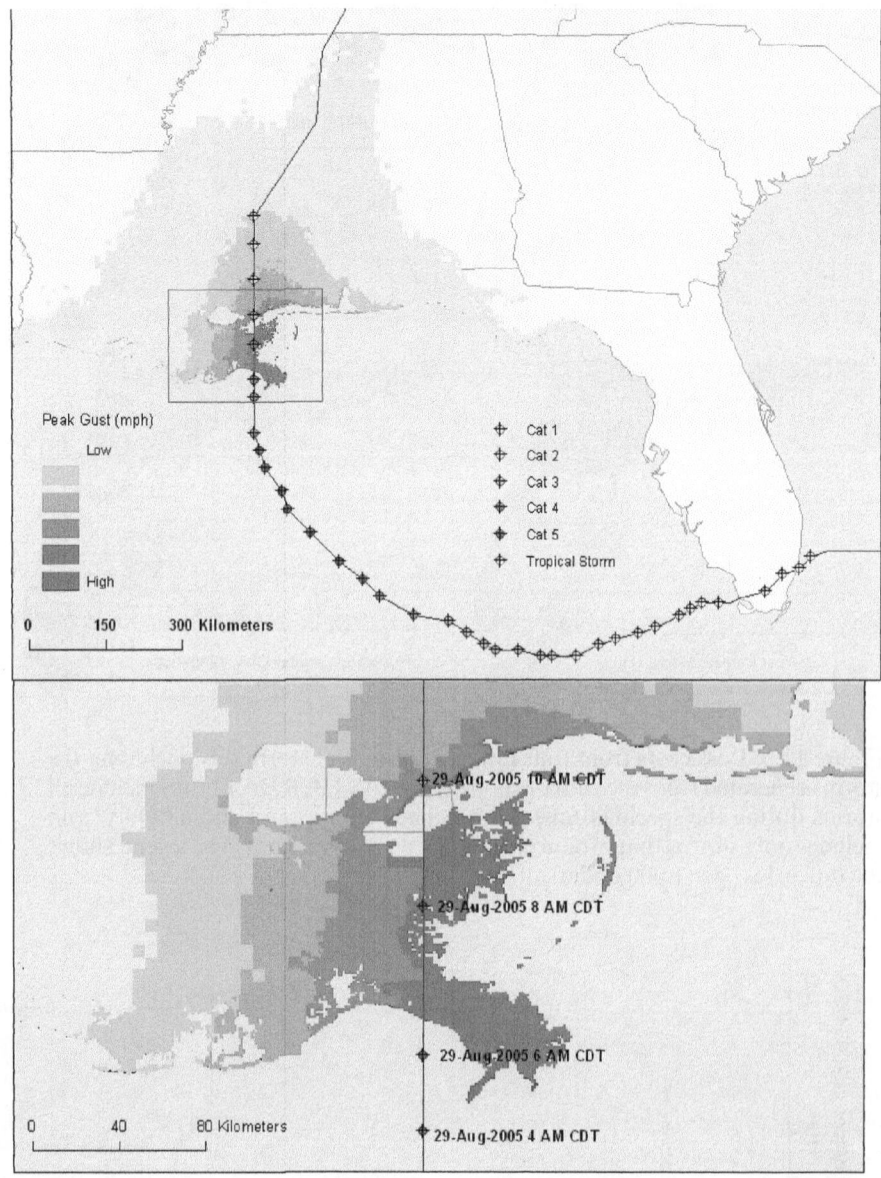

Figure 15.1. The path of Hurricane Katrina, showing first and second landfalls in Florida and Louisiana (above) and a close-up of the second landfall in Louisiana (below). (From RMS, 2005.)

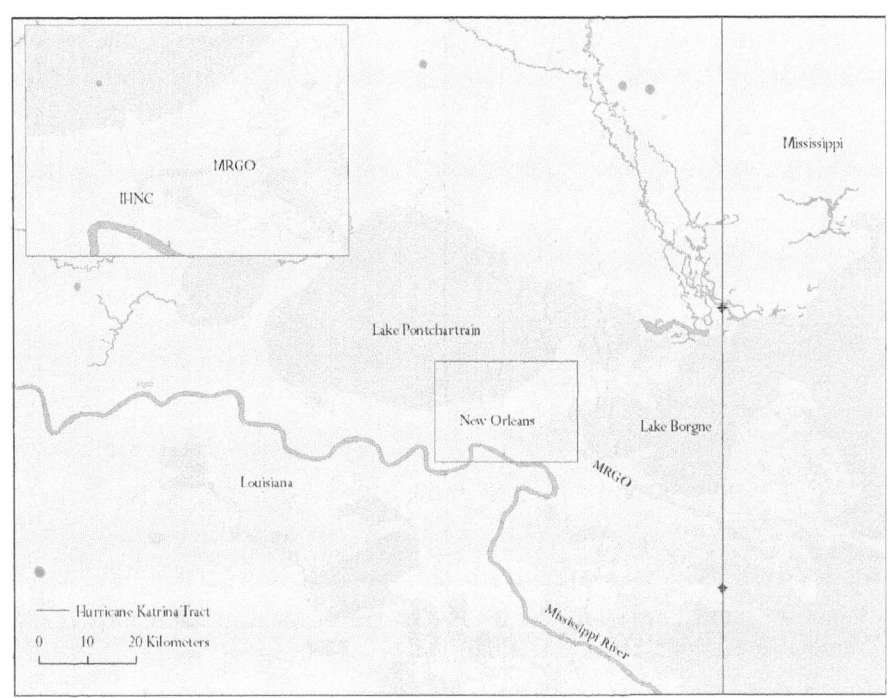

Figure 15.2. The region at landfall of Hurricane Katrina, with key locations labeled and close-up of New Orleans (in upper left).

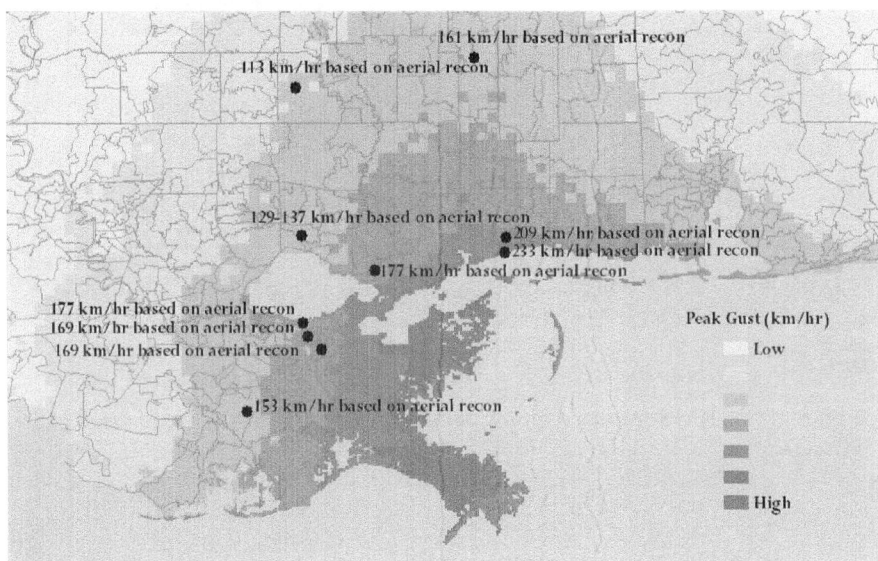

Figure 15.3. Map of the RMS-modeled wind field of the Gulf Coast landfall of Hurricane Katrina. (Based on RMS, 2005.)

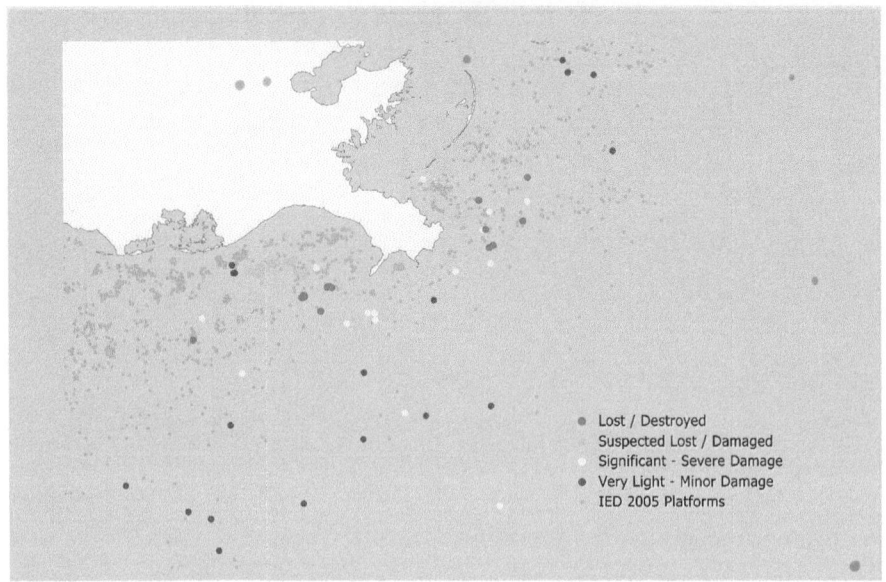

Figure 15.4. Offshore platforms affected by Hurricane Katrina: reports as of September 19, 2005. (In RMS, 2005.)

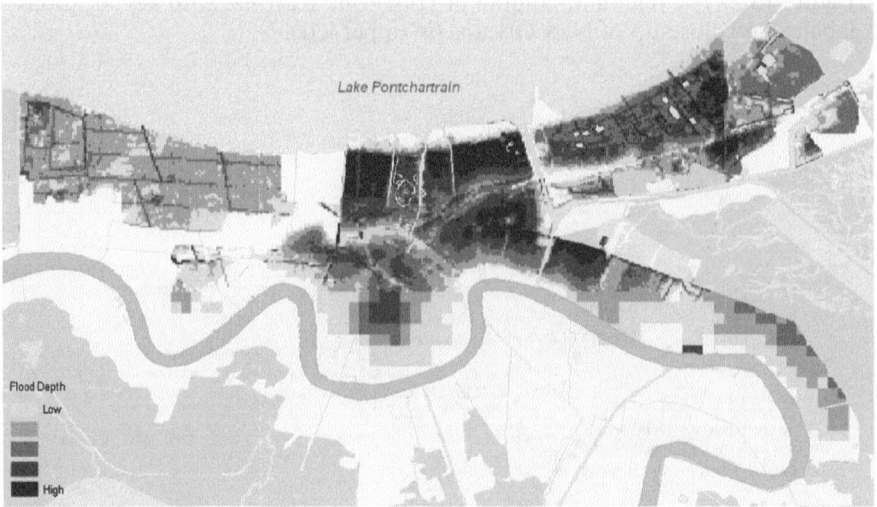

Figure 15.5. RMS-modeled flood depth in New Orleans. (Based on RMS, 2005.)

II

Impacts of weather and climate extremes

II

8

Extreme climatic events and their impacts: examples from the Swiss Alps

MARTIN BENISTON

Condensed summary

While changes in the long-term mean state of climate will have many important consequences on a range of environmental, social, and economic sectors, the most significant impacts of climate change are likely to be generated by shifts in the intensity and frequency of extreme weather events. Indeed, insurance costs resulting from extreme weather events have been steadily rising since the 1970s, essentially in response to increases in population pressures in regions that are at risk, but also in part because of recent changes in the frequency and severity of certain forms of extreme. Regions that are now safe from catastrophic windstorms, heat waves, and floods could suddenly become vulnerable in the future. Under such circumstances, the costs of the associated damage could be extremely high. This chapter provides an overview of certain climate extremes that in recent years have had very costly impacts in the Swiss Alps – namely, heat waves and strong convective precipitation – and how these events may change as climate warms in response to increased greenhouse gas concentrations.

8.1 Introduction

If climate warms as projected during the course of the twenty-first century, the thermal energy that drives many atmospheric processes will be enhanced and, as a consequence, many types of extreme event may increase in frequency and/or intensity. Although this intuitive reasoning has a physical basis, current climate trends do not unequivocally show that atmospheric warming in the past century has been accompanied by greater numbers of extreme events. This uncertainty is due in part to the fact that these are rare events that cannot be related in a statistically meaningful manner to changes in mean climatic conditions, as has been shown, for example, by Frei and Schär (2001). In

Climate Extremes and Society, ed. H. F. Diaz and R. J. Murnane. Published by Cambridge University Press. © Cambridge University Press 2008.

addition, while thermal energy is a prerequisite for generating and sustaining weather extremes, it is by no means a sufficient condition; if this were to be the case, then extreme events would be far more common. Counterintuitively, it has been shown for different parts of the world that climate variability and its propensity for generating extreme events has actually decreased during the twentieth century; such a decrease has been demonstrated, for example, by Beniston and Goyette (2007) for Switzerland, despite a clear warming signal during this time frame.

There is no single definition of what constitutes an extreme event. Extremes can be quantified *inter alia* on the basis of:

- how rare they are; this definition needs a statistical treatment based on frequency of occurrence. The Intergovernmental Panel on Climate Change (IPCC, 2001) has adopted the definition of an extreme as being an event that occurs below the 10% quantile or above the 90% quantile of a particular statistical distribution (probability distribution function, or PDF) of temperature, precipitation, pressure, and other parameters. The recent 2003 heat wave in Europe, for example, was estimated to have been the warmest since 1540 (Pfister *et al.*, 1999); i.e., probably a frequency of occurrence of 1 in 500.
- how intense they are; here, analyses of the exceedance of a particular threshold beyond which health, ecosystem, and economic damages occur are required. In Switzerland, for example, insurance companies are obliged to reimburse damages to infrastructure if wind speeds greater than 75 km/h are recorded by the weather services. This arbitrarily chosen figure typically defines the intensity threshold from the point of view of the insurance industry.
- how strong the impacts may be following a particular event or set of events; impact-based definitions of extremes are complex because not all sectors will necessarily be affected the same way. In addition, in many instances, damaging natural hazards can occur in the absence of a climatic extreme.

None of these definitions on its own is entirely satisfactory, however, and each definition corresponds to a particular situation but cannot necessarily be applied in a universal context.

Extreme climate events take the heaviest toll on human life and result in some of the highest damage costs of all natural hazards, with the exception of seismic-related hazards. The economic costs of extreme events have increased in the past few decades (e.g., Munich Re, 2005; Swiss Re, 2005), not necessarily because the frequency of extremes has changed per se but rather because there has been a substantial rise in the number of inhabitants and penetration of infrastructure in risk-prone areas. In the second half of the twentieth century, there were 71 "billion-dollar events" resulting from earthquakes, but more than 170 climate-related events: in particular, tropical cyclones, midlatitude winter storms, floods, droughts, and heat waves (Swiss Re, 2003). In Switzerland, extreme events have also placed a heavy burden on people and infrastructure in the relatively densely populated Alpine regions; the floods that affected many parts of the central and northern Swiss Alps in August 2005 are estimated to have been the costliest weather-related hazard to date in Switzerland, according to press releases and as yet unpublished statements from insurance firms.

Thus, there is an obvious incentive for the research community as well as the public and private sectors to focus on extreme climatic events and possible shifts in their frequency and intensity as climate changes in the course of the twenty-first century. Understanding the mechanisms underlying various forms of climatic extremes is important when assessing the manner in which they may evolve in the future, under changing climatic conditions. An improved understanding can in turn lead to the development of more accurate means of quantifying the costs associated with natural climate-related hazards and thereby provide the basis for strategies for adapting to climate change from an economic point of view.

In order to highlight some of these issues, the focus here will be on two forms of climatic extreme that have in the past resulted in serious health, environmental, and economic impacts in the Swiss Alps; namely, heat waves and heavy precipitation events. Recent trends and analyses of available climate model data will be summarized to demonstrate the sensitivity of a vulnerable and populated region to certain forms of extreme.

8.2 Observations and models

The Swiss weather service (MeteoSwiss) maintains a dense network of observation stations and manages the acquired daily data in the form of a digital database. There is a reasonable degree of confidence in the quality of climatological data for most Swiss locations, as the daily data have been homogenized for numerous sites (Begert *et al.*, 2003). The Swiss data have been used in many statistical studies of climate and climatic change in Switzerland (e.g., Beniston, 2004a), and the climatological stations span a range of altitudes from 300 to 3,600 m above sea level (ASL). Jungo and Beniston (2001) have shown through cluster analysis that, despite individual site heterogeneity, there is a close resemblance between climatological variables in different parts of Switzerland; this is true particularly on an altitudinal basis, such that even if some sites may be biased by local characteristics, the long-term trends are in general agreement among all stations.

When global and regional models are applied to climate change scenarios, they are powerful tools that allow insights into possible climate futures in

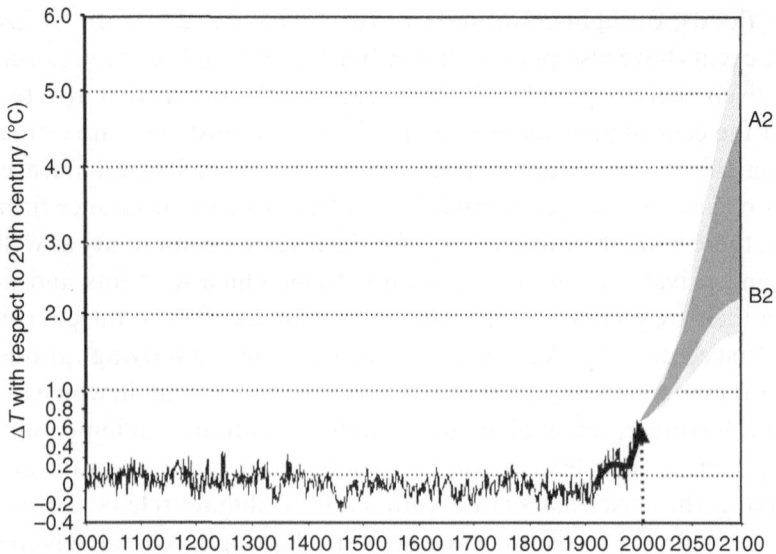

Figure 8.1. Range of possible climate futures according to various greenhouse gas emissions scenarios (IPCC, 2001), superimposed upon the climate reconstructions of Mann *et al.* (1999) for the past 1,000 years in the Northern Hemisphere. Solid black line for 1900–2000 represents the temperature curve obtained from the observational record.

response to various levels of greenhouse gas emissions and concentrations (IPCC, 2001). According to the scenarios used in the model, the response of climate is an increase in global mean temperatures ranging from 1.5 °C to 5.8 °C. These scenarios (Nakicenovic *et al.*, 2000) are based on various greenhouse gas emission pathways, which in turn depend on economic and population growth, and societal and technological choices; e.g., the rapidity with which the energy sector may reduce its dependency on fossil fuels, or deforestation and land use practices. Figure 8.1 illustrates the projected changes in global mean temperature to the year 2100 based on the full range of IPCC scenarios; the reconstructed time series of the past 1,000 years compiled by Mann *et al.* (1999) is provided to emphasize the totally different nature of the climate projected for the next decades. Even if the reconstructed curve is for the Northern Hemisphere only, and the model projections are global, the graph clearly highlights the fact that the future course of climate is completely out of the bounds of natural climate variability of past centuries. In this chapter, results for a low (B2 in Figure 8.1) and a high (A2 in Figure 8.1) emissions range will be given to evaluate the sensitivity of extremes to a particular level of greenhouse gases or if, on the contrary, no distinct thresholds can be detected. The A2 scenario leads to atmospheric CO_2 concentrations of about 800 parts per million by volume (ppmv) by 2100 (three times their pre-industrial values) and provides an estimate of the upper bound of climate futures discussed by the IPCC (2001); at 550 ppm, the B2 scenario corresponds to a doubling of greenhouse gas concentrations compared with their pre-industrial values. Refer to the IPCC (2001) and to Nakicenovic *et al.* (2000) for more information on the assumptions inherent in these scenarios.

The complexity and mutual interdependency of mountain environmental and socioeconomic systems pose significant problems for climate impact studies (Beniston, 2004a), essentially because the spatial resolution of general circulation models (GCMs) still remains too crude to adequately represent the topographic detail of most mountain regions. Impact research, on the other hand, requires information at fine spatial definition, where the regional detail of topography and land cover are important determinants in the response of natural and managed systems to change. Since the mid 1990s, the scaling problem related to complex topography has been addressed through regional modeling techniques, pioneered by Giorgi and Mearns (1991), and through statistical–dynamical downscaling techniques (e.g., Zorita and von Storch, 1999).

So-called "nested" approaches to regional climate simulations, whereby large-scale data or GCM outputs are used as boundary and initial conditions for regional climate model (RCM) simulations, have been applied to scenario computations for climate change in the twenty-first century (Giorgi and Mearns, 1999). The technique is applied to specific periods in time ("time windows") for which high-resolution simulations are undertaken over a given geographical area. The nested modeling approach represents a trade-off between decadal- or century-scale, high-resolution simulations that are expensive in terms of computational resources, and coarse-resolution results provided by long-term GCM integrations. Although the method has a number of drawbacks – in particular, the fact that the nesting is "one-way" (i.e., the climatic forcing occurs only from the larger to the finer scales and not vice versa) – RCMs have demonstrated skill in handling the regional detail of climate processes. This is an advantage for regions with complex topography, in particular where orographically enhanced precipitation often represents a significant fraction of seasonal or annual rainfall.

Since the mid 1990s, RCM spatial resolution has continually increased, partially as a response to the needs of the impacts community. Currently, simulations with 5 km or even 1 km grids are used to investigate the details of precipitation in relation to surface runoff, infiltration, and evaporation (e.g., Arnell, 1999; Bergström *et al.*, 2001), events such as extreme precipitation (Frei *et al.*, 1998), and damaging windstorms (Goyette *et al.*, 2003), thereby opening the way for studies on the impacts of extreme events. It should be emphasized, however, that very high-resolution simulations are possible for short-term case studies only and not for long-term climate simulations, which are possible only when the horizontal grid spacing of RCMs is 25 km or more.

Within the European Union, the PRUDENCE project (Prediction of Regional scenarios and Uncertainties for Defining EuropeaN Climate change risks and Effects; see http://prudence.dmi.dk) suite of regional climate models has been applied to the investigation of climate change over Europe for the last 30 years of the twenty-first century. Use of the models has allowed shifts in a number of key climate variables being assessed (Christensen *et al.*, 2002), including over the Alpine domain. In comparing model results for future temperature change influenced by enhanced greenhouse gas concentrations, there is generally close agreement between many of the models, as was reported by Déqué *et al.* (2005). The various models used in the PRUDENCE project operate at a 50 km horizontal grid resolution, and have completed two 30-year simulations; i.e., current climate or the control simulation for the period 1961–90, and the future greenhouse gas climate for the period 2071–2100. The fully coupled ocean–atmosphere GCM of the UK Hadley Centre, HadCM3 (Johns *et al.*, 2003), has been used to drive the higher-resolution atmospheric HadAM3 H model (Pope *et al.*, 2000), which in turn provides the initial and boundary conditions for the RCMs used in the PRUDENCE project.

8.3 Climate extremes in the Alpine region

The climate of the Alpine region is characterized by a high degree of complexity, because interactions between the mountains and the general circulation of the atmosphere result in features such as gravity wave breaking, blocking highs, and föhn winds. This complexity is exacerbated by the competing influences of a number of different climate regimes in the region, namely the Mediterranean, continental, Atlantic, and polar. Average temperatures following the end of the Little Ice Age (LIA) increased by up to 2 °C in many parts of Switzerland between 1901 and 2000 (e.g., Jungo and Beniston, 2001), which is well above the global average twentieth-century warming of about 0.6 °C reported by Jones and Moberg (2003).

Climate change in the Alps is a complex aggregate of short- to long-term forcings, such as those related to the North Atlantic Oscillation (NAO; e.g., Beniston and Jungo, 2002), the Atlantic Multidecadal Oscillation (AMO; see, for example, Beniston and Diaz, 2004), and atmospheric response to anthropogenic greenhouse gases (IPCC, 2001). Figure 8.2 shows the evolution of

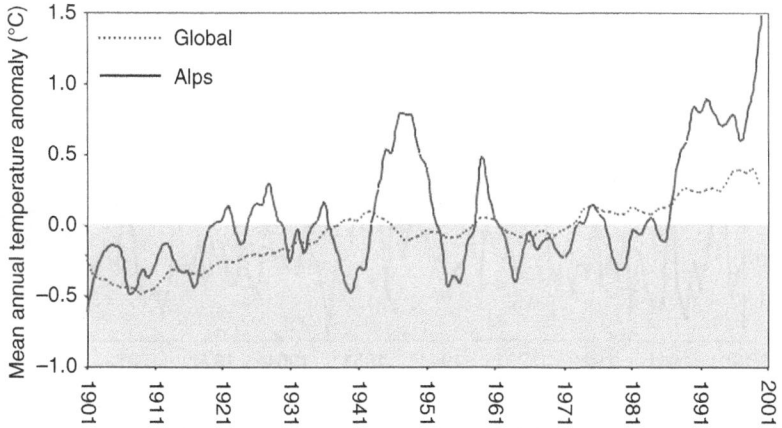

Figure 8.2. Departures of mean temperature from its climatological mean (1961–90 reference period) averaged over four Swiss sites (low elevations: Basel and Zürich; high elevations: Davos and Säntis). The global average change in temperature is also given here for comparison purposes.

temperature anomalies (i.e., departures of mean annual temperatures from the 1961–90 climatological reference period) during the course of the twentieth century at two low-elevation sites (Zürich, 569 m ASL, and Basel, 317 m ASL) and two high-elevation sites (Davos, 1,590 m ASL, and Säntis, 2,500 m ASL), where contamination by urbanization or the presence of temperature inversion layers is absent. The anomaly data have been merged into one single curve for purposes of clarity. Although synchronous with global mean temperature anomalies for the same period, the warming in the Alpine region is clearly stronger (e.g., Beniston, 2004a).

8.3.1 Summer heat waves

The record heat wave that affected many parts of Europe during the summer of 2003 has been seen by many as a "shape of things to come," reflecting the extremes of temperature that are projected for summers in the later decades of the twenty-first century (Beniston, 2004b; Schär *et al.*, 2004). The heat wave resulted in absolute maximum temperature records exceeding those that had stood since the 1940s and early 1950s in many locations in France, Germany, the United Kingdom, and Switzerland, according to information that was supplied by national weather agencies and that was highlighted in the annual report of the World Meteorological Organization (WMO, 2003). The human and environmental impacts were unprecedented in Europe: close to 35,000 excess deaths occurred during the summer of 2003 (with the greatest mortality at the height of the heat wave during August 2003), mostly in France but also

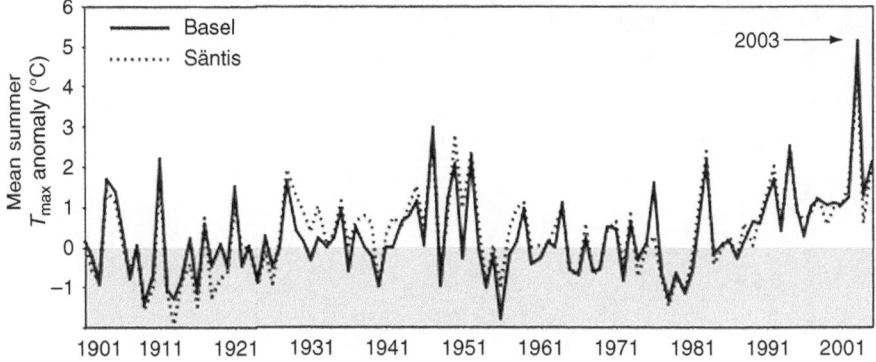

Figure 8.3. Departures of summer maximum temperatures from the 1961–90 means at Basel (317 m ASL) and Säntis (2,500 m ASL), Switzerland.

in Italy, Spain, Germany, and Switzerland (World Health Organization [WHO], 2003). Crop failures in France, Germany, and the United Kingdom cost a total of over 2 billion Euros. Rivers such as the Rhine in Germany and the Po in Italy ran almost dry; and some Alpine glaciers lost between 5% and 10% of their mass during this period.

Figure 8.3 illustrates the annual values of summer maximum temperature anomalies (i.e., T_{max}, the average of daily maximum temperatures recorded in June, July, and August [JJA]) from 1901 through 2003 at Basel, Switzerland, and the high-elevation site of Säntis. The 2003 event stands out as a "climatic surprise" in the sense that, for the first time ever, T_{max} exhibited positive monthly anomalies of up to 6 °C in August 2003. The heat wave came at the end of a 40-year period during which summers were markedly cooler than the warm summers of the mid-twentieth century, with the exception of some isolated (but noteworthy) events such as those in 1976, 1983, and 1994.

The RCM results mapped over Europe for maximum temperatures and threshold exceedances have implications for the future course of extreme events, such as an increase in heat waves and a reduction in cold spells and frost days. According to the baseline used, the very definition of heat wave could change in a future, systematically warmer climate, compared with today. The climate of southern Spain, for example, which is currently characterized by temperatures exceeding 30 °C for about 60 days per year on average, may in the future experience over 150 days or more at these warm temperatures. Under such circumstances, the notion of a heat wave may lose some of its impact value when a rare or exceptional feature of today's climate becomes commonplace in tomorrow's climate. Figure 8.4 shows the shift in JJA T_{max} between the 1961–90 reference period and 2071–2100 for the RCM grid point closest to Basel, for the IPCC A2 and B2 scenarios, for both the mean and the

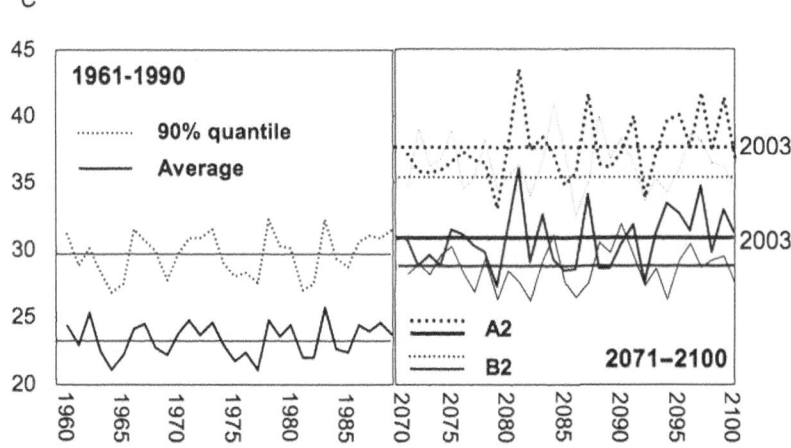

Figure 8.4. Comparisons between summer (JJA) mean maximum temperatures and their 90% quantiles for Basel, for each year of the reference 1961–90 climate and the A2 and B2 scenario climates. The horizontal lines depict the 30-year means for each time series. 2003 refers to the mean and 90% quantile values recorded during the summer of 2003 in Europe, in order to highlight the exceptional nature of that heat wave.

90% quantile. In the reference climate, mean temperatures in Basel are close to 23 °C and heat waves can be considered to occur when temperatures exceed 30 °C (which is the average level of the 90% quantile for this period). For the scenario climates, whatever the scenario chosen, temperatures are seen to rise on average between 5 °C and 7 °C over current values; the difference in temperature between the high- and low-emissions scenarios is less than that between the B2 scenario and the current climate. This result implies that, even with rather stringent policies to abate greenhouse gas emissions, the increase in temperatures as seen for the B2 scenario will potentially result in summer heat waves that are as intense as, or even stronger than, the 2003 heat wave; the potential for strong heat waves is even greater for the A2 scenario, as could be intuitively expected. The mean and 90% quantile statistics for the 2003 heat wave are provided in this diagram to highlight the fact that this event was exceptional and could be considered to be a "summer of the future." Indeed, statistically speaking, temperatures such as those experienced in the 2003 heat wave could occur one summer out of two in a future climate.

The 2003 heat wave, by mimicking quite closely the possible course of summers in the latter part of the twenty-first century, can thus be used within certain limits as an analog of what may occur with more regularity in the future. The physical processes that characterized the event, such as soil moisture depletion and the positive feedback on summer temperatures and the lack

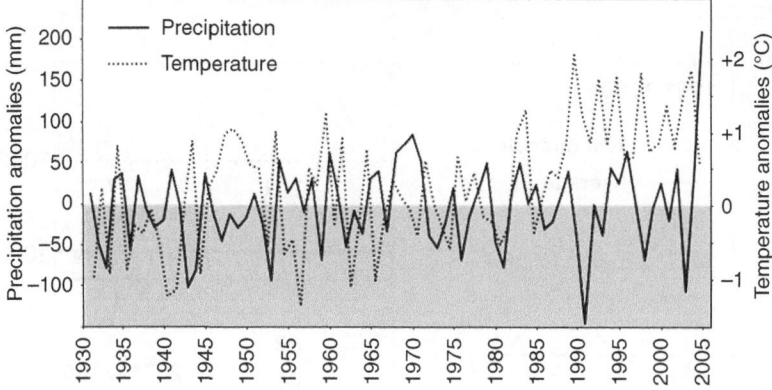

Figure 8.5. Anomalies of August precipitation and temperatures in Altdorf, in the central–northern part of the Swiss Alps, a region vulnerable to heavy precipitation and flood events.

of convective rainfall in many parts of the continent that generally occur from June through September, are projected to occur with greater frequency in the future. In view of the severity of the impacts related to the heat wave, it should help scientists in assessing the course of future climatic impacts, and decision makers in formulating appropriate response strategies.

8.3.2 Heavy precipitation events

The Alps particularly are exposed to both extremes of precipitation – i.e., heavy precipitation (including hail) and drought – according to the circulation patterns that are associated with extremes and their persistence. Many of the strong rainfall events result in flooding and geomorphologic hazards such as landslides, rock falls, and debris flows within regions of complex topography. If these events occur in the vicinity of populated regions, the impacts in human and economic terms can be enormous. Indeed, the August 2005 floods in Switzerland were estimated to be, on a gross domestic product (GDP) basis, as costly to the Swiss economy as the 2005 Katrina hurricane was to the US economy (unpublished figures from insurance firms).

Figure 8.5 shows the behavior of summer precipitation events, in the form of August precipitation totals recorded each year at Altdorf, a location in the central–northern Swiss Alps that is often subject to heavy precipitation events (Beniston, 2006). August is a prime month for strong convective downpours, especially when moist air converges into regions whose surfaces have been well heated during the summer months; in addition, explosive convection is exacerbated by forced uplift of air by the topography. Figure 8.5 shows that while

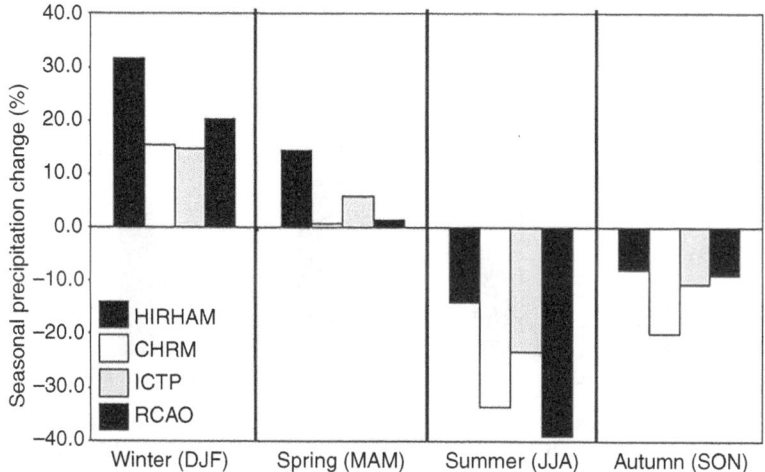

Figure 8.6. Seasonal precipitation change in the Swiss Alps as projected by four regional climate models for the IPCC A2 scenario. DJF, December–January–February; MAM, March–April–May; JJA, June–July–August; SON, September–October–November.

record precipitation was registered during the devastating flood of August 2005, intense events also occurred during other periods of the twentieth century, such as in the late 1960s. Superimposed on the precipitation data in Figure 8.5 are the August temperature anomalies, which show that there is no discernible trend that could link rainfall trends to increasing temperatures, even though the latter, warmer part of the record shows a possible increase in variability.

Numerical models of the climate system have greater difficulty in simulating precipitation as opposed to temperature, because of the complex microphysics involved and the fact that sub-grid-scale features such as topography or land use are often inadequate for assessing the correct location and intensity of precipitation. However, recent studies with RCMs applied to Europe show that they capture the broad features of observed precipitation and their seasonal shifts, even in the complex Alpine domain (e.g., Frei *et al.*, 2006; Christensen and Christensen, 2003; Beniston *et al.*, 2007). When they are applied to the A2 and B2 scenario climates for the period 2071–2100, most RCMs show a distinct shift in the seasonal precipitation totals compared with 1961–90, in regions of the Alps that are prone to strong rainfall and flooding. Figure 8.6 shows the shift in precipitation in the central–northern Swiss Alps for four of the RCMs used in the PRUDENCE project. Although there is some disagreement as to the absolute levels of change, the RCMs nevertheless agree on the signs of change; i.e., increases in winter and spring, and reductions in summer and autumn (Beniston, 2006). The annual precipitation amounts

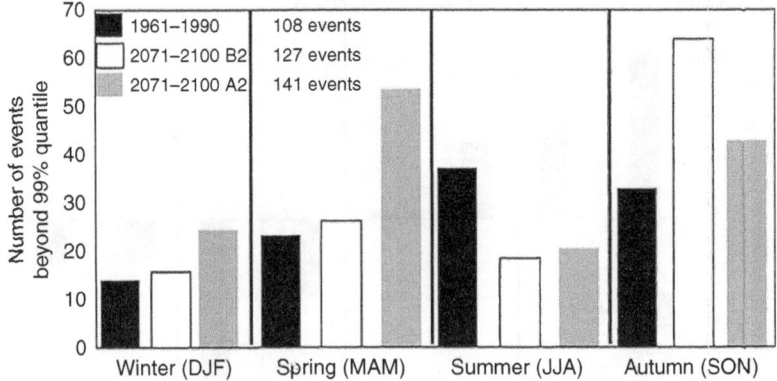

Figure 8.7. Changes in extreme precipitation events in the Swiss Alps as projected by the Danish HIRHAM model for the IPCC A2 and B2 scenarios.

remain remarkably similar between current and future climates (around 1,600 mm), but their seasonal distribution is markedly different. The principal cause of the change in seasonal patterns is related to strong summer warming and drying in the Mediterranean zone, which is likely to spread north to the Alps and beyond, and the distinctly rainier rainfall season that a milder climate is expected to bring to the region in winter.

As a result of the shifts in mean precipitation, the frequency of extreme events also changes in seasonality compared to the current climate. Figure 8.7 illustrates the seasonal change in the number of heavy daily precipitation events as simulated by one of the PRUDENCE RCMs, the HIRHAM (High-resolution Regional model with Hamburg physics; Christensen *et al.*, 2002), which is represented by the 99% quantile of precipitation in the northern part of the Swiss Alpine region. This model was selected because it has been used consistently over the past few years for investigating extremes in the Alpine region (e.g., Beniston, 2004b, 2005) and has proved to be quite reliable in reproducing observed climatic trends. For the scenario climates, the RCM projects the most important changes to occur in spring and autumn, with large increases in intense precipitation. Whatever the scenario considered, shifts between the current and scenario climates are large, and are certainly larger than those between the A2 and B2 scenarios. Summer rainfall extremes are projected to decline, but spring and autumn events are projected to increase substantially.

Heavy precipitation is a necessary but not always sufficient condition for floods, landslides, and other related damages. Hydrological systems are sensitive not only to heavy precipitation, but also to the permeability of soils, the prior history of rainfall (i.e., how saturated the soils are before an event), evaporation rates, land use, river channeling, and the buffering effect of snow.

As the freezing level rises, so does the potential for strong runoff since rainfall is captured over a greater surface area and thus more water may be channeled into river catchments. The same quantity of precipitation can thus result in very different responses according to the altitude level at which snow falls during a strong convective event. In the absence of the buffering effect of snow, a heavy downpour can lead to catastrophic impacts. Paradoxically, flood events and other rainfall-triggered hazards may not necessarily increase in a given scenario of future climate, because the snowfall levels in future springs and autumns are projected to remain below the freezing levels of current summers. However, although heavy summer precipitation is projected to decline sharply in the future, the risks related to such events will remain high because the snowline will be located at greater altitudes than it is under current climatic conditions (Beniston, 2006).

8.4 Impacts of extreme events

While many of the extreme events discussed here have been, and will increasingly be, costly in social, economic, and environmental terms, it needs to be emphasized once again that strong impacts are not necessarily or exclusively related to extreme weather events. Indeed, many of the geomorphologic hazards in the Alps are the result of long-term climatic and geological forcings that at some stage result in a particular threshold exceedance leading to sudden and intense slope instabilities.

Impacts of climate change in the Alps can be placed into five broad categories: hydrology, snow, and ice; plants, forest ecosystems, and mountain biodiversity; human health (this aspect will not be discussed further here); socioeconomic sectors such as tourism, agriculture, and hydropower; and financial services such as insurance. While extreme events may have a significant impact on one or more of these sectors, they can also be viewed as short-term and rapid "pulses of energy" into systems that are already experiencing decadal-scale climate change.

Snow and ice are key components of the hydrological cycle in the Alps, and the seasonal character and amount of runoff is closely linked to cryospheric processes, such as the timing of the spring snowmelt and the water that is added to rivers by seasonal glacier melt towards the end of the summer. Because of the sensitivity of mountain glaciers to temperature and precipitation, changes in climate have been shown to result in shifts in seasonal snowpack (Beniston *et al.*, 2003). In temperate mountain regions, glacier ice is often close to its melting point, so it may respond rapidly to apparently minor changes in temperature. The persistence and intensity of the 2003 heat wave had rapid and adverse effects on glacier mass balance, and the projected increases in the numbers of heat waves in the future will lead to an acceleration of glacier retreat. The consequences of changing snow patterns and reduced glacier volume for river runoff are likely to affect not only the watersheds within the mountains themselves, but also lowland regions in Germany, Italy, and France that are heavily dependent on the Alps for their water supply.

Biodiversity in mountain areas encompasses both natural and cultivated species; these systems are sensitive to climatic factors and are likely to have different vulnerability thresholds according to the species, the amplitude, and the rate of climate change. Plant life at high elevations is constrained primarily by direct and indirect effects of low temperatures, radiation, wind and storminess, or lack of water (Körner and Larcher, 1988). Plants respond to these climatological influences through a number of morphological and physiological adjustments. Adaptation to environmental change includes the progressive replacement of currently dominant species by more thermophilic species. Observations in the Alps (Grabherr *et al.*, 1994; Keller *et al.*, 2000) suggest that certain plants have already begun to respond in this manner to observed twentieth-century warming. A further mechanism is that the dominant species may be replaced by pioneer species of the same community that have enhanced adaptation capabilities (Pauli *et al.*, 1998). A third possibility is that environmental change may favor less-dominant species, which then replace the dominant ones through competition (Street and Semenov, 1990). Extreme events, especially heavy precipitation and also rare but severe winter windstorms, severely damage trees in particular and thereby reduce their anchoring capacity for soils on steep slopes, thus opening the way to enhanced slope erosion. Increasing loss of vegetation and forest cover under the long-term influence of climate change and the shock effects of repeated extreme weather would be detrimental to the environmental health of the Alps, which is strongly related to the quality of vegetation preservation in natural, seminatural, agricultural, and forest ecosystems. Maintaining Alpine biodiversity would represent an optimal strategy for averting natural hazards such as landslides, and for maintaining water quality in the numerous Alpine watersheds.

In the latter decades of the twentieth century, tourism and recreation was one of the fastest growing industries worldwide (Perry, 2000). Tourism has economic benefits for, and potential adverse effects on, mountain environments and local mountain communities. Changing climates may alter the seasonal patterns of tourism (for example, skiing in winter) and thus the environmental pressures associated with different forms of leisure activities. Lack of snow during some recent winters in the Alps has translated into serious economic shortfalls for many mountain resorts that have few alternate solutions for attracting tourists in winter. While the Alps may be seen as a cool haven during heat waves in the lowland regions, summer tourism rarely compensates for winter sports in terms of the income that skiing generates for many mountain communities. Extreme events, particularly floods and windstorms, can damage infrastructure such as housing, communication routes, or cable cars.

In the Alps, hydropower is the main source of energy for electricity production. Changes in the seasonal character of precipitation, and also the timing of the melting of the Alpine snowpack, will substantially change the periods during which reservoirs will be filled; they may occur towards the late spring and early summer, compared with autumn today. In this case, new water management techniques will be required to assess whether to produce more electricity during the off-peak summer months or wait until the peak-demand winter season before doing so, which would imply economic shortfalls as production is suspended or reduced. Hydropower infrastructure can partially buffer the effects of strong convective precipitation, by holding back floodwaters that would otherwise impact communities downstream in the valleys. This has not always been successful, however, as was seen during some of the flood events in the autumns of the early 1990s. In the central and southern Alps, where the dams were already full in preparation for the winter peak-demand period, they could not cope with a sudden and massive influx of water.

While long-term climate change will certainly contribute to environmental and economic adversity to many sectors of Alpine life, the additional effects of sudden and unexpected extreme events may compound these detrimental effects, sometimes beyond the threshold of economic viability. In the past, the negative impacts of an extreme weather event have been financially absorbed over time; if the return periods of floods, heat waves, storms, or drought were to be reduced, however, then even the robust economies of the Alpine countries would be stressed, and government subsidies and public and private insurance could have difficulties in coping with the increasing and recurrent events that many models are projecting for coming decades.

8.5 Conclusions

While changes in the long-term mean state of climate will have many important consequences for numerous environmental, social, and economic sectors, the most significant impacts of climate change are likely to come about from shifts in the intensity and frequency of extreme weather events. Indeed, insurance costs resulting from extreme weather events have been steadily rising since the 1970s, essentially in response to increases in population pressures in regions that are at risk, but also in part because of recent changes in the frequency and severity of certain forms of extremes. Regions that are now safe from catastrophic windstorms, heat waves, and floods could suddenly become vulnerable in the future. Under such circumstances, the costs of the associated damage could be extremely high.

This chapter has summarized certain types of weather extreme as they have affected the Alpine region in the recent past, and that may continue to do so in the future in a changing climate. It has been emphasized throughout that there are no simple links between the behavior of extremes and changes in mean climatic conditions, which adds complexity to our understanding of extreme events in midlatitude mountain regions. However, because of the repetitive nature of many of the impacts that are associated with heavy rain or high temperatures, it is possible and strongly recommended to plan for an increasing frequency of certain types of extreme events. Adaptation measures could certainly help to alleviate many of the negative impacts that are associated with these extremes. Some of these adaptation measures would in any case be beneficial in responding to long-term climate change, too.

References

Arnell, N. W. (1999). The effect of climate change on hydrological regimes in Europe: a continental perspective. *Global Environmental Change*, **9**, 5–23.

Begert, M., *et al.* (2003). Homogenization of climate time series and computation of the 1961–1990 norms. MeteoSuisse Publication, 67, Zürich, Switzerland (in German).

Beniston, M. (2004a). *Climatic Change and Its Impacts. An Overview Focusing on Switzerland*. Dordrecht, The Netherlands, and Boston, MA: Kluwer Academic Publishers.

Beniston, M. (2004b). The 2003 heat wave in Europe: a shape of things to come? *Geophysical Research Letters*, **31**, L02202.

Beniston, M. (2005). Warm winter spells in the Swiss Alps: strong heat waves in a cold season? *Geophysical Research Letters*, **32**, L01812.

Beniston, M. (2006). The August 2005 rainfall event in Switzerland: not necessarily an analog for strong convective events in a greenhouse climate. *Geophysical Research Letters*, **33**, L05701.

Beniston, M., and Diaz, H. F. (2004). The 2003 heat wave as an example of summers in a greenhouse climate? Observations and climate model simulations for Basel, Switzerland. *Global and Planetary Change*, **44**, 73–81.

Beniston, M., and Goyette, G. (2007). Changes in variability and persistence of climate in Switzerland: exploring twentieth-century observations and twenty-first-century simulations. *Global and Planetary Change* (in press).

Beniston, M., and Jungo, P. (2002). Shifts in the distributions of pressure, temperature and moisture in the Alpine region in response to the behavior of the North Atlantic Oscillation. *Theoretical and Applied Climatology*, **71**, 29–42.

Beniston, M., Keller, F., Koffi, B., and Goyette, S. (2003). Estimates of snow accumulation and volume in the Swiss Alps under changing climatic conditions. *Theoretical and Applied Climatology*, **76**, 125–40.

Beniston, M., Stephenson, D. B, Christensen, O. B., *et al.* (2007). Future extreme events in European climate: an exploration of regional climate model projections. *Climatic Change* (in press).

Bergström, S., Carlsson, B., Gardelin, M., Lindström, G., Pettersson, A., and Rummukainen, M. (2001). Climate change impacts on runoff in Sweden: assessments by global climate models, dynamical downscaling and hydrological modelling. *Climate Research*, **16**, 101–12.

Christensen, J. H., and Christensen, O. B. (2003). Severe summertime flooding in Europe. *Nature*, **421**, 805–6.

Christensen, J. H., Carter, T. R., and Giorgi, F. (2002). PRUDENCE employs new methods to assess European climate change. *Eos* (*American Geophysical Union Newsletter*), **83**, 13.

Christensen, O. B., Christensen, J. H., Machenhauer, B., and Botzet, M. (1998). Very high-resolution regional climate simulations over Scandinavia: present climate. *Journal of Climate*, **11**, 3204–29.

Déqué, M., Jones, R. G., Wild, M., *et al.* (2005). Global high resolution versus Limited Area Model climate change projections over Europe: quantifying confidence level from PRUDENCE results. *Climate Dynamics*, **25**, 653–70.

Frei, C., and Schär, C. (2001). Detection probability of trends in rare events: theory and application to heavy precipitation in the Alpine region. *Journal of Climate*, **14**, 1568–84.

Frei, C., Schär, C., Lüthi, D., and Davies, H. C. (1998). Heavy precipitation processes in a warmer climate. *Geophysical Research Letters*, **25**, 1431–4.

Frei, C., Schöll, R., Fukutome, S., Schmidli, J., and Vidale, P. L. (2006). Future change of precipitation extremes in Europe: intercomparison of scenarios from regional climate models. *Journal of Geophysical Research*, **111**, D06105, doi:10.1029/2005JD005965.

Giorgi, F., and Mearns, L. O. (1991). Approaches to the simulation of regional climate change. *Reviews of Geophysics*, **29**, 191–216.

Giorgi, F., and Mearns, L. O. (1999). Regional climate modeling revisited. *Journal of Geophysical Research*, **104**, 6335–52.

Goyette, S., Brasseur, O., and Beniston, M. (2003). Application of a new wind gust parameterization: multi-scale case studies performed with the Canadian RCM. *Journal of Geophysical Research*, **108**, 4371–89.

Grabherr, G., Gottfried, M., and Pauli, H. (1994). Climate effects on mountain plants. *Nature*, **369**, 448.

Intergovernmental Panel on Climate Change (IPCC) (2001). *Climate Change. The IPCC Third Assessment Report*. Volumes I (*The Scientific Basis*), II (*Impacts, Adaptation, and Vulnerability*), and III (*Mitigation*). Cambridge and New York: Cambridge University Press.

Johns, T. C., *et al.* (2003). Anthropogenic climate change for 1860 to 2100 simulated with the HadCM3 model under updated emission scenarios. *Climate Dynamics*, **20**, 583–612.

Jones, P. D., and Moberg, A. (2003). Hemispheric and large-scale surface air temperature variations: an extensive revision and an update to 2001. *Journal of Climate*, **16**, 206–23.

Jungo, P., and Beniston, M. (2001). Changes in the anomalies of extreme temperature in the twentieth century at Swiss climatological stations located at different latitudes and altitudes. *Theoretical and Applied Climatology*, **69**, 1–12.

Keller, F., Kienast, F., and Beniston, M. (2000). Evidence of the response of vegetation to environmental change at high-elevation sites in the Swiss Alps. *Regional Environmental Change*, **2**, 70–7.

Körner, C., and Larcher, W. (1988). Plant life in cold climates. In *Plants and Temperature*, ed. S. F. Long and F. I. Woodward. Cambridge, UK: The Company of Biologists Ltd., pp. 25–57.

Mann, M. E., Bradley, R. S., and Hughes, M. K. (1999). Global-scale temperature patterns and climate forcing over the past six centuries. *Nature*, **392**, 779–87.

Munich Re (2005). *Topics Geo, Annual Review: Natural Catastrophes 2004*. Munich, Germany: Munich Re Publications.

Nakicenovic, N., *et al.* (2000). *IPCC Special Report on Emission Scenarios*. Cambridge, UK, and New York: Cambridge University Press.

Pauli, H., Gottfried, M., and Grabherr, G. (1998). Effects of climate change on mountain ecosystems: upward shifting of alpine plants. *World Resources Review*, **8**, 382–90.

Perry, A. H. (2000). Impacts of climate change on tourism. In: *Assessment of Potential Effects and Adaptations for Climate Change in Europe*, ed. M. L. Parry. The ACACIA Report. Jackson Environment Institute, Norwich, and EU Publications, Brussels, pp. 217–26.

Pfister, C., *et al.* (1999). Documentary evidence on climate in sixteenth-century Europe. *Climatic Change*, **43**, 55–110.

Pope, D. V., Gallani, M., Rowntree, R., and Stratton, A. (2000). The impact of new physical parameterizations in the Hadley Centre climate model HadAM3. *Climate Dynamics*, **16**, 123–46.

Schär, C., Vidale, P. L., Lüthi, D., *et al.* (2004). The role of increasing temperature variability in European summer heat waves. *Nature*, **427**, 332–6.

Street, R. B., and Semenov, S. M. (1990). Natural terrestrial ecosystems. In *Climate Change: The First Impacts Assessment Report*, ed. W. J. McG Tegart, G. W. Sheldon, and D. C. Griffiths. Australian Government Publishing Service, Chapter 3.

Swiss Re (2003). *Natural Catastrophes and Reinsurance*. Zürich: Swiss Reinsurance Company Publications.

Swiss Re (2005). *Opportunities and Risk in Climatic Change*. Zürich: Swiss Reinsurance Company Publications.

World Health Organization (WHO) (2003). The health impacts of 2003 summer heat waves. Briefing note for the delegations of the fifty-third session of the WHO Regional Committee for Europe.

World Meteorological Organization (WMO) (2003). *The State of the World's Climate 2003 Report*. Geneva: WMO Publications.

Zorita, E., and von Storch, H. (1999). The analog method: a simple statistical downscaling technique: comparison with more complicated methods. *Journal of Climate*, **12**, 2474–89.

9

The impact of weather and climate extremes on coral growth

M. JAMES C. CRABBE, EMMA L. L. WALKER, AND DAVID B. STEPHENSON

M. JAMES C. CRABBE, EMMA L. L. WALKER, AND
DAVID B. STEPHENSON

Condensed summary

Coral reefs are complex underwater ecosystems that are particularly vulnerable to climate extremes. In this chapter, we review the meteorological processes that influence corals and their growth, illustrate a number of methods for growth rate modeling, and show how climate extremes can affect growth rates. We then provide two examples of detailed modeling of coral colony growth as a function of climate in the Caribbean, for reefs off the coasts of Jamaica and Curaçao. For the Jamaican reefs, non-branching coral recruitment was inversely correlated with storm severity. For the reefs off Curaçao, the only significant correlation, which was negative, was the maximum daily temperature with a 30-day moving average applied (p-value of 0.002), suggesting that during the measurement period, temperatures rose to values higher than optimum for growth, but not sufficiently high to cause bleaching.

Our results show that hurricanes and severe storms can limit the recruitment and survival of massive coral colonies, and that small changes in temperature can significantly influence branching coral growth rates. Even for the simple exponential growth models, it is possible to introduce parameters for climate variables and climate change that should be useful predictively. Future studies will link climate modeling with environmental genetics and studies on symbiont diversity.

9.1 Introduction

The deeps have music soft and low
When winds awake the airy spry,
It lures me, lures me on to go
And see the land where corals lie.
The land, the land where corals lie.
(Text by Richard Garnett (1835–1906). Set by Sir Edward Elgar (1857–1934),
Op. 37, first performance 1899, from Sea Pictures, no. 4.)

Climate Extremes and Society, ed. H. F. Diaz and R. J. Murnane. Published by Cambridge University Press. © Cambridge University Press 2008.

Coral reefs are ecosystems that are particularly vulnerable to climate extremes. In this chapter, we review the meteorological processes that influence corals and their growth, illustrate a number of methods for growth rate modeling, show how climate extremes can affect growth rates, and provide examples of detailed modeling of coral colony growth as it relates to climate for reefs in the Caribbean. Coral growth depends in a very complicated way on many weather and climate processes; it is not just a simple response to a single factor such as temperature. Coral growth needs certain conditions, such as sufficient heat and light and nutrients, and it is hindered by extreme events such as periods with high or low temperatures, and strong winds. Quantitative growth models can be used to study how corals respond to these different weather and climate events.

Coral reefs are marine ecosystems of great biodiversity: the "rain forests of the sea." They are found predominantly in the tropics, often in areas of great human poverty (Crabbe, 2006), but they also exist in other specialized locations (Figure 9.1). Coral reefs provide an environment in which one-third of all marine fish species and tens of thousands of other species are found, and from which 6 million tons of fish are caught annually. This abundant catch provides an income not only to national and international fishing fleets, but also for some local communities, which in addition rely on the local fish stocks

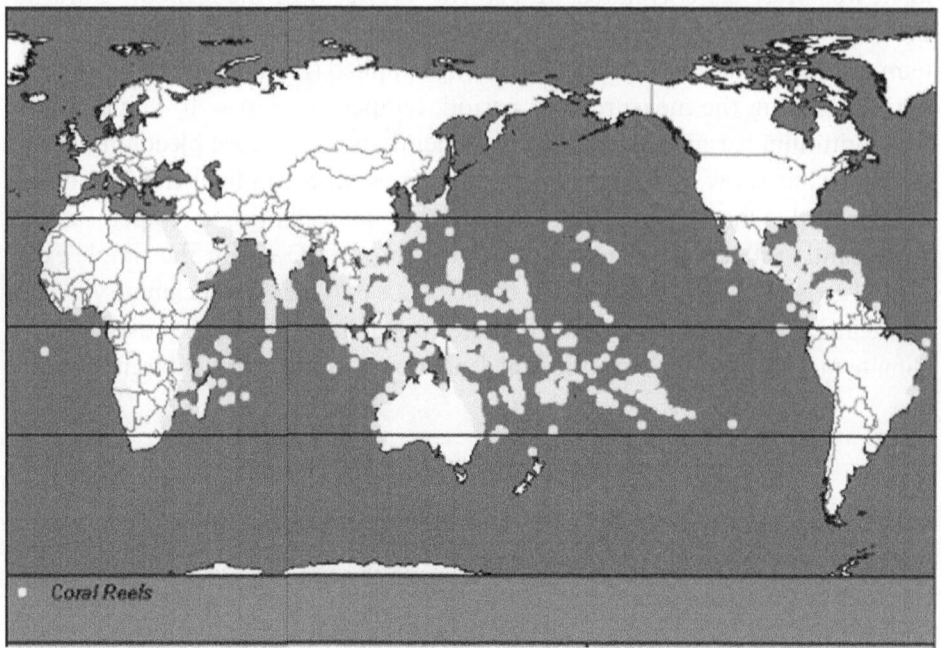

Figure 9.1. Worldwide distribution of corals and coral reefs. The dots indicate the positions of coral reefs. For color version, see plate section.

to provide nutritional sustenance. The reefs act as barriers to wave action and storms by reducing the incident wave energy through wave reflection, dissipation, and shoaling, protecting the land and an estimated half a billion people who live within 100 km of reefs. The annual global value of coral reefs has been estimated to be about US$375 billion (1997 values) (Costanza *et al.*, 1997).

Hermatypic corals are reef-building corals that belong to the order Scleractinia, and contain dinoflagellate symbiotic algae called zooxanthellae, located intracellularly in the gastrodermal (gut) layer of the coral. The predominant source of nutrition for corals comes in the form of photosynthetic products produced by the zooxanthellae. The symbiotic relationship between zooxanthellae and corals is that the zooxanthellae provide the coral with photosynthetic carbon, which is often enough to supply the coral's energy requirements (Muller-Parker and D'Elia, 1997); in turn, the coral provides protection and access to enough light for the zooxanthellae to photosynthesize. Zooxanthellae require only light, carbon dioxide, and inorganic nutrients, the last of which can be obtained from animal waste or the seawater. The ability of corals to obtain nutrients by the capture of zooplankton combined with the nutrients provided by the zooxanthellae allows corals to live in low-nutrient tropical waters.

9.2 Meteorological processes influencing corals

The growth and subsistence of coral depends on a number of requirements: temperature, irradiance, calcium carbonate saturation, turbidity, sedimentation, salinity, pH, and nutrients. These variables influence the physiological processes of photosynthesis and calcification as well as coral survival, and as a result coral reefs occur only in select areas of the world's oceans. Meteorological processes can alter these variables. Figure 9.2 summarizes the connections between different meteorological processes and coral requirements for growth and survival. These processes affect the distribution of corals on both global and synoptic scales (Walker, 2005).

9.2.1 Irradiance and coral growth rate

The abundance of corals in the tropics compared with the extratropics (poleward of latitudes 23°), which is apparent from Figure 9.1, is caused by differences in seasonal temperatures and the levels of annual solar irradiance, or insolation. The photosynthetic rates of zooxanthellae are dependent on both light and temperature (Muller-Parker and D'Elia, 1997). As a result, corals are found in warm, clear, shallow waters that have sufficient irradiance necessary for photosynthesis. Reef corals can grow from the surface to depths where there is between

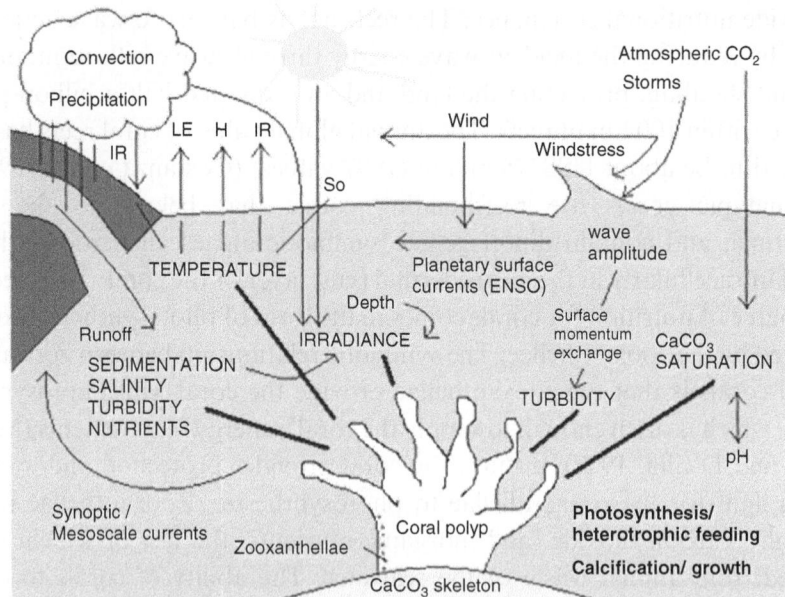

Figure 9.2. Schematic diagram summarizing key meteorological processes and coral requirements controlling calcification, photosynthesis, and survival.

1% and 10% of the surface irradiance (Chalker, 1981; Chalker et al., 1988). This means that most corals live within the top 100 m of the ocean. Deep-sea corals inhabit the colder deep waters of continental shelves and offshore canyons in waters at depths ranging from 50 to 1,000 m. Where current and substrate conditions are suitable, these corals form thickets, or groves, of high complexity. Also, under some conditions, such as for corals living at greater depths, the energy provided for calcification may be supplemented through heterotrophic feeding so that the growth rate would not be entirely dependent upon the intensity of irradiance (see, e.g., Kleypas, 1997; Crabbe and Smith, 2005, 2006).

The dominance of irradiance as a factor affecting coral reef growth has been demonstrated by the simulation of the growth of corals as a function of light only, and comparison of these results to measured growth rates (Bosscher and Schlager, 1992). This study covered reefs dominated by framework corals in regions where there was little lateral sediment transport and accumulation.

For reefs where lateral sediment transport and accumulation are important, the effects of lagoons, wave action, erosion, and resedimentation on growth rates would need to be included. In studies on the growth rates of seven eastern Pacific corals in the upwelling region of the Golfo de Papagayo, Costa Rica (Jiménez and Cortés, 2003), no significant correlation between the growth rate of the species *Pavona clavus* and the number of sunshine hours was found.

However, a negative growth anomaly, which coincided with the entry of heavy sediment loads to the cove where the samples were collected, was found. The heavy loads of sediment were caused by intense land removal activities for a road construction approximately 100 m from the site. The growth rates of *Pavona clavus* were also lower in El Niño years, when the sea surface temperatures (SSTs) were above normal, than in non–El Niño years (17.0 ± 2.8 mm yr^{-1} vs. 21.1 ± 3.0 mm yr^{-1}). Along the central and south Pacific coast of Costa Rica, an increase in seawater temperatures associated with El Niño produced bleaching and mortality of several species of coral.

Total solar radiation fluctuates much less during a year in the tropics (about 400 cal cm^{-2}day^{-1} at 20° S) than in the extratropics (about 1,000 cal cm^{-2} day^{-1} at 60° S) as a result of the combination of high intensity and constant day length. The even distribution of high-intensity insolation throughout the year in the tropics provides the coral's zooxanthellae with enough light to photosynthesize all year round.

Diurnal and seasonal changes in radiation are predictable, and large-scale shifts are not likely to occur over the next century (Kleypas *et al.*, 1999a, b). However, the future changes in the levels of irradiance on coral reefs are hard to predict, because the effects of cloud cover and water transparency cannot be predicted at the global scale (Guinotte *et al.*, 2003). In addition, the mesoscale effects of cloudiness and storms, which can reduce surface irradiance and also increase turbidity and phytoplankton blooms, are highly unpredictable (Brown, 1997).

9.2.2 Sea level and coral growth

Sea level is predicted to rise over the next 100 years from the thermal expansion of seawater and the melting of continental ice (Intergovernmental Panel on Climate Change [IPCC], 2001, 2007). This rise will increase the depth of the water column above a coral reef, and as the depth is related to the level of irradiance by the Beer–Lambert law, the intensity of irradiance will decrease. The stability of sea level for the past few thousand years has led to reefs growing up to a point where they are limited by the level of the sea; therefore, an increase in sea level may be beneficial to some reefs, allowing an increase in upward growth (Buddemeier *et al.*, 2004). Fast-growing corals such as members of the genus *Acropora*, which add up to 20 cm year^{-1} to their branch tips, should not have any problem keeping up with changing sea levels (Done, 1999). However, slower-growing species of corals, such as *Diploria*, are likely to be drowned as the predicted sea level rise rate overtakes the radial growth rate of the coral of around 1 cm yr^{-1} (Lough and Barnes, 2000). The IPCC (2001) reports that coral reefs on small island states such as in the Caribbean Sea and the Indian Ocean are less likely to be able to keep pace due to additional severe stress caused by anthropogenic factors. The next IPCC report (2007) concludes that small island states are already experiencing negative effects of global warming, resulting in changes to their ecosystems and socioeconomic conditions.

If growth rates are reduced due to increasing sea surface temperatures or decreased carbonate saturation states, then it will become increasingly difficult for reefs to keep up with rising sea levels. In addition, by being forced to keep up with rising sea levels, the reefs may have an increasing growth rate; but as the calcification rate would be unlikely to increase, this would result in a decreasing skeletal density that would make reefs more vulnerable to storms and other erosional forces (Hoegh-Guldberg, 1999). An increase in sea level rise could also lead to increased erosion of shorelines, resulting in higher levels of sedimentation and lower levels of irradiance.

9.2.3 Temperature and coral growth rate

A number of studies have examined the impact of air temperature variations on coral growth rates. For example, coral skeletal density, determined by using banding patterns (see, e.g., Dodge and Vaisnys, 1975) from a colony of *Porites lutea* in the Great Barrier Reef in the district of Haapiti, Australia (Bessat and Buiges, 2001), for the period 1958–1990, was linearly correlated with air temperature 25 km from where the core was taken with a value of $r = 0.37$ ($r = 0.56$ with a 5-year filter). For the same period, the correlation between air temperature and annual calcification rate was lower, at $r = 0.28$. These results indicated that a 1 °C rise in temperature would lead to an increase in the density of about 10.5%; however, this result has to be tempered by predicted decreased carbonate saturation rates, as well as by temperature rises above the optimum for coral growth, as was mentioned above. The effect of sea surface temperature and solar radiation on growth rates of the massive coral *Porites lutea* on 29 reefs on the Great Barrier Reef, Australia, was studied by Lough and Barnes (2000). As in the previous study (Bessat and Buiges, 2001), the annual density banding patterns from coral cores were measured, but were compared to annual sea surface temperature rather than to air temperature, and also to annual solar radiation. Measurements were taken over a range of 9° latitude from north to south along the Great Barrier Reef. Both average annual sea surface temperature and average incoming solar radiation significantly decreased with latitude ($r = -0.98$, $r = -0.65$, respectively.) The correlation with solar radiation was not as strong because the solar radiation is reduced in the northern Great Barrier Reef owing to greater than average cloud cover. The findings were combined with those from Grigg (1981, 1997) on corals in Hawaii, and those from

Table 9.1. *Correlations between annual sea surface temperature and calcification and extension rates, and correlations between annual incoming solar radiation and calcification rates and extension rates*[a]

	Correlation of average annual sea surface temperature and:		Correlation of average annual incoming solar radiation and:	
	Calcification rate	Extension rate	Calcification rate	Extension rate
29 reefs from the Great Barrier Reef	0.84	0.72	0.59	0.58
44 reefs from Great Barrier Reef, Hawaii, and Thailand	0.91	0.91	0.72	0.71

[a] Data are for 29 reefs from the Great Barrier Reef (Lough and Barnes, 2000) and for 44 reefs from the Great Barrier Reef, Hawaii (Grigg, 1981, 1997), and Thailand (Scoffin *et al.*, 1992). All correlation values are statistically significant.

Scoffin *et al.* (1992) on corals in Thailand. The results of the correlations (Table 9.1) show that the calcification rates and extension rates were highly correlated with sea surface temperature and to a lesser extent with incoming solar radiation. The extrapolation of the regression line for the minimum sea surface temperature indicates that the average calcification ceased at approximately 17.5 °C.

Nie *et al.* (1997) quantified the relationship between sea surface temperature and coral growth rate in the northern part of the South China Sea. The coral *Porites lutea* was studied by using five coral cores from the Xisha Islands and southern Hainan Island waters. The r-values for the positive correlation between the growth band width and the sea surface temperature were 0.88, 0.77, 0.89, and 0.85, values similar to those calculated by Lough and Barnes (2000) (Table 9.1).

Buddemeier *et al.* (2004) suggested that the observed increase in coral reef calcification rate with ocean warming over the latter half of the twentieth century (Lough and Barnes, 2000; Bessat and Buiges, 2001) was most likely due to increased metabolism and photosynthetic rates of the zooxanthellae (Leclerq *et al.*, 2000).

Growth rates also depend on minimum seasonal temperatures (Slowey and Crowley, 1995). Two sets of growth data were used to produce a single record of average growth rates of the coral species *Montastrea annularis* at the Flower Garden Banks in the Gulf of Mexico. Changes in average winter air temperature were found to correspond to changes in coral growth rate ($r = 0.53$, and $r = 0.78$ for smoothed data). Interdecadal changes in the growth rates of the corals corresponded to changes in average minimum winter season air temperatures at New Orleans (contemporaneous correlations $r = 0.42$ and $r = 0.86$ for smoothed data). Slowey and Crowley (1995) acknowledged that the correspondence between the changes in the two was not one to one because the influence of air temperature on water temperature depends on a number of meteorological and oceanographic factors. The minimum temperatures over the Gulf of Mexico can be caused by the passage of fronts bringing cold, dry air from Canada; Slowey and Crowley (1995) suggested that this process is primarily responsible for stressing corals at the Flower Gardens and reducing their winter growth rates. There was a major shift towards colder winters during the 1950s, and this shift coincided with the decline of coral growth at the Flower Gardens.

Coral bleaching

Most of the pigmentation within corals is within the zooxanthellae. Coral bleaching is caused by corals losing their zooxanthellae. The coral appears white, or bleached, because the white calcium carbonate coral skeleton shows through the translucent living tissue. Thermal bleaching occurs when the coral is exposed to prolonged above-normal temperatures, resulting in additional energy demands on the coral, depleted reserves, and reduced biomass. Under these circumstances, the coral is unable to house the zooxanthellae and so becomes bleached (Muller-Parker and D'Elia, 1997). The effect of high temperatures can be aggravated by high levels of irradiance. Gleason and Wellington (1993) reported that corals tend to bleach first on their upper, most sunlit surfaces. However, the absence of mass bleaching events occurring in the presence of high ultraviolet (UV) radiation intensity and normal temperatures indicates that high UV radiation is not a primary factor in causing mass bleaching (Hoegh-Guldberg, 1999).

Corals can die as a result of bleaching, though they may partially or fully recover from bleaching events (Lough, 2000). Bleaching causes a decrease in the growth rate of corals, and the time taken for a coral to recover from a bleaching event may be several years or decades. If the frequency of bleaching increases, then the capacity for coral reefs to recover is diminished (Done, 1999).

Bleaching tends to occur in regions where high temperatures are the norm (Muller-Parker and D'Elia, 1997). There is no single bleaching threshold for all locations, times, and species, but most bleaching events occur when the temperature is at least 1 °C higher than seasonal maximum temperatures (Winter et al., 1998; Hughes et al., 2003). There have been six major episodes of coral bleaching since 1979, affecting reefs in every part of the world (Figure 9.3). Secondary peaks in SST occurred in 1993 and 1994 following the 1991–92 El Niño, which led to conjectures that climate change was affecting ENSO. For a widely used index of El Niño/Southern Oscillation (ENSO) events, see www.cdc.noaa.gov/people/klaus.wolter/MEI/#ElNino.

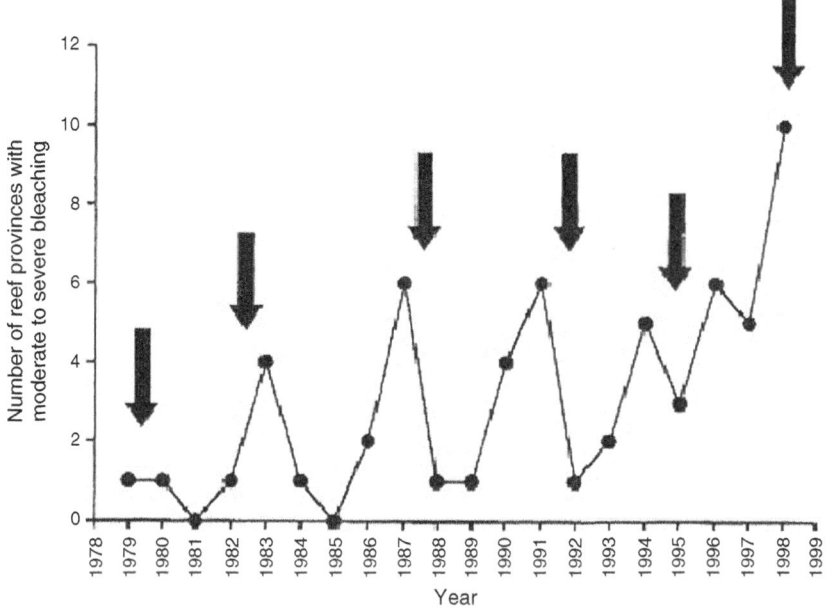

Figure 9.3. Number of reef provinces showing bleaching since 1979. Arrows indicate 12-month periods when there were El Niño events. Adapted from Hoegh-Guldberg (1999).

The 1998 El Niño event was the largest, killing an estimated 16% of the world's corals (Hughes *et al.*, 2003). There are virtually no reports of coral bleaching prior to 1979. The lack of reports may be due to a lack of reef observers then, but it is more likely that there were few or no bleaching events, since neither tourist resorts at the time nor indigenous fishers are aware of any events occurring before this time (Hoegh-Guldberg, 1999).

Forty-seven sites were studied where bleaching occurred during 1997–98, in the Indian Ocean/Middle East, Southeast Asia, Pacific Ocean, and Caribbean/ Atlantic (Lough, 2000). All these regions showed a significant warming trend during 1903–1999 of about 0.05 °C per decade, with regional variations in the periods of greatest warming. Extreme sea surface temperature maxima (defined as the top 20%) increased in frequency since 1979 compared with the previous 76 years in all the regions apart from the Atlantic/Caribbean (Table 9.2). The sea surface temperatures in 1998 were the highest at most of the sites in the previous 97 years. This study indicates that warm season sea surface temperature anomalies associated with mass bleaching events in the Indian Ocean, Southeast Asia, and parts of the Pacific Ocean are more likely to occur during an El Niño event.

While ENSO has a significant impact on coral bleaching in those areas of the tropics directly affected by the associated rise in SST, bleaching events have

Table 9.2. *Frequency in years[a] of extreme maxima in SST[b],* 1903–78 *and* 1979–99

	SST maximum index	
	1903–78	1979–99
47 sites	8	2
Indian Ocean/Middle East	8	2
Southeast Asia	13	2
Pacific	11	2
Caribbean/Atlantic	5	4

[a] Mean number of years between extremes.
[b] Based on top 20% of values over 97 years.
Adapted from Lough (2000).

occurred in all regions of the Pacific. This is likely the result of the general rise in ocean surface temperature documented over the past several decades (e.g., IPCC, 2001).

Projected ocean temperatures and bleaching events

The frequency with which corals will be bleached in the future has been estimated by using projections of future sea surface temperatures from four different general circulation models (GCMs) forced by the IPCC IS92a emission scenario (Hoegh-Guldberg, 1999). The SST projections were combined with thermal thresholds for corals, derived by using the Integrated Global Ocean Services System (IGOSS) dataset provided by the Joint World and Scientific Meteorological Organization (WMO) and United Nations Educational, Scientific and Cultural Organization (UNESCO), the Joint Intergovernmental Oceanographic Commission's (JCOMM) Technical Commission for Oceanography and Marine Meteorology, and from literature and Internet reports of bleaching events. The key assumption made was that reef-building corals and their zooxanthellae are unable to adapt or acclimatize to sporadic thermal stress. All SST projections indicated that the frequency of bleaching events is set to rise rapidly, with the highest rates in the Caribbean, Southeast Asia, and Great Barrier Reef, and the lowest rates in the central Pacific. The frequency of bleaching events was predicted to become annual in most oceans by 2040, and the Caribbean and Southeast Asia are projected to reach this point by 2020, triggered by seasonal changes in seawater temperature rather than by El Niño events.

The geographical patterns and the timing of probable repeat occurrences of coral mortality in the Indian Ocean have been estimated (Sheppard, 2003).

Forecast sea surface temperatures at 33 sites in the Indian Ocean were blended onto historical sea surface temperatures. The forecast temperatures were estimated by using the IS92a scheme, which follows a median path (IPCC, 2001). The probability of repeat critical sea surface temperatures was then estimated by combining the sea surface temperature at a site, the rate of rise, and the temperature that was lethal to more than 90% of the shallow-water corals in 1998. The probabilities of the warmest month reaching the lethal 1998 temperatures over time, for four sites, are shown in Figure 9.4, for different scenarios. The curves indicate a 50% probability of SSTs being warm enough by 2030 for the occurrence of coral bleaching events at Sites 1 and 2 (Comoros and Chagos in the Indian Ocean), and by 2070 in the Saudi Arabian Gulf (Site 3): see Sheppard (2003).

There is some evidence that corals can adapt to climate change. Corals can contain different clades (subspecies) of zooxanthellae (Stat *et al.*, 2006; Van Oppen and Gates, 2006). Baker *et al.* (2004) studied corals in Panama, the Persian (Arabian) Gulf, and the western Indian Ocean, following the 1997–98 El Niño-induced bleaching event, and compared these to corals in areas that were relatively unaffected by El Niño. The surviving corals were found to contain a higher percentage of zooxanthellae of the genus *Symbiodium* in clade D than the unaffected corals. For example, 62% of coral colonies in Panama had clade D symbionts compared with just 1.5% of colonies in the Red Sea

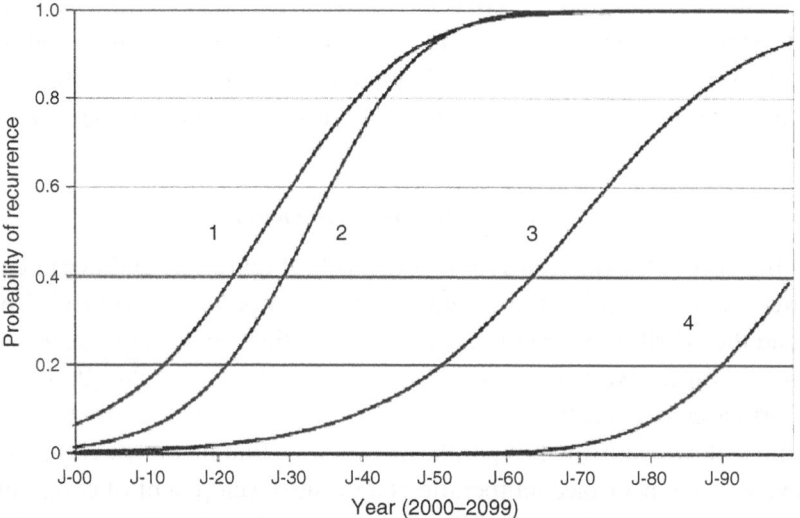

Figure 9.4. Probabilities of the warmest months reaching the 1998 temperatures lethal to coral at four sites in the Indian Ocean: 1, Comoros; 2, Chagos; 3, Saudi Arabian Gulf; 4, Minicoy, North Lakshadweep. The warmest month was March in Comoros and Chagos, May in North Lakshadweep, and September in the Saudi Arabian Gulf. Adapted from Sheppard (2003).

(which was relatively unaffected by El Niño and does not experience such high seasonal temperatures). This result suggested that clade D *Symbiodium* were more thermally tolerant than other clades. Rowan (2004) found responses that were similar to those of Baker *et al.* (2004), where *Symbiodinium* clade D was found to be more resistant to high temperatures than clade C. Rowan (2004) proposed that corals may be able to adapt to global warming by recombination with temperature-resistant zooxanthellae.

Hughes *et al.* (2003) did not support the stance that bleaching is a mechanism for corals to adapt to climate change by expelling susceptible zooxanthellae in order to uptake more resistant ones. Instead, Hughes *et al.* (2003) described bleaching as "a stress response, which is often followed by high mortality, reduced growth rates, and lower fecundity."

The IPCC (2001) stated that increases in CO_2 and temperature over the next 50 years will exceed those that the corals have previously thrived in over the last half-million years. This accelerating rate of environmental change may be too fast for the evolutionary capacity of corals (Hughes *et al.*, 2003). While we have found clade D zooxanthellae and multiple zooxanthellae clades in corals from Ningaloo reef in Australia, we have yet to determine whether this indicates "symbiont switching" (introduction of a new clade from exogenous sources) or "symbiont shuffling" (host contains multiple clades and a shift in dominance is introduced) in response to environmental or climate factors, and whether this phenomenon extends to other coral reefs around the world (Crabbe and Carlin, 2004; Carlin *et al.*, 2006; Crabbe and Carlin, 2007). It has been suggested that the majority of corals may not change their symbionts (Goulet, 2006), but whether this is a matter of sensitivity of analysis remains to be seen (Mieog *et al.*, 2007).

9.2.4 *Winds and ocean currents*

A comparison of the distribution of coral reefs (Figure 9.1) and ocean currents indicates a correlation between regional ocean temperatures controlled by currents and the distribution of corals. For example, there are no coral reefs south of the equator on the west coast of South America; the Humboldt–Peru Current, which brings cold water from the higher latitudes towards the equator, together with winds blowing northward parallel to the coast that favor upwelling, result in the lowering of sea surface temperatures and the development of cool, nutrient-rich waters off the west coast of South America. These waters are depleted of nutrients by phytoplankton by the time they reach the central Pacific. However, as corals can thrive in warm, nutrient-poor water, they are abundant in the central Pacific. There are few coral reefs on the west coast of Africa, where the southward flowing Canary Current and northerly upwelling, favorable winds result in nutrient-rich, cooler water at these latitudes. Finally, the northward-flowing Benguela Current, off the southwest coast of Africa, and longshore winds also result in nutrient-rich, cooler water in this region. The Benguela Current is weaker than the Humboldt–Peru Current and weakens considerably well before reaching the equatorial Atlantic; this may account for the few coral reefs occurring on the equator in this region. Coral reefs are found on both the east coast of Africa and to a lesser extent on the east coast of South America, perhaps due to the warm water driven by the easterly equatorial currents.

9.2.5 Wave energy

Reef corals are adapted to withstand a typical range of wave energy in their environment. However, extreme wind events from winter storms or hurricanes can produce waves with unusually extreme size and energy that can damage or destroy corals (e.g., Jokiel, 2006). The increased wave energy can also enhance turbidity and abrasion and remove juveniles that are needed for coral regrowth. Massive corals, in particular, provide important protection of coastlines from high wave energy. While the tsunami in December 2004 damaged a number of reefs – notably in Indonesia, Thailand, the Andaman and Nicobar Islands, and Sri Lanka – damage was patchy, and was generally overshadowed by major anthropogenic threats such as overfishing and deforestation (Wilkinson *et al.*, 2006).

9.2.6 Precipitation, salinity, and sedimentation

Corals are affected by sedimentation, often caused by discharge from rivers. There are no obvious links between mean rainfall and occurrence of corals. The areas with the highest rainfall are the northeast and northwest coasts of South America, the Equatorial Guinea coast, Africa, Madagascar, and the Maritime Continent. Coral reefs occur in all these areas, suggesting that perhaps sedimentation affects corals on a more local scale or that the reefs in these areas are not adjacent to rivers. It may also be that it is the rainfall intensity rather than the mean rainfall that affects the distribution of corals, causing large amounts of sedimentation from rivers during tropical storms or wet seasons.

Turbidity and sedimentation can both affect the growth rate of corals. Suspended sediment can reduce the amount of irradiance reaching the coral by shading it; this is a problem in areas of high hydrodynamic energy, where sediments tend to remain in suspension. Sedimentation can also lead to smothering, in which large reefs are buried by sedimentation from the bank-top (region above the coral); this is a greater problem in areas of low hydrodynamic energy (Hubbard and Scaturo, 1985; Crabbe and Smith, 2005; Crabbe *et al.*, 2006).

Salinity is another variable that determines where corals can grow. Reefs are limited to areas of normal marine salinity (33‰–36‰) (Hubbard, 1997). Reefs that are affected by changes in salinity are those situated near rivers and areas of high precipitation. Lirman (2003) tested the hypothesis that both salinity and sedimentation influence growth and survivorship of corals. Small (<5 mm in diameter) polyp colonies of *Siderastrea radians* were taken from hard-bottom communities of Biscayne Bay, Florida, and glued to ceramic tiles. The combined affects of sedimentation and salinity were then measured by exposing the polyp colonies to a variety of experimental conditions, including different salinity and different amounts of sedimentation. Sixteen colonies were exposed to low salinity (20‰) and sediment-burial treatments for periods of 24 hours at weekly intervals, and 16 colonies were kept under the same conditions as those in water in Florida's Biscayne Bay (35‰) as a control. After 1 month, the corals exposed to low salinity had mean radial growth rates slower than those of the controls, although this difference was not statistically significant at the 5% level, and no colony mortality was observed within either group. Colonies buried under sediments at weekly intervals had radial extension rates significantly slower than those in the controls. In this case, 30% of colonies exposed to burial treatments experienced total mortality at the end of 1 month.

The Intertropical Convergence Zone (ITCZ) results in cloudiness and rainfall, which may affect irradiance, temperature, sedimentation, and salinity. Large-scale convection caused by converging air masses and warm sea surface temperatures transports moisture upwards into the atmosphere, forming clouds and rain. Cumulous clouds formed by the ITCZ are highly reflective, reducing the downwelling shortwave radiation by approximately 10% (McFarlane and Evans, 2004), and therefore the amount of photosynthetic active radiation for photosynthesis. Thick clouds can also reduce the amount of ultraviolet radiation through reflection (ultraviolet radiation may be harmful to corals at high intensities; Chadwick-Furman, 1996).

Corals are abundant underneath the most intense cloudy regions of the ITCZ: the Maritime Continent, central Pacific, and Central America. However, they are also abundant in the southern Pacific, the Red Sea, the Persian Gulf, and the East and South China Seas, which the ITCZ does not reach. This suggests that either the cloud cover and precipitation from the ITCZ does not affect the distribution of corals, or that corals have adapted to the radiation conditions in each area.

9.2.7 *Atmospheric CO_2 and calcium carbonate chemistry*

Corals grow by combining calcium ions with carbonate ions and depositing a calcium carbonate skeleton (calcification) in the form of aragonite. The concentration of calcium ions in seawater is much higher than the concentration of the carbonate ion; therefore, the rate of calcification is partly controlled by the concentration of carbonate ions in seawater (Kleypas *et al.*, 1999a). Aragonite and calcite are more soluble in cold, pressurized water and less soluble in warm, non-pressurized water. As a result, aragonite and calcite are super-saturated in warm, shallow tropical waters and under-saturated in deep, cold water. Aragonite supersaturation allows corals to precipitate calcium carbonate skeletons in warm tropical waters (Kleypas *et al.*, 1999a).

An increase in the partial pressure of CO_2 lowers the pH of seawater and decreases the concentration of carbonate ion. As corals use the carbonate ion to form their skeletons, a decrease in the levels of carbonate ion will lead to a reduction in the calcification rate, less carbonate accumulation on average, and probably lower extension rates or weaker skeletons in some corals (Caldeira and Wickett, 2003; Feely *et al.*, 2004). The results of these changes would be a reduction in the ability of the coral to calcify and to withstand erosion (Guinotte *et al.*, 2003). As atmospheric CO_2 concentrations rise in the future, the pH and carbonate ion concentrations in surface seawater will fall, perhaps to the point where aragonite becomes under-saturated and corals are unable to form their skeletons.

Nevertheless, the greatest potential threat to corals is likely to be an increase in seawater temperature, rather than the calcium carbonate saturation state, as many corals already live precariously near their temperature thresholds (IPCC, 2001, 2007). The IPCC (2001) document also suggests that as well as an increase in the mean sea surface temperature, there will also be an increase in the variability of temperature, possibly causing more bleaching events than would occur under a smooth and gradual increase.

9.3 Coral colony growth rates and models

9.3.1 Introduction

Modeling coral colony growth with high precision has value not so much as a predictive tool but as a tool for monitoring real-time changes to growth from anthropogenic or other effects. There are many ways to model the growth of coral colonies, from simple exponential functions to more complex logistic functions, von Bertalanffy functions, and polynomial functions (see, e.g., Crabbe *et al.*, 2002; Crabbe and Smith, 2002; 2003). A study of coral growth in Discovery Bay, Jamaica, provides an example of a polynomial function approach (Crabbe *et al.*, 2002). In this sampling study, the authors took pains to measure corals that were widely separated.

Figure 9.5 shows numbers of colonies (from a total of 1,650) estimated to have been recruited per year (which assumes their survival, or retention, from that date), for Discovery Bay, Jamaica. Arrows refer to the severe storms encountered

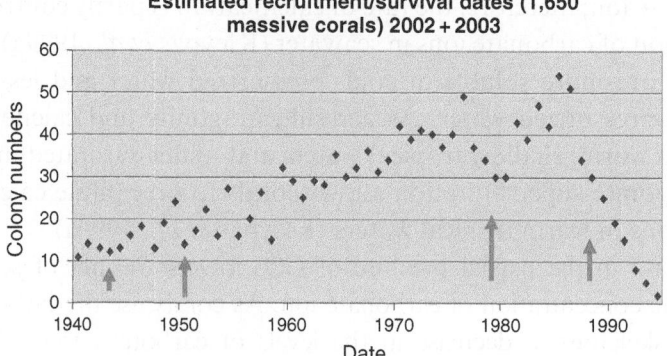

Figure 9.5. Numbers of colonies estimated to have been recruited per year (which assumes their survival, or retention, from that date), for Discovery Bay, Jamaica. The total number of massive corals measured was 1,650. Arrows refer to the severe storms encountered in the area in 1944, 1951, 1980, and 1988. Adapted from Crabbe *et al.* (2004).

in the area in 1944, 1951, 1980, and 1988. Interestingly, after the most severe recent storm, Hurricane Allen in 1980, massive coral recruitment took a number of years to return to former levels. This was probably due to the changes in the substrate topography, and to the stress on coral colonies. Stress has been shown to reduce the reproductive output of corals (Baird and Marshall, 2002; Ward, 1995). The fall after 1988 reflects a combination of storm effects, pollution, and small colonies. Figure 9.6 shows the relationship between numbers of colonies recruited and storm severity for the years since 1940 when there were tropical storms or hurricanes near Discovery Bay. There was a significant negative correlation ($r = 0.72$; $p < 0.01$) between recruitment estimates and storm severity. Intermediate storm severity resulted in variable levels of recruitment of non-branching corals, while the severest storms resulted in significantly ($p < 0.002$; Student's t-test) lower recruitment estimates. We have also shown that hurricanes and severe storms limit the recruitment and survival of massive nonbranching corals of the Mesoamerican barrier reef and on patch reefs near the Belize coast in the Caribbean, and suggest that marine park managers may need to assist coral recruitment in years where there are hurricanes or severe storms (Crabbe *et al.*, 2008).

Hurricane Allen in 1980 caused major damage to the Discovery Bay reefs (Woodley *et al.*, 1981), with great destruction of branching corals (Woodley, 1992) and delayed mortality (Knowlton *et al.*, 1981). We now know that the storm also severely limited non-branching coral recruitment. This also happened for the severe storms of 1951 and 1944. It had been suggested that as the hurricane exposed large amounts of substratum, thus greatly increasing the area for recruitment of sessile organisms, new reef development would

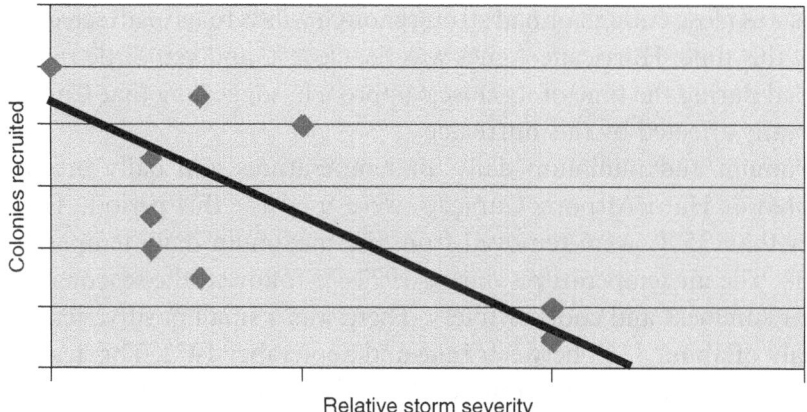

Figure 9.6. Graph of coral colonies recruited against storm severity in Discovery Bay, Jamaica. Storm intensity was calculated according to Crabbe *et al.* (2002). Adapted from Crabbe *et al.* (2002).

follow (e.g., Graus *et al.*, 1984). This did not happen, no doubt because survival is greater in the shade or under surfaces, although growth is greater on upper surfaces (Birkeland, 1977, 1997). Predation (Knowlton *et al.*, 1988) and algal growth would also interfere with recruitment; Hurricane Allen resulted in significant increases in algal settling on the substratum (Woodley *et al.*, 1981).

9.3.2 Modeling coral growth in Curaçao

Curaçao is completely surrounded by fringing reefs, which provide protection, food, and income to the island population of approximately 150,000. There are several documented natural and anthropogenic threats to the Curaçao reefs. Massive coral bleaching impacted the Curaçao coral reef in 1987, 1990, and 1995; hurricanes and disease have caused considerable damage; and massive coastal development has led to increased sedimentation and sewage, which are also harmful to the reefs (Burke and Maidens, 2004).

Curaçao is situated outside of the southern edge of the Atlantic hurricane belt. Hurricanes pass within 100 miles of Curaçao on average once every 4 years. The Curaçao coral reef has been damaged by hurricanes in the past. The most recent hurricanes causing considerable damage to Curaçao were José and Lenny in 1999, Georges in 1998, Luis and Marilyn in 1995, Hugo in 1989, Donna in 1960, and Dog in 1950.

The coral weights measured by Bak (1976), and reviewed here, were measured over the period May 25, 1972, to June 25, 1973, on the Curaçao coral reef. During this period, three hurricanes, one tropical storm, and three subtropical storms occurred in the Caribbean (UNISYS, 2006). None of these storms was close enough or had strong enough winds to have affected Curaçao during this time; Hurricane Agnes was the closest, and very little rainfall was recorded during the time of its closest approach, suggesting that Curaçao was not greatly affected by this hurricane.

Maximum and minimum daily air temperatures and daily precipitation measured at Hato Airport, Curaçao, were used for this period. The values greater than 35 °C were removed from the maximum daily temperatures as outliers. The air temperatures during 1972–73 followed the seasonal trend of warmer summers and cooler winters. There was a small positive temperature anomaly of about 1 °C, between June and September 1972. This temperature anomaly could have been a result of the El Niño occurring at this time.

A 30-day moving average was applied to the maximum and minimum daily temperatures and daily precipitation, as is shown in Figures 9.7 and 9.8, respectively.

The underwater masses of the branching coral *Acropora palmata*, on the Curaçao reef, were measured *in situ* (under water) at approximately monthly intervals between May 25, 1972, and June 2, 1973, and between December 13, 1973, and June 14, 1974. Data were kindly provided by Professor Bak. These are some of the longest ever studies of *in situ* coral mass measurements.

The corals measured during the latter period showed a higher growth rate than during the earlier period, suggesting that large colonies accumulate calcium carbonate more rapidly than smaller colonies (Bak, 1976).

The exponential models gave good fits to the data ($R^2 = 0.9967$, 0.9937). However, the log of the corrected measured coral mass, Y', as a function of time did not lie on a straight line, indicating that the fit could not be entirely described by an exponential function. A more complicated model may be needed to describe coral growth, such as the rational polynomial function used by Crabbe *et al.* (2002) and Crabbe and Smith (2003). However, for this study we used a smoothing spline to produce a non-parametric fit to the data on total coral mass as a function of time and to determine a time-dependent growth rate. We then examined the correlation between the time-dependent growth rate and the 30-day moving average values for daily maximum temperature, minimum temperature, and precipitation. The results suggest that the 30-day averaged maximum daily temperature could explain about 3% of the variability in the time-dependent growth rate and that there was no correlation between coral growth rate and minimum daily temperature or daily precipitation. Interestingly, the temperature correlation was negative. This result suggests that during the measurement period, temperatures rose to values higher than optimum for growth, thus inhibiting coral growth, but were not sufficiently high to cause bleaching of this species. Thus what would normally be a positive correlation became a slight but significant negative correlation. This result points to the very

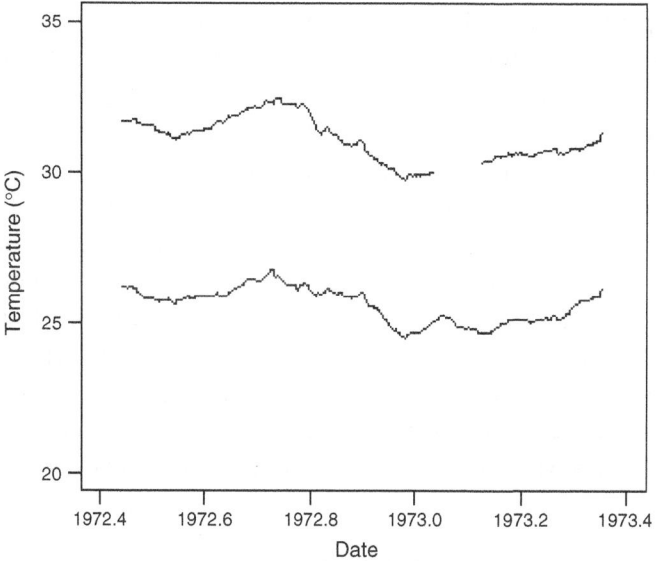

Figure 9.7. Daily maximum and minimum temperatures (degrees Celsius) with a 30-day moving average applied in Curaçao from May 25, 1972, through June 25, 1973. Missing values in the maximum daily temperatures are due to the removal of doubtful data points.

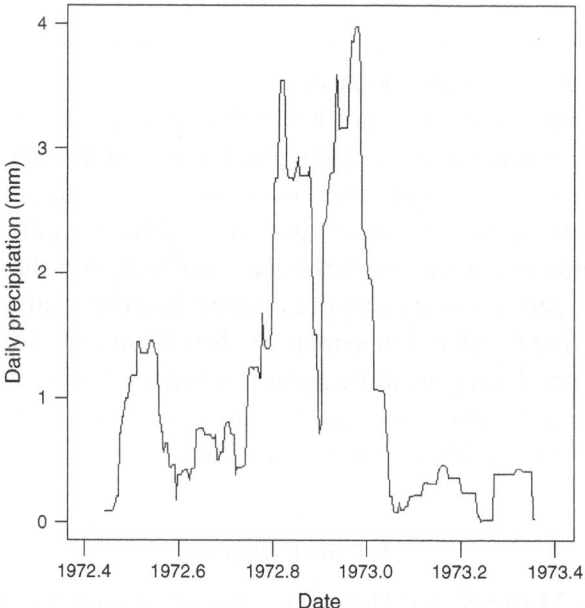

Figure 9.8. Daily precipitation (millimeters) in Curaçao, Netherlands Antilles from May 25, 1972, through June 25, 1973. Data were smoothed with a 30-day moving average filter.

narrow temperature range for coral growth, and to how sensitive corals are to fluctuations in temperature (Dodge and Lang, 1983; Crabbe, 2007).

In this example, there are several sources of uncertainty in the measurement and analysis of the data. In order to calculate the coral growth rate, the initial coral mass was estimated by assuming an exponential growth model over the first few data points; this procedure will have caused an error in all subsequent calculations based on the total coral mass to some degree. Another error was due to the estimation of the coral growth rate after applying a smoothing spline to the data. However, the overall correlation will not have been affected to a great degree by these errors, as these are systematic errors that will increase or decrease the whole coral growth rate dataset. In future studies, we will be looking in more detail at temperature effects on growth rate modeling in different coral reef areas, and to correlations with zooxanthellae clades (Crabbe, 2003; Carlin *et al.*, 2006).

9.4 Conclusions

In this chapter, we have illustrated how climate processes and extremes can influence the physiological processes responsible for the growth of coral reef colonies. Coral growth takes place within narrow limits of temperature, irradiance, salinity, pH, and turbidity, all variables that are influenced by climate and weather. In periods of pronounced climate change in the future, it appears unlikely that corals or their symbiont algae could evolve to keep pace; the expected rates of change are too large.

We have shown that on reefs off the north coast of Jamaica, hurricanes and severe storms can limit the recruitment and survival of massive coral colonies. We have studied a number of empirical models for coral growth, and have shown that small changes in temperature can significantly influence branching coral growth rates in coral reefs off the coast of Curaçao in the Caribbean. We are currently undertaking further work on modeling the coral reefs of Jamaica, as well as studying the clades of symbiotic algae from reefs around the world, with a view to developing accurate predictive models for coral growth dependent upon climate variables, as well as to understanding the functional genomics of coral–algal symbiosis that can be influenced by climate extremes.

Acknowledgments

We thank Rick Murnane and Henry Diaz for organizing the climate extremes meeting in Bermuda in 2005, which provided the inspiration for this chapter; we also thank them for their work in editing the chapter for inclusion in this volume. The Earthwatch Institute and the Royal Society kindly provided essential funding for the coral measurements made on the reefs of Jamaica. Professor Rolf Bak provided some essential data from his study of coral growth in Curaçao, the anonymous referees helped in improving the manuscript, and Dr Albert Maris of the Antilles Meteorological Service provided meteorological data for the period during which coral growth was being measured on Curaçao.

References

Baird, A. H., and Marshall, P. A. (2002). Mortality, growth and reproduction in Scleractinian corals following bleaching on the Great Barrier Reef. *Marine Ecology Progress Series*, **237**, 133–41.

Bak, R. P. M. (1976). The growth of coral colonies and the importance of crustose coralline algae and burrowing sponges in relation with carbonate accumulation. *Netherlands Journal of Sea Research*, **10**, 285–337.

Baker, A. C., Stargert, C. J., McClanahan, T. R., and Glynn, P. W. (2004). Corals' adaptive response to climate change. *Nature*, **430**, 741.

Bessat, F., and Buiges, D. (2001). Two centuries of variation in coral growth in a massive *Porites* colony from Moorea (French Polynesia): a response of ocean-atmosphere variability from south central Pacific. *Palaeogeography, Palaeoclimatology, Palaeoecology*, **175**, 381–92.

Birkeland, C. (1977). The importance of rate of biomass accumulation in early successional stages of benthic communities to the survival of coral recruits. *Proceedings of the Third International Coral Reef Symposium*, pp. 15–21.

Birkeland, C. (1997). Introduction. In *Life and Death of Coral Reefs*, ed. C. Birkeland. New York: Chapman & Hall, pp. 1–12.

Bosscher, H., and Schlager, W. (1992). Computer simulation of reef growth. *Sedimentology*, **39**, 503–12.

Brown, B. E. (1997). Adaptations of reef corals to physical environmental stress. *Advances in Marine Biology*, **31**, 221–99.

Buddemeier, R. W., Kleypas, J. A., and Aronson, R. B. (2004). *Coral Reefs and Global Climate Change: Potential Contributions of Climate Change to Stresses on Coral Reef Ecosystems*. Report, Pew Center on Global Climate Change, Arlington, VA.

Burke, L., and Maidens, J. (2004) *Reefs at Risk in the Caribbean*. Washington, D.C.: World Resources Institute.

Caldeira, K., and Wickett, M. E. (2003). Anthropogenic carbon and ocean pH. *Nature*, **425**, 365.

Carlin, J. P., Humphries, S., Crabbe, M. J. C., and Mazet, F. (2006). Zooxanthellae diversity in the Ningaloo Reef tract, western Australia. *Abstracts of the 7th Reef Conservation UK Meeting*, 6.

Chadwick-Furman, N. E. (1996). Reef coral diversity and global change. *Global Change Biology*, **2**, 559–68.

Chalker, B. E. (1981). Simulating light-saturation curves for photosynthesis and calcification by reef-building corals. *Marine Biology*, **63**, 135–41.

Chalker, B. E., Barnes, D. J., Dunlop, W. C., and Jokiel, P. L. (1988). Light and reef building corals. *Interdisciplinary Science Review*, **13**, 222–37.

Costanza, R., D'Arge, R., De Groot, R., *et al.* (1997). The value of the world's ecosystem services and natural capital. *Nature*, **387**, 253–60.

Crabbe, M. J. C. (1988). Simple graphical methods for use with complex ligand binding and enzyme mechanisms. *Federation of European Biochemical Societies Letters*, **235**, 183–8.

Crabbe, M. J. C. (2003). A novel method for the transport and analysis of genetic material from polyps and zooxanthellae of Scleractinian corals. *Journal of Biochemical and Biophysical Methods*, **57**, 171–6.

Crabbe, M. J. C. (2006). Challenges for sustainability in cultures where regard for the future may not be present. *Sustainability: Science, Practice and Policy*, **2**, http://ejournal.nbii.org/progress/2006fall/communityessay.crabbe.html.

Crabbe, M. J. C. (2007). Global warming and coral reefs: modeling the effect of temperature on *Acropora palmata* colony growth. *Computational Biology and Chemistry*, **31**, 294–7.

Crabbe, M. J. C., and Carlin, J. P. (2004). Zooxanthellae clade populations in corals. *Abstracts of the 38th Population Genetics Group*, 6.

Crabbe, M. J. C., and Carlin, J. P. (2007). Industrial sedimentation lowers coral growth rates in a turbid lagoon environment, Discovery Bay, Jamaica, *International Journal of Integrative Biology*, **1**, 37–40.

Crabbe, M. J. C., and Smith, D. J. (2002). Comparison of two reef sites in the Wakatobi Marine National Park (SE Sulawesi, Indonesia) using digital image analysis. *Coral Reefs*, **21**, 242–4.

Crabbe, M. J. C., and Smith, D. J. (2003). Computer modeling and estimation of recruitment patterns of non-branching coral colonies at three sites in the Wakatobi Marine Park, SE Sulawesi, Indonesia: implications for coral reef conservation. *Computational Biology and Chemistry*, **27**, 17–27.

Crabbe, M. J. C., and Smith, D. J. (2005). Sediment impacts on growth rates of Acropora and Porites corals from fringing reefs of Sulawesi, Indonesia. *Coral Reefs*, **24**, 437–41.

Crabbe, M. J. C., and Smith, D. J. (2006). Modelling variations in corallite morphology of *Galaxea fascicularis* coral colonies with depth and light on coastal fringing reefs in the Wakatobi Marine National Park (SE Sulawesi, Indonesia). *Computational Biology and Chemistry*, **30**, 155–9.

Crabbe, M. J. C., Karaviotis, S., and Smith, D. J. (2004). Preliminary comparison of three coral reef sites in the Wakatobi Marine National Park (SE Sulawesi, Indonesia): estimated recruitment dates compared with Discovery Bay, Jamaica. *Bulletin of Marine Science*, **74**, 469–76.

Crabbe, M. J. C., Mendes, J. M., and Warner, G. F. (2002). Lack of recruitment of non-branching corals in Discovery Bay is linked to severe storms. *Bulletin of Marine Science*, **70**, 939–45.

Crabbe, M. J. C., Wilson, M. E. J., and Smith, D. J. (2006). Quaternary corals from reefs in the Wakatobi Marine National Park, SE Sulawesi, Indonesia, show similar growth rates to modern corals from the same area. *Journal of Quaternary Science*, **21**, published online: 12 May 2006. doi:10.1002/jqs.1011.

Crabbe, M. J. C., Martinez, E., Garcia, C., Chub, J., Castro, L., and Guy, J. (2008). Growth modeling indicates hurricanes and severe storms are linked to low coral recruitment in the Caribbean. *Marine Environmental Research* (in press).

Dodge, R. E., and Lang, J. C. (1983). Environmental correlates of hermatypic coral (*Montastrea annularis*) growth on the East Flower Gardens Bank, northwest Gulf of Mexico. *Limnology and Oceanography*, **28**, 228–40.

Dodge, R. E., and Vaisnys, J. R. (1975). Hermatypic coral growth banding as an environmental recorder. *Nature*, **258**, 706–8.

Done, T. J. (1999). Coral communities adaptability to environmental change at the scales of regions, reefs and reef zones. *American Zoologist*, **39**, 66–79.

Feely, R. A., Sabine, C. L., Lee, K., Berelson, W., Kleypas, J., Fabry, V. J., and Millero, F. J. (2004). Impact of anthropogenic CO_2 on the $CaCO_3$ system in the oceans. *Science*, **305**, 362–6.

Gleason, M. G., and Wellington, G. M. (1993). Ultraviolet radiation and coral bleaching. *Nature*, **365**, 836–8.

Goulet, T. L. (2006) Most corals may not change their symbionts. *Marine Ecology Progress Series*, **321**, 1–7.

Graus, R. R., Macintyre, I. G., and Herchenroder, B. E. (1984). Computer simulation of the reef zonation at Discovery Bay, Jamaica: hurricane disruption and long-term physical oceanographic controls. *Coral Reefs*, **3**, 59–68.

Grigg, R. W. (1981). Coral reef development at high latitudes in Hawaii. *Proceedings of the Fourth International Coral Reef Symposium, Manila*, **1**, 687–93.

Grigg, R. W. (1997). Paleoceanography of coral reefs in the Hawaiian-Emperor Chain – revisited. *Coral Reefs*, **16**, S33–8.

Guinotte, J. M., Buddemeier, R. W., and Kleypas, J. A. (2003). Future coral reef habitat marginality: temporal and spatial effects of climate change in the Pacific basin. *Coral Reefs*, **22**, 551–8.

Hoegh-Guldberg, O. (1999). Climate change, coral bleaching and the future of the world's coral reefs. *Marine Freshwater Research*, **50**, 839–66.

Hubbard, D. K. (1997). Reefs as dynamic systems. In *Life and Death of Coral Reefs*, ed. C. Birkeland. New York: Chapman and Hall, pp. 43–67.

Hubbard, D. K., and Scaturo, D. (1985). Growth rates of seven species of Scleractinian corals from Cane Bay and Salt River, St. Croix, U.S. Virgin Islands. *Bulletin of Marine Science*, **36**, 325–38.

Hughes, T. P., Baird, A. H., Bellwood, D. R., *et al.* (2003). Climate change, human impacts, and the resilience of coral reefs. *Science*, **301**, 929–33.

Intergovernmental Panel on Climate Change (IPCC) (2001). *Third Assessment Report of the Intergovernmental Panel on Climate Change, 2001: The Scientific Basis*. Cambridge, UK: Cambridge University Press.

Intergovernmental Panel on Climate Change (IPCC) (2007). Summary for Policymakers from Working Group 1. http://www.ipcc.ch/SPM2 feb 07.pdf

Jiménez, C. and Cortés, J. (2003). Growth of seven species of Scleractinian corals in an upwelling environment of the eastern Pacific (Golfo De Papagayo, Costa Rica). *Bulletin of Marine Science*, **72**, 187–98.

Jokiel, P. L. (2006). Impact of storm waves and storm floods on Hawaiian reefs. *Proceedings of the 10th International Coral Reef Symposium*, pp. 390–8.

Kleypas, J. A. (1997). Modelled estimates of global reef habitat and carbonate production since the last glacial maximum. *Paleoceanography*, **12**, 533–45.

Kleypas, J. A., Buddemeier, R. W., Archer, D., Gattuso, J.-P., Langdon, C., and Opdyke, B. N. (1999a). Geochemical consequences of increased atmospheric carbon dioxide on coral reefs. *Science*, **284**, 118–20.

Kleypas, J. A., McManus, J. W., Lambert, A. N., and Meñez, A. (1999b). Environmental limits to coral reef development: Where do we draw the line? *American Zoologist*, **39**, 146–59.

Knowlton, N., Lang, J. C., and Keller, B. D. (1988). Fates of staghorn coral isolates on hurricane-damaged reefs in Jamaica: the role of predators. *Proceedings of the Sixth International Coral Reef Symposium*, **2**, 83–8.

Knowlton, N., Lang, J. C., Rooney, M. C., and Clifford, P. (1981). Evidence for delayed mortality in hurricane-damaged Jamaican staghorn corals. *Nature*, **294**, 251–2.

Leclerq, N., Gattuso, J.-P., and Jaubert, J. (2000). CO_2 partial pressure controls and the calcification rate of a coral community. *Global Change Biology*, **6**, 329–34.

188 *M.J.C. Crabbe et al.*

Lirman, D. (2003). A simulation of the population dynamics of the branching coral *Acropora palmata*: effects of storm intensity and frequency. *Ecological Modelling*, **161**, 167–80.

Lough, J. M. (2000). Unprecedented thermal stress to coral reefs? *Geophysical Research Letters*, **27**, 3901–4.

Lough, J. M., and Barnes, D. J. (2000). Environmental controls on growth of the massive coral *Porites*. *Journal of Experimental Marine Biology*, **245**, 225–43.

McFarlane, S. A., and Evans, K. F. (2004). Clouds and shortwave fluxes at Nauru. Part II: Shortwave flux closure. *Journal of Atmospheric Sciences*, **61**, 2602–15.

Mieog, J. C., van Oppen, M. J. H., Cantin, N. E., Stam, W. T., and Olsen, J. L. (2007). Real-time PCR reveals a high incidence of *Symbiodinium clade D* at low levels in four scleractinian corals across the Great Barrier Reef: implications for symbiont shuffling. *Coral Reefs*, **26**, 449–57.

Muller-Parker, G, and D'Elia, C. F. (1997). Interactions between corals and their symbiotic alga. In *Life and Death of Coral Reefs*, ed. C. Birkeland. New York: Chapman and Hall, pp. 96–112.

Nie, B., Chen, T., Liang, M., Wang, Y., Zhong, J., and Zhu, Y. (1997). Relationship between coral growth rate and sea surface temperature in the northern part of the South China Sea during the past 100 years. *Science in China* (Series D), **40**, 173–82.

Rowan, R. (2004). Thermal adaptation in reef coral symbionts. *Nature*, **430**, 742.

Scoffin, T. P., Tudhope, A. W., Brown, B. E., Chansang, H., and Cheeney, R. F. (1992). Patterns and possible environmental controls of skeletogenesis of *Porites lutea*, South Thailand. *Coral Reefs*, **11**, 1–11.

Sheppard, C. R. C. (2003). Predicted recurrences of mass coral mortality in the Indian Ocean. *Nature*, **425**, 294–297.

Slowey, N. C., and Crowley, T. J. (1995). Interdecadal variability of Northern Hemisphere circulation recorded by Gulf of Mexico corals. *Geophysical Research Letters*, **22**, 2345–8.

Stat, M., Carter, D. and Hoegh-Guldberg, O. (2006). The evolutionary history of Symbiodinium and Scleractinian hosts: symbiosis, diversity, and the effect of climate change. *Perspectives in Plant Ecology Evolution and Systematics*, **8**, 23–43.

UNISYS (2006). http://weather.unisys.com/hurricane/index.html.

Van Oppen, M. J. H., and Gates, R. D. (2006). Conservation genetics and the resilience of reef-building corals. *Molecular Ecology*, **15**, 3863–83.

Walker, E. L. L. (2005). The role of weather and climate processes in coral growth. M.S. thesis, University of Reading.

Ward, S. (1995). The effect of damage on the growth, reproduction and storage of lipids in the Scleractinian coral *Pocillopora damicornis* (Linnaeus). *Journal of Experimental Marine Biology and Ecology*, **187**, 193–206.

Wilkinson, C., Souter, D., and Goldberg, J., eds. (2006). *Status of Coral Reefs in Tsunami Affected Countries: 2005*. Townsville, Queensland: Australia Institute of Marine Sciences.

Winter, A., Appledorn, R. S., Bruckner, A., Williams, E. H., Jr., and Goenaga, C. (1998). Sea surface temperatures and coral reef bleaching off La Parguera, Puerto Rico (northeastern Caribbean Sea). *Coral Reefs*, **17**, 377–82.

Woodley, J. D. (1992). The incidence of hurricanes on the north coast of Jamaica since 1870: are the classic reef descriptions atypical? *Hydrobiologia*, **247**, 133–38.

Woodley, J. D., Chornesky, E. A., Clifford, P. A., *et al.* (1981). Hurricane Allen's impact on Jamaican coral reefs. *Science*, **214**, 749–55.

10

Forecasting US insured hurricane losses

THOMAS H. JAGGER, JAMES B. ELSNER, AND MARK A. SAUNDERS

Condensed summary

Coastal hurricanes generate huge financial losses within the insurance industry. The relative infrequency of severe coastal hurricanes implies that empirical probability estimates of the next big loss will be unreliable. Hurricane climatologists have recently developed statistical models to forecast the level of coastal hurricane activity based on climate conditions prior to the season. Motivated by the usefulness of such models, in this chapter we analyze and model a catalog of normalized insured losses caused by hurricanes affecting the United States. The catalog of losses dates back through the twentieth century. The purpose of this work is to demonstrate a preseason forecast tool that can be used for insurance applications. Although wind speed is directly related to damage potential, the amount of damage depends on both storm intensity and storm size. As anticipated, we found that climate conditions prior to a hurricane season provide information about possible future insured hurricane losses. The models exploit this information to predict the distribution of likely annual losses and the distribution of a worst-case catastrophic loss aggregated over the entire US coast.

10.1 Introduction

Coastal hurricanes are a serious social and economic concern for the United States. Strong winds, heavy rainfall, and storm surge kill people and destroy property. The destructive power of hurricanes rivals that of earthquakes. On August 28, 2005, Hurricane Katrina's winds reached 78 meters per second $(m\,s^{-1})$ in the central Gulf of Mexico, making it one of the strongest Atlantic hurricanes ever recorded. Early morning on the next day, Katrina struck Plaquemines Parish, Louisiana, with winds estimated near 65 $m\,s^{-1}$. Katrina caused an estimated US$38 billion (bn) in insured losses as it roared across Louisiana, Mississippi, and Alabama.

Climate Extremes and Society, ed. H. F. Diaz and R. J. Murnane. Published by Cambridge University Press. © Cambridge University Press 2008.

It is important to know the return periods for losses incurred from storms of Katrina's magnitude or stronger and how the return periods vary when the climate fluctuates or changes (Elsner et al., 2006a). It is also valuable to be able to forecast the probability of a large loss before the hurricane season. Skillful forecasts of insured losses at lead times (forecast horizons) of 6 months or more would certainly benefit risk managers and others who are interested in acting on these forecasts. The rarity of severe hurricanes implies that empirical estimates of return periods likely will be unreliable. Fortunately, extreme value theory provides models for rare events and a justification for extrapolating to levels that are much greater than have already been observed. Moreover, statistical theory combined with knowledge of climate variability and its connection to regional storminess allows forecasts of seasonal hurricane activity.

Probability estimates of extreme hurricanes are available in the literature (Darling, 1991; Rupp and Lander, 1996; Heckert et al., 1998; Chu and Wang, 1998), but these studies do not address the question of how hurricane probabilities change with climate. This is done in Jagger et al. (2001), but the focus is on the probability of hurricanes of any intensity and not on the probability of the most extreme winds. Jagger and Elsner (2006) model the most extreme hurricane winds along the US coast and show how the probability of winds exceeding extreme thresholds changes with climate factors, including the North Atlantic Oscillation (NAO) and the El Niño/Southern Oscillation (ENSO).

Predictions of basin-wide Atlantic hurricane activity have been around since the middle 1980s (Gray, 1984). Studies focusing on climate factors that influence hurricane frequency regionally (Lehmiller et al., 1997; Bove et al., 1998; Maloney and Hartmann, 2000; Elsner et al., 2000a; Murnane et al., 2000; Saunders et al., 2000; Jagger et al., 2001; Larson et al., 2005) are more recent. Insights into climate conditions that affect regional hurricane activity are used to help predict landfall activity (Lehmiller et al., 1997; Elsner and Jagger, 2004, 2006; Saunders and Lea, 2005). Preseason forecasts of the number of hurricanes expected to affect the coast are useful especially if they are issued with significant lead time.

Saunders and Lea (2005) were the first to link predictions of US hurricane activity to skillful seasonal forecasts of loss. Here we present forecast models that can be used to directly predict the probability of a significant US financial loss from July 1. The models combine the strategy of Jagger and Elsner (2006) to estimate return periods with the strategy of Elsner and Jagger (2006) to forecast US hurricane activity before the start of the hurricane season. We begin with an examination of the normalized insured loss data and the data associated with climate fluctuations. We then describe our modeling strategy and show results from a preseason model that predicts the annual expected loss and a model that predicts the worst-case scenario over a 100-year time horizon.

10.2 Normalized insured losses: 1900–2005

The work presented in this chapter was motivated by Katz (2002), who modeled total annual economic damage associated with hurricanes with a compound Poisson process. The process is compound since the total number of damaging hurricanes per year is fitted with a Poisson distribution, while the monetary amount of damage for individual hurricanes is fitted by the log-normal distribution. Damage totals are thus represented as a "random sum," with variations in total damage being decomposed into two sources, one attributable to variations in the frequency of events and another to variations in the damage from individual events. Results from Katz (2002) indicate a dependence of both hurricane occurrence and damage amount on the state of ENSO. Our idea is similar but with the following differences. First, we use preseason covariates to represent the climate rather than a contemporaneous above/below normal factor. Second, we use a threshold for dividing the loss data into small and large loss events, and third, we use simulation (random samples) to generate the distribution of losses.

We obtained insured loss data from Collins and Lowe (2001), who have produced a normalized record of insured losses for all hurricanes affecting the United States between 1900 and 1999. The normalization adjusts the damage from each hurricane to match what it would be if the storm had struck in the year 2000. This normalization is achieved by allowing for changes in inflation, wealth, and population, plus an additional factor, which represents a change in the number of housing units that exceeds population growth between the year of the loss and 2000. We extend the original Collins and Lowe (2001) data to 2005 using insured losses provided by the US Property Claims Service and inflate all losses to reflect 2005 US dollar values. The insured loss data for 1900–2005 comprise 178 loss events. The Collins and Lowe (2001) insured loss dataset is similar to the loss dataset of Pielke and Landsea (1998), who estimated total economic losses attributable to hurricanes since 1900. The rank correlation between the two annual hurricane loss time series is high, at 0.90 (1900–99).

Figure 10.1 shows the distribution and time series of insured losses over the period 1900–2005. The histogram bars indicate the percentage of events with losses in groups of US$1bn. The distribution is highly skewed, with 34% of the events having losses exceeding US$1bn and 19% of the events having losses

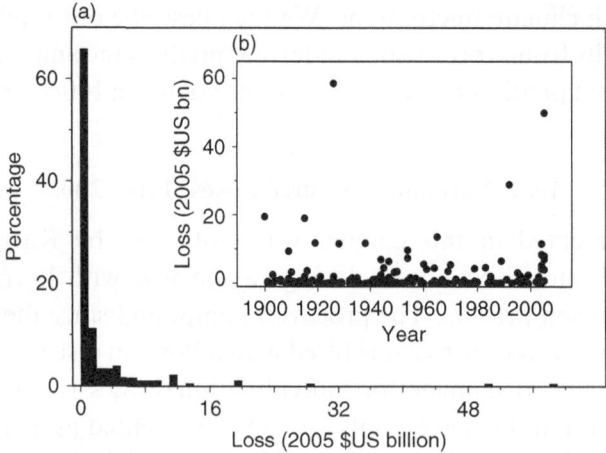

Figure 10.1. (a) Distribution of insured losses from hurricanes in the United States (excluding Hawaii). The distribution is highly skewed, with a few events generating very large losses. (b) Time sequence of the losses. Individual years may have more than one loss event.

exceeding US$3bn. The worst loss occurred with the 1926 hurricane that struck southeast Florida, creating an estimated insured loss adjusted to 2005 dollars of US$58.5bn. Hurricane Katrina in 2005 comes in second, with an estimated total loss of US$38.1bn. The time series of event losses is shown as an insert to Figure 10.1. Years with more than one loss have more than one dot. The data display large year-to-year variability but no obvious long-term trend, although here the data are not disaggregated into loss amount and number of loss events. The insured loss exceedances are shown in Table 10.1. Of the 178 loss events since 1900, 113 exceeded US$100 million (mn) in losses and 10 of these exceeded US$10bn. The geographic distribution of losses is shown in Figure 10.2. Plots are made for losses in four sizes, ranging from less than US$100 mn to more than US$10 bn. There does not appear to be a large geographic variation in loss locations with loss amount, with the exception of the largest loss amounts confined to southern exposures.

Because of the large skewness in loss values, we transform the data by using logarithms. A logarithmic transformation of the loss data is also used in Katz (2002). Here we use the base 10 logarithm for ease of interpretation. The logarithm to base 10 of a US$1bn loss is equal to 9. Figure 10.3 shows the logarithm of insured losses. The time series of log transformed annual losses shows no significant trend, although two of the highest yearly totals occurred in 2004 and 2005. The distribution of the logarithm of annual losses approximates a normal distribution, although there is some asymmetry in the tails. A quantile–quantile plot of the logarithm of losses against a normal distribution

Table 10.1. *Insured loss exceedances (US dollars adjusted to 2005)*

Values are the number of events exceeding various loss thresholds.

US$ (2005)	Exceedance number	Events
1 mn[a]		177
10 mn		172
100 mn		113
1 bn[a]		61
10 bn		10

[a] Abbreviations: mn, million; bn, billion.

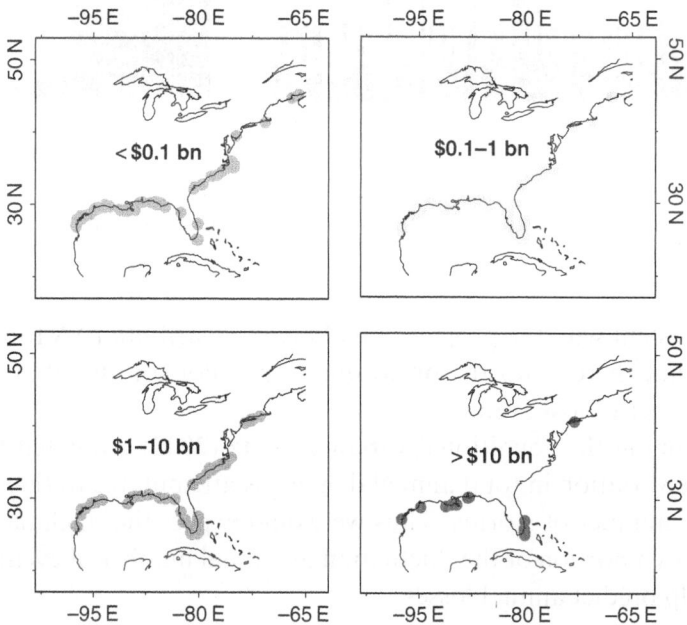

Figure 10.2. Geographic distribution of normalized insured losses from hurricanes striking the United States between 1900 and 2005. For color version, see plate section.

indicates a reasonable fit and provides evidence that the distribution of individual losses is log normal. Figure 10.3 also shows the annual number of loss events and the distribution. Again we see no obvious trend over time. There were three years with six loss events, with the most recent being 2004. The observed mean rate of loss events is 1.68, with a variance that is nearly equal, at

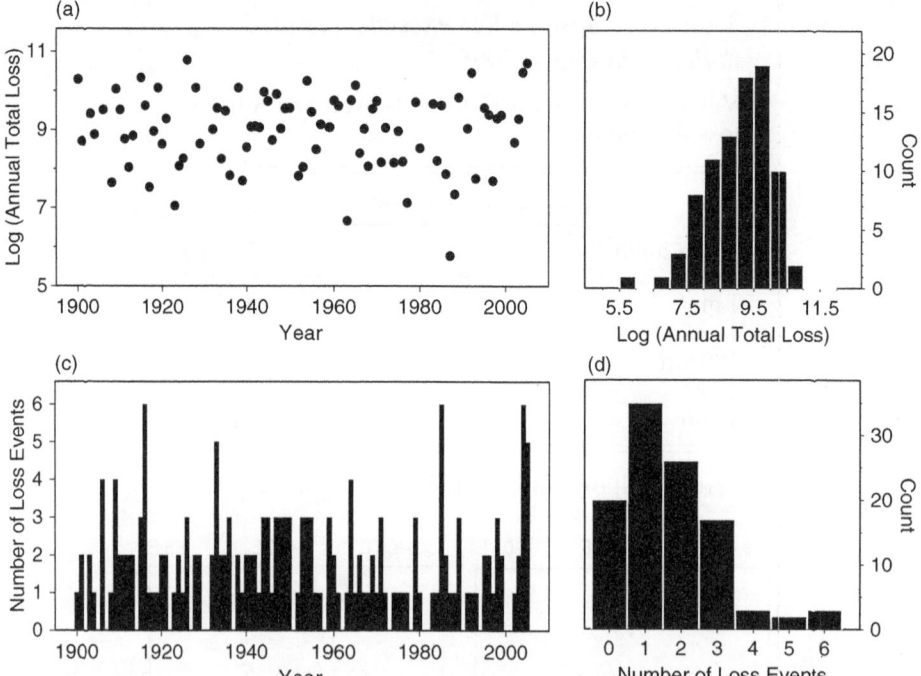

Figure 10.3. (a) Time series and (b) distribution of the logarithm of annual total insured losses and the (c) time series and (d) distribution of the number of loss events.

1.896, consistent with the property of a Poisson distribution. A formal χ^2 test indicates that there is no reason to question a Poisson distribution for the annual number of loss events.

By examining the conditional variance, Katz (2002) estimated that about 17% of the variation in total annual damage is attributable to fluctuations in the annual number of storms. Thus we would expect that a climate variable that explains a portion of the fluctuation in annual number of events could be used to help predict annual losses.

10.3 Climate variations

We argue that the annual distribution of insured hurricane losses depends to some extent on preseason climate factors. This conclusion is reasonable given that statistical relationships between US hurricane activity and climate are well established (Bove *et al.*, 1998; Elsner and Kara, 1999; Elsner *et al.*, 1999; Saunders *et al.*, 2000; Elsner *et al.*, 2000a, b, 2001; Elsner, 2003; Elsner *et al.*, 2004; Saunders and Lea, 2005). More importantly for the present work, Jagger *et al.* (2001) and Jagger and Elsner (2006) modeled the wind speeds of hurricanes at or near landfall and showed that the exceedance probabilities (e.g., wind speeds in excess of 100 knots) vary appreciably with the phase of the ENSO, the NAO, and Atlantic sea surface temperatures (SSTs). Similarly, Murnane *et al.* (2000) modeled the probability of coastal hurricanes conditioned on ENSO. A study by Goldenberg *et al.* (2001) suggests that the number and strength of Atlantic hurricanes follow a multidecadal cycle of changes in North Atlantic Ocean currents. This cycle, called the Atlantic Multidecadal Oscillation (AMO), might be related to changes in radiative forcing and/or changes in the thermohaline circulation.

The ENSO is characterized by basin-scale fluctuations in sea level pressure (SLP) between Tahiti and Darwin. Although noisier than equatorial Pacific SSTs, pressure values are available back to 1900. The Southern Oscillation Index (SOI) is defined as the normalized sea level pressure difference between Tahiti and Darwin. The SOI is strongly anti-correlated with equatorial Pacific SST, with an El Niño warming event associated with negative SOI values. Units are standard deviations. The relationship between ENSO and hurricane activity is strongest during the hurricane season, but we are interested in a predictive relationship, so we use a May–June average of the SOI as our predictor. The monthly SOI values (Ropelewski and Jones 1997) are obtained from the Climatic Research Unit (CRU) of the University of East Anglia, UK.

The NAO is characterized by fluctuations in sea level pressure differences. Index values for the NAO (NAOI) are calculated as the difference between the SLP for Gibraltar and for a station over southwest Iceland, and are obtained from the CRU (Jones *et al.*, 1997). The values are averaged over the pre- and early-hurricane season months of May and June (Elsner *et al.*, 2001). We speculate that the relationship might result from a communication between the middle latitudes and the tropics (Tsonis and Elsner, 1996) whereby below normal values of the NAO during the spring lead to dry conditions over the continents and to a tendency for greater summer/fall middle tropospheric ridging (enhancing the dry conditions). In turn, tropospheric ridging over the eastern and western sides of the North Atlantic basin during the hurricane season tends to keep the middle tropospheric trough of low pressure, responsible for hurricane recurvature, farther to the north and away from the westward tracking tropical cyclones (Elsner and Jagger, 2006).

The AMO is characterized by fluctuations in SST over the North Atlantic Ocean. Modeled SST and US National Oceanic and Atmospheric Administration (NOAA) optimal interpolated SST datasets were used to compute Atlantic SST anomalies north of the equator (Enfield *et al.*, 2001). Anomalies (in degrees centigrade, °C) are computed by month for the base period 1951–2000. Data are obtained from the NOAA–Cooperative Institute for Research in Environmental Sciences' Climate Diagnostics Center (NOAA–CIRES CDC) back to 1871. For this study we average the Atlantic SST anomalies over the hurricane preseason months of May and June.

In summary, the distribution of US insured losses from hurricane winds is statistically modeled by using covariate (predictor) data for the period 1900–2005. We are interested in the preseason values (May–June averaged) of SOI, NAO, and Atlantic SST as predictors for the distribution of likely losses during the US hurricane season, which runs principally from July through October. Figure 10.4 shows time series of the covariate values used in the model. All three series display variability from year to year with a distinct nonlinear trend in the late springtime values of Atlantic SST.

The upper and lower quartile values of the SOI are 0.60 and −0.75 standard deviation (s.d.), respectively, with a median (mean) value of −0.16 (−0.10) s.d. Years of below (above) normal SOI correspond to El Niño (La Niña) events and thus to a lower (higher) probability of hurricanes. The upper and lower quartile values of the NAO are 0.42 and −1.08 s.d., respectively, with a median (mean) value of −0.39 (−0.32) s.d. Years of below (above) normal values of the NAO correspond to a weak (strong) NAO phase and thus to a higher (lower)

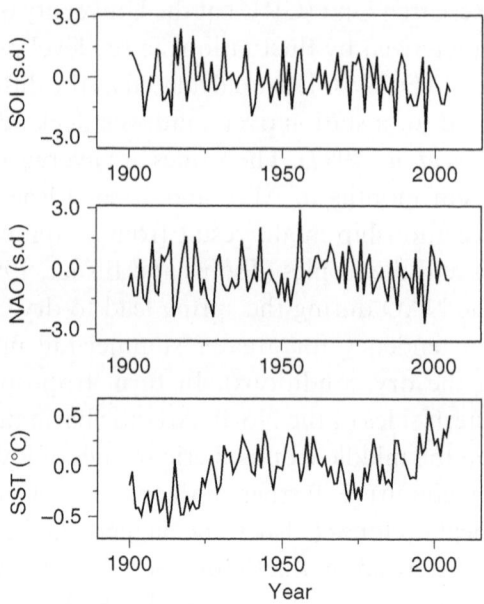

Figure 10.4. Time series of the three covariates used to predict insured wind losses from hurricanes before the start of the hurricane season. The values are averaged over the months of May and June. A linear detrended version of the Atlantic SST is sometimes referred to as the AMO.

probability of US hurricanes. The upper and lower quartile values of the Atlantic SST are 0.13 °C and −0.23 °C, respectively, with a median (mean) value of −0.04 (−0.04) °C. Years of above (below) normal values of SST correspond to a higher (lower) probability of hurricane activity. The linear correlation between the SOI and the NAO (SST) is a negligible + 0.03 (−0.04). The linear correlation between the NAO and Atlantic SST is a marginally significant value of −0.21.

10.4 Large and small losses

The total amount of insured losses calibrated to 2005 US dollars from the 178 events (1900–2005) is estimated at US$421bn. The large skewness in the insured losses per event and per annum suggests that it might be a good strategy to separate large losses from small losses for the purpose of prediction. It is often quoted that 80% of the total damage from all hurricanes is caused by the top 20% of the strongest storms. Figure 10.5 shows that the distribution of loss data is even a bit more skewed than that. In fact, we find that the top 30 loss events (less than 17% of the total number of loss events) account for more than 80% of the total losses.

The relative infrequency of the largest loss events argues for a split that favors including more data for modeling. Here we use a cutoff of US$100 mn and find that 113 of the 178 events (63.5%) exceeded this threshold. The

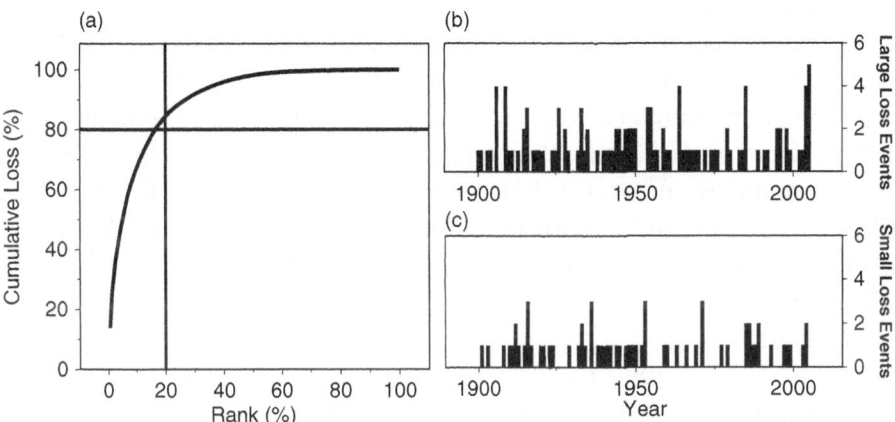

Figure 10.5. (a) Cumulative percent of total losses as a function of percent ranking. The reference lines indicate the oft-cited 80%/20% relationship, whereby the top 20% of the strongest hurricanes account for 80% of the losses. A split of the event counts into (b) large loss events and (c) small loss events based on losses exceeding US$100 mn is shown as annual time series. For reference, 2004 experienced four large and two small loss events.

remaining 65 events (36.5%) account for only 0.6% of the total losses. Thus it might be reasonable to assume that the small loss events are at the "noise" level. Time series of the annual number of large and small loss events are shown in Figure 10.5. The rank correlation between the two series is a negligible 0.06.

Next we examine the influence of the covariates, discussed in the previous section, on both the magnitude of annual loss and the number of annual loss events. For the number of loss events, we consider small and large loss events separately. Using the preseason Atlantic SST, we are able to explain 13% of the variation in the logarithm of loss values exceeding US$100 mn using an ordinary least squares regression model. The relationship is positive, indicating that warmer Atlantic SSTs are associated with larger losses as expected. The rank correlation between the amount of loss (exceeding US$100 mn) and the May–June Atlantic SST is $+0.31$ (p-value $= 0.0086$) over all years in the dataset and is $+0.37$ (p-value $= 0.0267$) over the shorter 1950–2005 period.

We also examine models for the number of loss events using the covariates. We find that the NAO is useful in predicting both the number of large loss events and the number of small loss events. The relationship is negative, indicating that when the preseason value of the NAO decreases, the probability of a loss event increases. The rank correlation between the total number of loss events and the preseason NAO is -0.29 (p-value $= 0.0032$) over all years and is -0.12 (p-value $= 0.3812$) over the shorter 1950–2005 period. Interestingly, we find no significant preseason relationship between event counts and SST or the SOI.

The analysis confirms that it is reasonable to model small and large loss events separately. However, it should be noted that it might be more appropriate to add measurement error to the data so as to reduce the weight of the smaller measurements rather than separate the data as is done here. Our final strategy combines a model for the loss amount with two models for the number of loss events: one for large losses and the other for small losses. We use the NAO for predicting the number of loss events (both large and small) and the SST for predicting the amount of damage given a loss event. We find that including the preseason SOI covariate does not help in forecasting the upcoming season's losses either for the amount of loss or for the number of loss events. This result is consistent with those for the models developed in Elsner and Jagger (2006) and Elsner et al. (2006b) for predicting coastal hurricane activity based on preseason data. Since it is well known that ENSO has an influence on shear during the hurricane season, it might be advantageous to include a predicted value of the SOI for the hurricane season rather than a preseason value as is done here.

10.5 Predicting annual losses

Results from the previous section provide the needed background for building
a preseason model capable of predicting the annual expected loss. The model
uses a hierarchical Bayesian specification. The final form of the model was
based on comparison of the deviance information criterion (DIC) using several
different models involving the three covariates. The DIC is a generalization of
the Akaike information criterion (AIC) and Bayesian information criterion
(BIC). It is useful in Bayesian model selection where the posterior distributions
of the models are obtained by Markov chain Monte Carlo (MCMC) simulation. Like the AIC and BIC, it is an asymptotic approximation as the sample
size becomes large. It is only valid when the posterior distribution is close to
multivariate normal. We chose the model with the lowest value of DIC.

A schematic of the hierarchical model is shown in Figure 10.6. The predicted annual loss (TL) is the sum of the individual loss amounts (both large
[LL_L] and small [LL_S] amounts) multiplied by the respective number of large
(N_L) and small (N_S) loss counts. Given the mean (μ_L) and standard deviation
(σ_L) of the logarithm of large losses, the logarithm of large loss follows a
truncated normal distribution. Small loss amounts are also specified by using
a truncated normal distribution, although the mean is not a function of any of

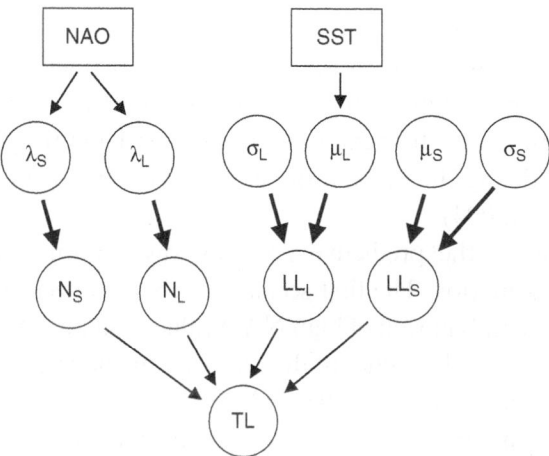

Figure 10.6. Hierarchical graph illustrating our strategy for simulating
annual insured losses based on preseason values of the NAO and Atlantic
SST. The connection between nodes is either stochastic (thick arrow) or
logical (thin arrow). Node λ_L (λ_S) is the mean annual rate of large (small)
losses, N_L (N_S) is the annual count of large (small) loss events, μ_L (μ_S) is the
mean amount of large (small) loss on a log scale, σ_L (σ_S) is the standard
deviation of large (small) loss amounts, LL_L (LL_S) is the logarithm of large
(small) loss amount, and TL is the total loss.

the covariates. Given a mean annual rate of large losses (λ_L), the annual number of large losses follows a Poisson distribution with the natural logarithm of the rate given as a linear function of the NAO. Similarly, given a mean annual rate of small losses (λ_S), the annual number of small losses follows a Poisson distribution with the natural logarithm of the rate given as a separate function of the NAO.

Samples of the annual losses are generated using WinBUGS (Windows version of Bayesian inference Using Gibbs Sampling) developed at the Medical Research Council in the United Kingdom (Gilks *et al.*, 1996; Spiegelhalter *et al.*, 1996). WinBUGS chooses an appropriate MCMC sampling algorithm based on the model structure. In this way, annual losses are samples conditional on the model coefficients and the observed values of the covariates. The cost associated with a Bayesian approach is the requirement to formally specify prior beliefs. Here we take the standard route and assume noninformative priors, which as the name implies provide little information about the parameters of interest. Markov chain Monte Carlo analysis, in particular Gibbs sampling, is used to sample the parameters given the data, since no closed form distribution exists for the truncated normal (or for the generalized Pareto distribution [GPD] used in the next section).

We check for mixing and convergence by examining successive sample values of the parameters. Samples from the posterior distributions of the parameters indicate relatively good mixing and quick settling as two different sets of initial conditions produce sample values that fluctuate around a fixed mean. Based on these diagnostics, we discard the first 10,000 samples and analyze the output from the next 10,000 samples. The utility of the Bayesian approach for modeling the mean number of coastal hurricanes is described in Elsner and Jagger (2004).

Figure 10.7 shows the predictive posterior distributions of losses for two different climate scenarios. The first scenario is characterized by preseason conditions featuring a combination of high NAO values and low SST values. To offer a strong contrast, we set the values to their maximum and minimum, respectively, over the 106-year period (1900–2005; NAO $= +2.9$ s.d. and SST $= -0.61\,°C$). This situation is unfavorable for hurricane activity along the US coast (Elsner and Jagger, 2006). Simulation results show that the probability of no loss (47%) is close to the probability of at least some loss (53%). This result contrasts with those from the second scenario, which is characterized by conditions favorable for hurricane activity (NAO $= -2.7$ s.d. and SST $= +0.55\,°C$). Here the probability of at least some loss is 94%.

Perhaps more useful is the predictive distribution of losses, given that at least some loss occurs. Here the distributions are shown for the logarithm

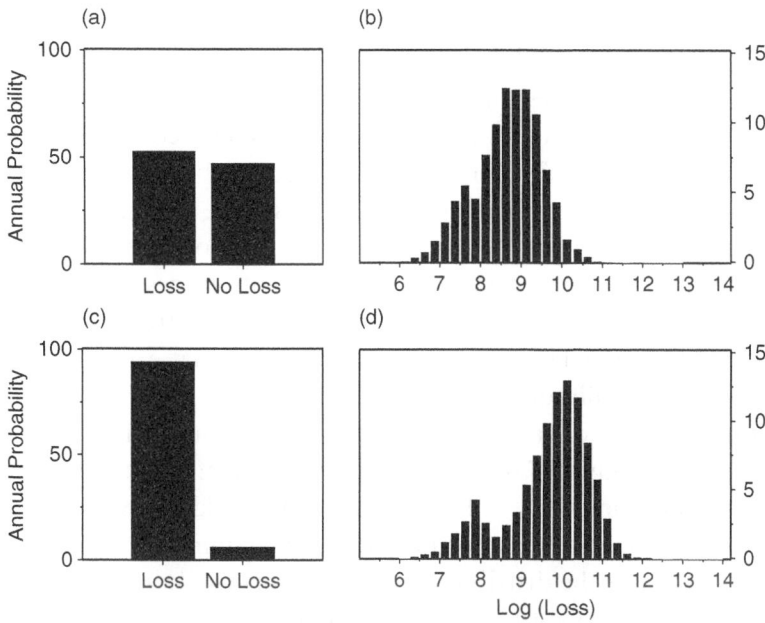

Figure 10.7. Simulated annual losses for two different climate scenarios. (a) The probability of at least some loss and (b) the probability distribution of loss amounts given at least one loss event under preseason climate conditions foreshadowing an inactive US landfall season. Plots (c) and (d) are the same as plots (a) and (b), respectively, except that they are based on preseason conditions foreshadowing an active US landfall season.

of total annual loss from both scenarios and, as expected, the results are divergent. In the case of favorable late springtime conditions for hurricane activity, the loss distribution is shifted toward substantially higher loss amounts relative to the case of unfavorable conditions. Converting to 2005 dollars, the expected yearly loss in a year with at least one loss when conditions are favorable for hurricanes is estimated at US$25.2bn. This compares with US$2.1bn when conditions are unfavorable. The overall expected loss (taking into account the non-zero probability of no losses) is US$23.7bn under favorable climate conditions and US$1.1bn under unfavorable conditions. Therefore, assuming the model is correct and the future will be the same as the past, the model is useful in portending the amount of insured losses before the start of the season. The interesting side hump in the distribution of losses is likely an artifact of using a truncated normal distribution. Both the 2004 and 2005 hurricane seasons featured late springtime negative NAO values and above normal Atlantic SST values, which combined to produce a forecast (hindcast) of above normal insured loss probabilities.

10.6 Predicting extreme losses

While the modeling strategy described above makes sense for forecasting the distribution of average losses associated with climate conditions before the start of the season, for financial planning it might be of greater interest to know the maximum possible loss. In this case, the normal distribution is replaced by an extreme value distribution for the logarithm of losses. For example, the family of generalized Pareto distributions describes the behavior of individual extreme events. Consider observations from a collection of random variables in which only those observations that exceed a fixed value are kept. As the magnitude of this value increases, the GPD family represents the limiting behavior of each new collection of random variables. This property makes the family of GPDs a good choice for modeling extreme events involving large insured losses. The choice of threshold, above which we treat the values as extreme, is a compromise between retaining enough observations to properly estimate the distributional parameters (scale and shape), but few enough that the observations follow a GPD family. A negative value for the shape parameter implies an upper limit to the maximum possible loss.

The GPD describes the distribution of losses that exceed a threshold u but not the frequency of losses at that threshold. As we did with the annual loss model, we specify that, given a rate of loss events above the threshold, the number of loss events follows a Poisson distribution. Here there is no need to consider small loss events, as we are interested only in the large ones. Combining the GPD for the distribution of large loss amounts with the Poisson distribution for the frequency of loss events allows us to obtain return periods for given levels of annual losses.

We determine the particular threshold value for the set of insured losses by examining the plots shown in Figure 10.8. The mean residual life (MRL) plot shows the value of the mean excess as a function of threshold. The MRL plot is produced by averaging the difference (residual) in the observed logarithm of loss above a threshold as a function of the threshold. For example, at a log loss of 9 we subtract 9 from each observed log loss and average only the positive values (excesses). We repeat this operation for all thresholds. The mean excess is the expected value of the amount that the observations exceed the threshold. The standard error on the mean excess allows us to compute confidence levels for the estimate. A nearly straight-line negative relationship between the mean excess and the loss above some threshold indicates the set of extreme losses. In other words, if extreme values follow a GPD, then their expected value is a linear function of the threshold. From the plot we see that a straight-line relationship is noted for losses at about 9 (US$1bn). The other two plots

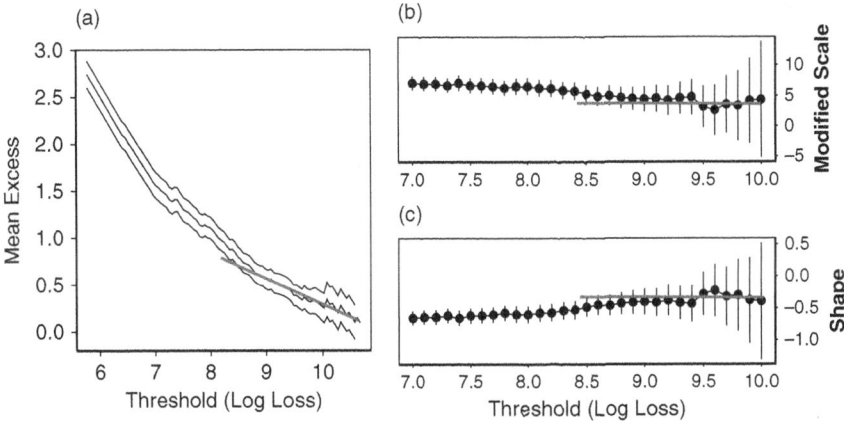

Figure 10.8. (a) Mean residual life plot for the logarithm of insured hurricane losses. The outside lines are the 95% confidence limits. An approximate linear decrease of the mean excess occurs after a threshold of about 9. The value of the (b) scale and (c) shape parameters from the GPD at various thresholds. The systematic variation is not detectable for thresholds exceeding about 9.

show the GPD parameters as a function of threshold. The systematic variation in the scale and shape parameters with threshold appears to end at a threshold value between 8.5 and 9, suggesting that only events with losses exceeding this level are extreme. Taken together, the diagnostic plots suggest that a threshold value is US$1bn in losses.

As with the annual loss model, we use a Bayesian hierarchical specification for the model of extreme losses. Markov chain Monte Carlo samples are used to generate posterior predictive distributions. Here we are interested in the return level as a function of return period. A schematic of the hierarchical model is shown in Figure 10.9. The annual return level (RL_y) is determined by the return level of individual extreme events (RL_E) and the annual frequency of such events above a threshold rate (λ). The annual number of extreme events follows a Poisson distribution, with the natural logarithm of the rate specified as a linear function of the NAO. Given values for the scale (σ) and shape (ξ) parameters, the return level of individual extreme events follows a GPD. The logarithm of the scale parameter is a linear function of the NAO, and the shape parameter is a linear function of the SOI.

As before, samples of the return levels are generated by using WinBUGS and we use non-informative prior distributions. Samples from the posterior distribution of the model parameters indicate good mixing and convergence properties. We discard the first 10,000 samples and analyze the output from the next 10,000 samples. Applications of Bayesian extremal analysis are

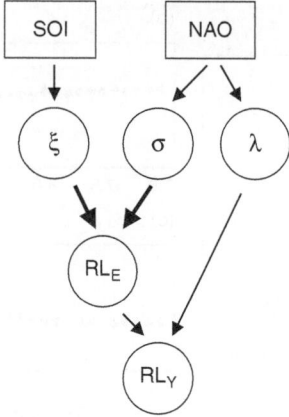

Figure 10.9. Hierarchical graph illustrating our strategy for simulating return levels for extreme losses conditional on the preseason values of the NAO and SOI. Nodes ξ and σ are the shape and scale parameters of the GPD, respectively; λ is the mean rate of extreme losses; RL_E is the return level for a particular loss event; and RL_Y is the return level for total losses over the year.

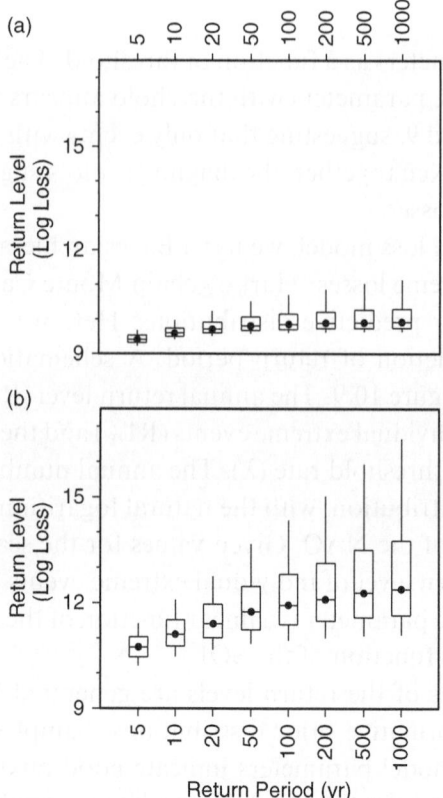

Figure 10.10. Simulated extreme losses for two different climate scenarios. (a) The distribution of return levels in the logarithm of insured losses for the case of unfavorable conditions for US hurricane landfalls, and (b) the distribution of return levels for the case of favorable conditions for US hurricane landfalls.

relatively rare (Coles and Tawn, 1996; Katz *et al.*, 2002; Coles *et al.*, 2003). In the context of local hurricane winds, Casson and Coles (1999) used a Bayesian analysis to estimate parameters of spatial regression models. They showed that including the spatial characteristics of extremes provides a substantial reduction in the width of the confidence intervals for high quantiles. Bayesian approaches to modeling extreme wind behavior are given in Walshaw (2000) and Jagger and Elsner (2006).

Figure 10.10 shows the predictive posterior distributions of extreme losses for two different climate scenarios. The first scenario is characterized by preseason conditions featuring a combination of high NAO and high SOI values. Again, to offer a strong contrast, we set the values to their maximum and minimum over the 106-year period (1900–2005). Box and whisker plots are used to illustrate the variation in simulated extreme loss amounts for increasing return periods.

Results show the clear difference in expected extreme losses for the different climate conditions. Under the unfavorable scenario for US hurricanes, we find the expected return level of a 50-year extreme event at less than US$10bn; this compares with a return level of a 50-year extreme event loss of approximately US$630bn under favorable scenarios for US hurricanes. Thus the model is also useful for projecting extreme losses over longer time horizons given the preseason values of the climate covariates.

10.7 Summary

Coastal hurricanes are capable of generating large financial losses for the insurance industry. The rarity of large losses in the historical record implies that empirical estimates of next year's loss will have large errors. Annual loss totals are directly related to the size and number of hurricanes affecting the coast. Since some skill exists in forecasts of coastal hurricane activity, it makes sense to investigate the potential of predicting losses directly. This chapter demonstrates a strategy for making forecasts of annual insured losses by July 1 using preseason values for the NAO, Atlantic SST, and the SOI. Models are specified by using hierarchical Bayesian technology, and predictive posterior distributions are generated by using MCMC sampling. Markov chain Monte Carlo sampling provides a method of generating future loss projections. According to the model of expected annual loss, the probability of incurring a loss is higher compared to the climatological average when the NAO is negative. Also, the amount of loss is greater when Atlantic SST is above normal. Both conditions were met before the 2004 and 2005 hurricane seasons. While we did not perform an out-of-sample test of model skill, a similar hierarchical Bayesian model for US hurricane activity using the same covariates is cross-validated and shown to have skill above climatology in Elsner and Jagger (2006).

These results are consistent with current understanding of hurricane climate variability. Forecasts of extreme loss amounts are also possible using a somewhat different model specification and including a preseason value of the SOI. Return level loss amounts exceed those of climatology under conditions characterized by a negative NAO. It might be possible to develop a similar model using data from as early as February 1 (see Elsner *et al.*, 2006b). While the models here are developed from aggregate loss data for the entire United States due to Atlantic hurricanes, it would be possible to apply the techniques to model data representing a subset of losses, capturing, for example, a particular reinsurance portfolio. Moreover, since the models use MCMC sampling, they can be extended quite easily to include measurement error as well as missing data. The models and data are available on our web site (Google key words: hurricane climate).

Acknowledgments

We thank Richard Murnane, Henry Diaz, and Charles King for organizing the Assessing, Modeling, and Monitoring the Impacts of Extreme Climate Events workshop in 2005 at the XL House in Hamilton, Bermuda. Support for this study was provided by the National Science Foundation (ATM-0086958, ATM-0435628, and BCS-0213980) and by the Risk Prediction Initiative (RPI-05001). The views expressed here are those of the authors and do not reflect those of the funding agencies.

References

Bove, M. C., Elsner, J. B., Landsea, C. W., Niu, X., and O'Brien, J. J. (1998). Effect of El Niño on U.S. landfalling hurricanes, revisited. *Bulletin of the American Meteorological Society*, **79**, 2477–82.

Casson, E., and Coles, S. (1999). Spatial regression models for extremes. *Extremes*, **1**, 449–68.

Chu, P. -S., and Wang, J. (1998). Modeling return periods of tropical cyclone intensities in the vicinity of Hawaii. *Journal of Applied Meteorology*, **37**, 951–60.

Coles, S., and Tawn, J. A. (1996). A Bayesian analysis of extreme rainfall data. *Applied Statistics*, **45**, 463–78.

Coles, S., Pericchi, L. R., and Sisson, S. (2003). A fully probabilistic approach to extreme rainfall modeling. *Journal of Hydrology*, **273**, 35–50.

Collins, D. J., and Lowe, S. P. (2001). A macro validation dataset for U.S. hurricane models. *Casualty Actuarial Society, Winter Forum*, 217–52.

Darling, R. W. R. (1991). Estimating probabilities of hurricane wind speeds using a large-scale empirical model. *Journal of Climate*, **4**, 1035–46.

Elsner, J. B. (2003). Tracking hurricanes. *Bulletin of the American Meteorological Society*, **84**, 353–6.

Elsner, J. B., and Jagger, T. H. (2004). A hierarchical Bayesian approach to seasonal hurricane modeling. *Journal of Climate*, **17**, 2813–27.

Elsner, J. B., and Jagger, T. H. (2006). Prediction models for annual U.S. hurricane counts. *Journal of Climate*, **19**, 2935–52.

Elsner, J. B., and Kara, A. B. (1999). *Hurricanes of the North Atlantic: Climate and Society*. New York: Oxford University Press.

Elsner, J. B., Bossak, B. H., and Niu, X.-F. (2001). Secular changes to the ENSO-U.S. hurricane relationship. *Geophysical Research Letters*, **28**, 4123–6.

Elsner, J. B., Jagger, T. H., and Niu, X. (2000a). Shifts in the rates of major hurricane activity over the North Atlantic during the twentieth century. *Geophysical Research Letters*, **27**, 1743–6.

Elsner, J. B., Jagger, T. H., and Tsonis, A. A. (2006a). Estimated return periods for Hurricane Katrina. *Geophysical Research Letters*, **33**, L08704, doi:10.1029/2005GL025452.

Elsner, J. B., Kara, A. B., and Owens, M. A. (1999). Fluctuations in North Atlantic hurricanes. *Journal of Climate*, **12**, 427–37.

Elsner, J. B., Liu, K.-b., and Kocher, B. (2000b). Spatial variations in major U.S. hurricane activity: statistics and a physical mechanism. *Journal of Climate*, **13**, 2293–305.

Elsner, J. B., Murnane, R. J., and Jagger, T. H. (2006b). Forecasting U.S. hurricanes 6 months in advance. *Geophysical Research Letters*, **33**, L10704, doi:10.1029/2006GL025693.

Elsner, J. B., Niu, X. -F., and Jagger, T. H. (2004). Detecting shifts in hurricane rates using a Markov chain Monte Carlo approach. *Journal of Climate*, **17**, 2652–66.

Enfield, D. B., Mestas-Nuñez, A. M., and Trimble, P. J. (2001). The Atlantic Multidecadal Oscillation and its relation to rainfall and river flows in the continental U.S. *Geophysical Research Letters*, **28**, 2077–80.

Gilks, W. R., Richardson, S., and Spiegelhalter, D. J. (1996). *Markov Chain Monte Carlo in Practice*. Boca Raton, FL: Chapman & Hall/CRC.

Goldenberg, S. B., Landsea, C. W., Mestas-Nuñez, A. M., and Gray, W. M. (2001). The recent increase in Atlantic hurricane activity: causes and implications. *Science*, **293**, 474–9.

Gray, W. M. (1984). Atlantic hurricane frequency. Part I: El Niño and 30 mb quasi-biennial oscillation influences. *Monthly Weather Review*, **112**, 1649–68.

Heckert, N. A., Simiu, E., and Whalen, T. (1998). Estimates of hurricane wind speeds by "peaks over threshold" method. *Journal of Structural Engineering*, ASCE, **124**, 445–9.

Jagger, T. H., and Elsner, J. B. (2006). Climatology models for extreme hurricane winds near the United States. *Journal of Climate*, **19**, 3220–36.

Jagger, T. H., Elsner, J. B., and Niu, X. (2001). A dynamic probability model of hurricane winds in coastal counties of the United States. *Journal of Applied Meteorology*, **40**, 853–63.

Jones, P. D., Jónsson, T., and Wheeler, D. (1997). Extension to the North Atlantic Oscillation using early instrumental pressure observations from Gibraltar and Southwest Iceland. *International Journal of Climatology*, **17**, 1433–50.

Katz, R. W. (2002). Stochastic modeling of hurricane damage. *Journal of Applied Meteorology*, **41**, 754–62.

Katz, R. W., Parlange, M. B., and Naveau, P. (2002). Statistics of extremes in hydrology. *Advances in Water Resources*, **25**, 1287–304.

Larson, J., Zhou, Y., and Higgins, R. W. (2005). Characteristics of landfalling tropical cyclones in the United States and Mexico: climatology and interannual variability. *Journal of Climate*, **18**, 1247–62.

Lehmiller, G. S., Kimberlain, T. B., and Elsner, J. B. (1997). Seasonal prediction models for North Atlantic basin hurricane location. *Monthly Weather Review*, **125**, 1780–91.

Maloney, E. D., and Hartmann, D. L. (2000). Modulation of hurricane activity in the Gulf of Mexico by the Madden-Julian Oscillation. *Science*, **287**, 2002–4.

Murnane, R. J., *et al.* (2000). Model estimates hurricane wind speed probabilities. *Eos, Transactions of the American Geophysical Union*, **81**, 433–8.

Pielke, R. A., Jr., and Landsea, C. W. (1998). Normalized hurricane damages in the United States, 1925–95. *Weather Forecasting*, **13**, 621–31.

Ropelewski, C. F., and Jones, P. D. (1997). An extension of the Tahiti-Darwin Southern Oscillation Index. *Monthly Weather Review* **115**, 2161–5.

Rupp, J. A., and Lander, M. A. (1996). A technique for estimating recurrence intervals of tropical cyclone-related high winds in the tropics: results for Guam. *Journal of Applied Meteorology*, **35**, 627–37.

Saunders, M. A., and Lea, A. S. (2005). Seasonal prediction of hurricane activity reaching the coast of the United States. *Nature*, **434**, 1005–8.

Saunders, M. A., Chandler, R. E., Merchant, C. J., and Roberts, F. P. (2000). Atlantic hurricanes and Northwest Pacific typhoons: ENSO spatial impacts on occurrence and landfall. *Geophysical Research Letters*, **27**, 1147–50.

Spiegelhalter, D. J., Best, N. G., Gilks, W. R., and Inskip, H. (1996). Hepatitis B: a case study in MCMC methods. In *Markov Chain Monte Carlo in Practice*, ed. W. R. Gilks, S. Richardson, and D. J. Spiegelhalter. London: Chapman & Hall/CRC, pp. 45–58.

Tsonis, A. A., and Elsner, J. B. (1996). Mapping the channels of communication between the tropics and midlatitudes. *Physica D*, **92**, 237–44.

Walshaw, D. (2000). Modelling extreme wind speeds in regions prone to hurricanes. *Applied Statistics*, **49**, 51–62.

11

Integrating hurricane loss models with climate models

CHARLES C. WATSON, JR., AND MARK E. JOHNSON

Condensed summary

Hurricane loss modeling has become an important if not vital aspect of many elements of hurricane planning, especially in the financial sector. The insurance industry has provided financial motivation for the development of complex hurricane damage and loss models. These models rely on a number of databases and model subcomponents that interact in complex ways, the most critical of which is hurricane climatology. The required hurricane climatology is developed through various analyses of the historical record. Unfortunately, there are numerous issues with the historical record that make detailed analysis of that record problematic, such as the length of the record, the quality of the observations, and the potential that the record is complicated by natural or anthropogenic climate signals. As climate modeling continues to advance, there is increasing potential for the use of these models to drive loss models, overcoming many of the limitations of the existing historical record. Here we describe the results of a study conducted for the Florida Commission on Hurricane Loss Projection Methodology (FCHLPM), a part of which was an assessment of historical hurricane climatology, and the potential for the use of general circulation climate models in driving loss models for both existing and future climates. The Community Climate System Model (CCSM) was used to drive a mesoscale model, which in turn was used to create inputs to an ensemble of loss models. The climate model/mesoscale model combination appears to do a reasonable job of reproducing Atlantic basin hurricane climatology, with the resulting loss costs being comparable to those obtained by traditional methods. While further work remains to fully establish the validity of this approach, the potential for the use of climate models in estimating the impact of future climate change scenarios seems promising.

Climate Extremes and Society, ed. H. F. Diaz and R. J. Murnane. Published by Cambridge University Press. © Cambridge University Press 2008.

11.1 Introduction

Hurricane loss models are essential for estimating loss costs to support the insurance, financial, and government sectors that must cope with these natural disasters. Hurricanes Hugo (1989), Andrew (1992), and Iniki (1992) made it apparent that the use of econometric models to establish insurance premiums was woefully inadequate. Computer hurricane loss models, which had been in existence within the insurance industry for some time (Friedman, 1984), were consequently promoted to greater public prominence. In Florida, the legislature established the Florida Commission on Hurricane Loss Projection Methodology (FCHLPM) to formally evaluate models used for residential insurance purposes through a comprehensive evaluation process (Florida State Statutes, 627.0628 F. S.; see also www.sbafla.com/methodology). The commission developed a process to formally evaluate these models, and in so doing it addressed the skepticism of state insurance regulators. The mission of the FCHLPM is as follows:

The mission of the Florida Commission on Hurricane Loss Projection Methodology is to assess the efficacy of various methodologies which have the potential for improving the accuracy of projecting insured Florida losses resulting from hurricanes and to adopt findings regarding the accuracy or reliability of these methodologies for use in residential rate filings.

This commission is thus charged to explore opportunities for improving hurricane loss projection methodologies within the context of risk models submitted for their consideration and with methods emerging in the scientific literature. The authors of this chapter have been involved with the commission since 1996 as members of an advisory group consisting of specialists in statistics, engineering, meteorology, and actuarial and computer sciences known as the Professional Team. Initially, the commission proceeded by developing a set of standards that computer hurricane models needed to pass in order to be approved by the commission. Upon approval by the commission, results from the model could be used in the context of an insurance rate filing.

The commission standards were originally driven by consideration of existing modeling efforts and were not necessarily designed to encourage completely different approaches. The models initially approved by the commission and all models approved since contain meteorological, vulnerability, and actuarial components that, while unique in implementation, are basically similar in approach. Nevertheless, non-traditional hurricane risk models are not precluded, and to the extent that these models would comply with the commission's mission, they would likely be welcomed. The commission has funded the Professional Team to examine areas with potential for improving both the standards and the state of the art of hurricane modeling. For example, the commission funded work by the Professional Team on sensitivity and uncertainty analysis (Iman *et al.*, 2005a,b) that became the basis of two commission statistical standards. This chapter focuses on a recently funded effort to investigate general circulation models (GCMs, or climate models; here we refer to coupled ocean–atmosphere models) and their possible use in future hurricane loss models that would come before the commission. Of course, this research is of independent interest as a means to reduce the uncertainty in hurricane loss projections. This chapter documents the effort by describing climate models and our approach to using them in the context of hurricane loss projection estimation.

Classical or legacy hurricane models are inherently constrained by the historical record. The record can be mined to extract critical characteristics and distributions to model the extreme phenomena. Simulations of the modeled systems in essence smooth the irregularities in the historical record, filling in spatial gaps and allowing for slightly stronger events given a sufficient time frame. Ultimately, the generated results are a reflection of the historical record, although one that has been perturbed in statistically reasonable fashions. In fact, in the commission deliberations, it is recognized that some expert judgment is necessary to ensure that the match between model and historical results is consistent with the scientific understanding of tropical cyclones.

Depending on the point of view, there are 50–150 years of viable historical Atlantic hurricane data, which are only marginally sufficient to develop probabilistic models for forecasting extreme tropical cyclone activity. Return period estimation and probable maximum loss (PML) are particularly sensitive to the length of the historical record available for analysis and extrapolation. Restrictions to the historical record where all events have been consistently observed with satellite, aircraft, or side view radar coverage, or some specified combination of these assessments, reduces the total time extent of the available dataset (to achieve more reliable data) but hinders the extrapolation to future extreme events. The emphasis in this chapter is on average annual losses, which are the basis of premiums for residential exposures. Analyses of PMLs play a critical role for reinsurance and risk assessment and can benefit from the results of hurricane loss models.

Statistically based models forecast storms that should be reasonably consistent with the basic characteristics of storms that have appeared over the same time frame. This characteristic is both an advantage and a disadvantage. The advantage is that the results will pass a basic reality check in that

unrealistic superstorms will not be generated and the basin frequency and
intensities will approximate the overall record. However, a disadvantage is
that, unless explicitly modeled, trends in frequency and/or intensity will be
averaged out in the estimation process. The recent flurry of hurricane activity
during the 2004 and 2005 seasons spawned consideration of short-term high-
frequency rates for models. Such short-term "corrections" to the long-term
frequency are not without considerable controversy. One perspective is that
the current generation of hurricane risk models does not consider short-term
cycles in activity caused by global climate signals such as the El Niño/Southern
Oscillation (ENSO), much less decadal or longer cycles or trends. In 2006, the
proprietary modelers communicated to the commission about the possibility
of submitting models that contain short-term adjustments.

Given the interest in conducting specific forecasts of short-term changes
to long-term trends, as well as the debates over climate change and hurricane
frequencies, the commission was concerned how the standards would apply
to innovative techniques such as full GCM-based approaches. Although no
such submissions appear imminent, the possibility is intriguing, thus trigger-
ing funding for the present study. The clear advantage of the GCM approach
is that the initialization of the model does not require explicit specification
of hurricane formation locations, track distributions, dynamic intensity
functions, and so forth. The details of individual storms in the historical
record (e.g., maximum wind speed) and possible climate signals in that record
have been the focus of the contentious discussions by Emanuel (2005a,b),
Pielke (2005), and Landsea (2005). The vagaries of the measurement systems
over time are not central to the initialization of the GCM models. Historical
storms provide rough validation values but are not critical to the performance of GCMs. With the GCM operating correctly, vortices naturally
form in the course of the hurricane season and behave in reaction to
basin-wide conditions. Moreover, this approach allows the interesting possibility of being run into future years under various greenhouse gas emission
scenarios.

The remainder of this chapter documents our experiments in coupling loss
models to GCM outputs. The basic structure of hurricane loss models is
reviewed in Section 11.2. This overall structure has been employed by every
commercial model that has come before the commission to date, as well as the
Florida Public Model developed under the aegis of Florida International
University. The core of this chapter is Section 11.3, which documents how
GCMs can provide the engine to generate synthetic storms, central pressures,
and maximum winds (the information included in the "best-track" HURDAT
files) files in support of hurricane loss models.

11.2 Overview of loss models

Hurricane loss models have traditionally consisted of an input set of historical or synthetic storms that constitute a frequency or occurrence model, and additional meteorological, vulnerability, and actuarial components. In support of these components, databases on historical events and their detailed characteristics are necessary. For average annual loss cost estimation, probability distributions governing the stochastic generation of events are also necessary. For a given, fixed exposure, the hurricane loss model would then be executed to simulate tens of thousands of years in order to produce loss cost estimates and attendant uncertainties in the estimates. This overview of hurricane loss model construction pertains both to the first loss model approved by the commission (the AIR model; Clark, 1986, 1997) and to the current public domain model (Powell *et al.*, 2005), as well as a model that has garnered ongoing funding from the Federal Emergency Management Agency's (FEMA's) HAZUS project (Vickery *et al.*, 2000). The Risk Management Solutions, Inc. (RMS) and EQECAT models also fit into this framework, as has been evidenced by their public submissions to the commission.

Although in principle the hurricane loss models as shown diagrammatically in Figure 11.1 are straightforward to construct, the "devil is in the details." In particular, databases on historical hurricanes (e.g., HURDAT (with updates); Schwerdt *et al.*, 1979; Ho *et al.*, 1987) must be mined to develop probability distributions of frequency of occurrence by location, of intensity by location, of forward speed, and so forth. A wind field model needs to be selected that can incorporate hurricane characteristics, such as central pressure, radius of maximum winds, forward speed, etc. A parametric wind field provides the wind speed at the sites of the exposures of interest (e.g., residential housing stock). In addition, some (but not all) published hurricane models consider further

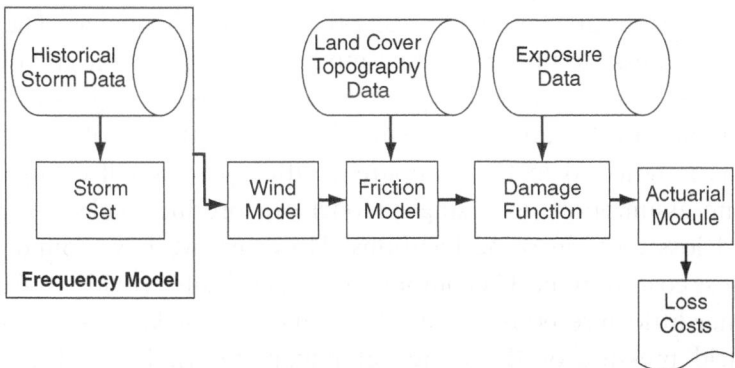

Figure 11.1. Traditional structure of hurricane loss models.

adjustment to the wind speed due to frictional effects and overland weakening. Finally, the developer of a model would need to determine a damage function that converts wind forces on structures to actual physical damages to the structure (roof, cladding, windows, and so forth, but converted through the function to a proportion of structure damage). Vulnerability functions produce "ground-up losses," which need additional refinement in the actuarial component of the model to emulate financial losses. The vulnerability functions have historically been the part of the model that distinguishes the various submissions, and they are considered highly proprietary. For vulnerability functions derived from insurance claims data (Friedman, 1984), an enormous amount of data processing, curve fitting, and engineering expert judgment take place, especially to handle mitigation impacts (shutters, tie-downs, roof-to-wall connectors, and so forth). Loss data for major events must be obtained from insurance companies or other sources to develop vulnerability curves or to validate existing curves against those with more recent events.

Watson and Johnson (2004) and Watson *et al.* (2004) investigated 324 combinations of wind field models, friction models, and damage functions drawn from the open literature, and they found that their range of average annual loss costs bracketed those reported by the proprietary modelers. Perhaps this is not surprising given that there is a finite set of parametric wind field models and damage functions. On the other hand, our implementations followed the descriptions given in the open literature and did not necessarily conform perfectly to the proprietary versions. The multi-model platform allowed the examination of the sources of variation driving the loss costs among model combinations (Watson and Johnson, 2006). The choice of wind field – the least sensitive (i.e., proprietary) part of the model – was shown to be the main driver of the variation.

Some politicians have gasped at the occasional two- to three-factor disparity of loss costs in specific counties among the proprietary models. They rightly ask how it is possible that individual models developed with great care and expense by teams of professionals produce such widely varying results. At first, one is tempted to despair in reviewing the state of the art of hurricane loss models as is reflected in these apparent differences. Small changes in key underlying assumptions in a single model can produce wide swings in the generated loss costs in some locations. However, we have found that the median loss cost from the 324 combinations at each site provides stable results that change little in response to variations in otherwise key assumptions such as far field pressure or the radius of maximum winds – values that are inaccurate or incompletely known for historical storms. We refer to this scheme as an ensemble median approach. This approach was further analyzed in a technical report to the commission (Watson and Johnson, 2006). Several variants of the HURDAT dataset (Jarvinen *et al.*, 1984, as updated annually by the Tropical Prediction Center) were considered, as well as various other model inputs such as land cover, and while individual model results varied widely, the results from the median ensemble were again much more stable. Thus the median ensemble allows us to more fully explore the impact of various alternative methodologies, such as GCM methods, for studying the underlying hurricane climatology without the risk of an interaction with a specific modeling technique (wind model, friction model, etc.) biasing the results.

11.3 Climate models as drivers of loss estimation models

Linking climate models to loss models can potentially solve two major short-comings associated with existing loss models. One shortcoming is that the historical record of hurricanes is short relative to the frequency of the most extreme events and incomplete with respect to key parameters required for loss modeling. The Atlantic basin has the best records, and the official HURDAT dataset starts in 1851. However, other areas of the world have significantly less temporal coverage. In particular, the best-track record for the South Pacific basin has barely 30 years of data – far too limited for a comprehensive analysis. The other shortcoming is that the existing loss models assume that climate is stationary. This assumption, of course, makes it difficult to evaluate the potential impact of either human or natural climate change. A climate model could be used to generate a set of synthetic storms consistent with a variety of climate conditions. However, a key question must be answered before the present generation of climate models can be usefully linked to loss models: How realistically can climate models reproduce observed tropical cyclone frequencies and storm characteristics?

The present generation of global climate models produces low-pressure systems that resemble tropical cyclones (hurricanes). These tropical cyclone-like vortices (TCLVs) occur with approximately the correct frequencies and tracks of observed tropical cyclones, but they do not directly reproduce key characteristics of tropical cyclones (Camargo *et al.*, 2005). In particular, while tracks are realistic, TCLVs produced by climate models are too large and the peak wind speeds are too low. As model resolutions increase, TCLVs begin to bear a closer resemblance to observed hurricanes, but at a massive and debil-itating cost in simulation time and data storage requirements. One method of overcoming these costs is the use of regional grids, wherein a lower-resolution global model is used to drive a higher-resolution regional grid. High-resolution (from a climate modeling standpoint) regional climate models (RCMs) have been applied to tropical cyclone modeling (Oouchi *et al.*, 2006). While they produce more realistic storms, they fail to reproduce the observed frequencies of intense storms with peak winds above $40\,\mathrm{m\,s^{-1}}$. It must be noted that even if it had produced accurate wind speeds, a relatively high-resolution simulation (from a climate modeling standpoint) using a 20 km grid is still inadequate for direct use in loss modeling owing to the spatial distribution of exposures and the inability to reproduce a realistic eye wall structure (Johnson and Watson, 1999; Watson and Johnson, 2004). As was noted in Johnson and Watson (1999), a 4 or 5 km grid seems to be the minimum grid size necessary to accurately reproduce losses from intense storms. Therefore, it seems reasonable to conclude that generating realistic hurricanes suitable for loss modeling directly from the present generation of GCMs and RCMs is not practicable at this time.

Given the coarse resolution of the current generation of GCMs and RCMs, and the need for finer resolutions for loss modeling, it is tempting to despair over the present potential to link loss models to GCM outputs. However, further consideration indicates that what is needed is not a fully accurate simulation of all of the details of tropical cyclones, but rather realistic generation of the minimum required parameters used in the parametric loss models: track and forward speed, the maximum wind speed, radius of maximum winds, and radius to environmental pressure. An alternative approach that was implemented in the context of the Florida project (Watson and Johnson, 2006) used a GCM to generate basic data files that can be used as surrogates for the Automated Tropical Cyclone Forecast (ATCF) system (Sampson and Schrader, 2000) storm track files, which in turn are transformed into a HURDAT-equivalent data file for use in existing loss models. Figure 11.2 shows a block diagram of our approach. A fully coupled ocean–atmosphere model (the National Center for Atmospheric Research's [NCAR's] Community Climate System Model [CCSM]) was run at the T42 resolution in order to generate TCLVs. A moving subgrid was established by using a subroutine based on a mesoscale nested model (the Penn State University/NCAR MM5 model; Grell *et al.*, 1993) to generate a 5 km analysis in the region bounding a detected vortex.

Finally, the characteristics of the tropical cyclone as modeled by the MM5-based subgrid were then interpreted to create a "synthetic ATCF file," which was processed in turn to create a synthetic HURDAT file of hurricane climatology that is analogous to that created from the observed historical record. The following sections describe in more detail each step in this process.

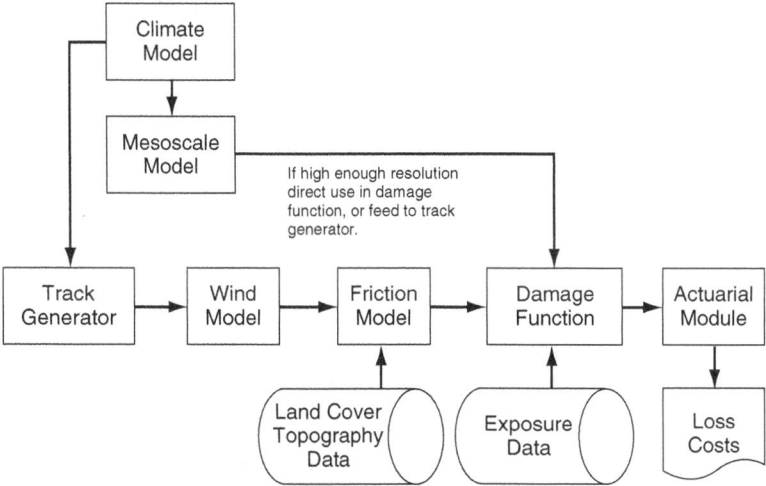

Figure 11.2. Climate model-based loss model.

11.3.1 Modifications to the CCSM and nesting methodology

The standard configuration of the CCSM was used for these simulations. A subroutine was added to the CCSM to scan the Atlantic basin every 6 hours of model time to find potential vortices. The following criteria, modified from Wu and Lau (1992), were used to decide whether or not to implement a nested grid over a given surface low:

- Positive vorticity at 850 millibar (mb)
- Negative vorticity at 200 mb
- Upward vertical motion at 500 mb
- Warm core (18 °C or higher)
- High relative humidity at the surface (70% or higher)
- Positive temperature anomaly from surface to 300 mb

If these criteria are met, a subgrid process to generate a two-way moving nest is spawned centered over the low. Every 6 hours of model time, the subgrid process is evaluated to determine whether the nest is needed. In addition to the above criteria, winds in the low must be above 15 m s^{-1} to maintain the grid for the next cycle.

The code for the nest was a straightforward implementation of the MM5 model as modified by Tenerelli and Chen (2001) for tropical cyclone modeling. The nest is moved at 6 hour intervals, in keeping with the reevaluation process noted above. The GCM outputs are applied to the MM5 nest as boundary conditions by using the code provided with MM5. The reverse nesting (MM5 to CCSM) is applied by averaging the MM5 fields to the appropriate CCSM levels and resolutions at each CCSM time step and inserting these fields back into the appropriate grid cell. Computationally, two Linux clusters were used: a 64-node cluster for the CCSM and a 16-node cluster for the subgrids.

11.3.2 Generation of simulated ATCF "A Deck"

Parametric wind models used in loss estimation use a few key parameters to define the storm (Iman *et al.*, 2006). For our real time and recent (after 1990) analyses, we used the Automated Tropical Cyclone Forecast System (Sampson and Schrader, 2000) "A Deck" records to extract the appropriate parameters. To ensure compatibility with our previously developed models and techniques, we inserted code into the CCSM/MM5 model to generate a simulated ATCF format file. At each synoptic time during a model run (0, 6, 12, and 18 h GMT, the nested grid was analyzed to extract data to create a simulated ATCF "A Deck" comprehensive archive (CARQ), record. Even when a 5 km grid was used, the resolution was such that some additional interpolation of the grid was needed. This was achieved by extracting the wind profile perpendicular to the storm motion and fitting it to the Air Force Global Weather Command (AFGWC) wind profile (Brand *et al.*, 1977). The maximum velocity and radius to maximum wind values that gave the best fit to that profile were saved in the records. All other variables (such as wind radii, far field pressure, minimum pressure, etc.) were obtained from the grid.

11.3.3 Atlantic basin results

The number of storms generated by the coupled CCSM/MM5 moving grid compared favorably with that in the historical record. Figure 11.3 plots the average number of tropical storms (34 knots or higher) from the 1851–2005 HURDAT records versus a 130-year integration of the nested subgrid model. Overall, the model was slightly low in most months, but not significantly so. The model did not produce a storm in February during the 130-year run, although this could have been due to the criteria used to select TCLVs for subgrid runs. Given the few historical storms in that month, this is not considered a major deficiency.

Once it has been established that the model-based climatology is producing storms with the correct frequencies, the next logical issue is to assess the intensities of the generated storms. Figure 11.4 shows the number of storms produced per 100 years of observed time (from the 1851–2005 HURDAT dataset) compared to simulated storms from the 130-year CCSM/MM5

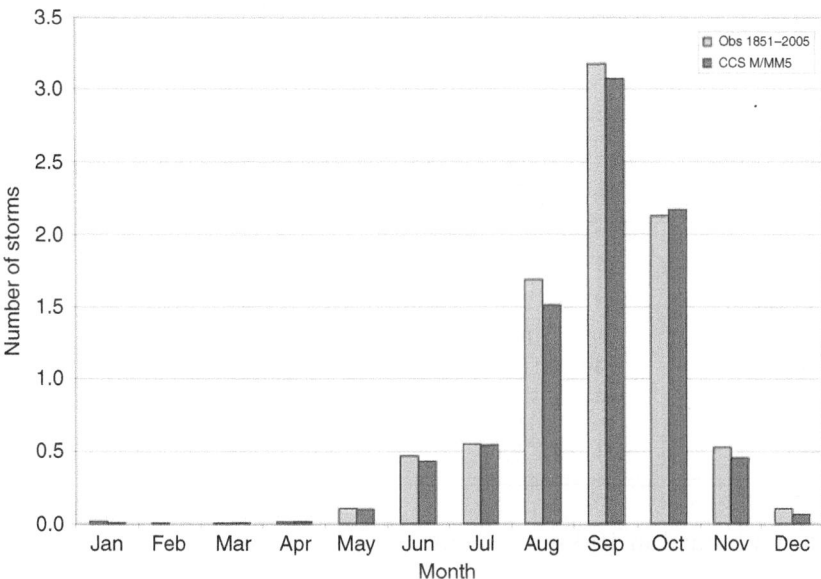

Figure 11.3. Average number of storms per month: observed versus modeled.

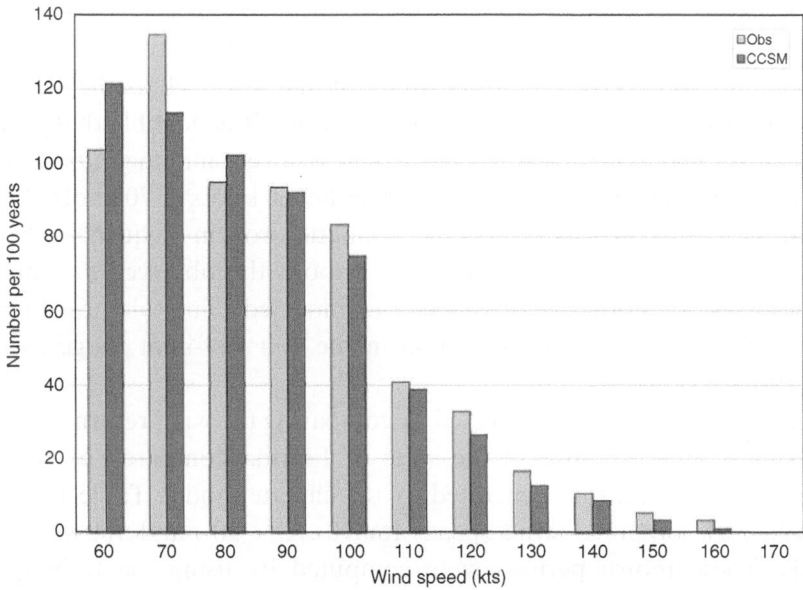

Figure 11.4. Comparison of the distribution of observed and climate model-simulated wind speeds (in knots).

Table 11.1. *Selected return period comparison (knots) for selected sites in Florida*

Return period (yr)	Observed				CCSM/MM5			
	10	25	50	100	10	25	50	100
Site								
Jacksonville	61	79	88	92	61	80	89	99
Cape Canaveral	63	76	79	84	63	76	79	89
Miami	90	102	109	114	78	100	102	113
Tampa Bay	64	83	91	93	63	79	88	93
Apalachicola	68	78	83	90	69	79	83	89
Pensacola	70	92	96	98	70	87	93	98
Orlando	61	75	77	88	61	74	76	83

model run binned in 10-knot increments. For example, there were 41 storms observed with peak winds between 110 and 120 knots, while the climate model produced 38 per 100 years. As would be suspected based on previous studies, the number of extreme storms is slightly below the observed, but unlike studies such as Oouchi *et al.* (2006), not seriously so. An interesting feature of the plot is that for the model-based runs, the number of events per bin decreases monotonically, whereas the observed (or, more accurately, reported) winds peak in the 70–80-knot bin. Recall that the US National Hurricane Center (NHC) real-time wind estimates as well as the HURDAT dataset winds are rounded to the nearest 5 knots. It is possible that there is a bias against the 65-knot value, as this would be a very minimal hurricane; it is more likely that a storm would be rated at either 60 knots or 70 knots. Another interesting feature of this plot is the dramatic drop in frequencies between the 100-knot bin and the 110-knot bin in both the observed and modeled climatologies. In both the observed and modeled climatologies, there are roughly twice as many storms falling in the 100–109-knot range as in the 110–119-knot range.

A final verification was conducted by comparing the wind return periods for wind speeds at seven sites in the state of Florida computed by using the historical record and that generated by the climate model. Table 11.1 shows observed and computed wind speeds for 10-, 25-, 50-, and 100-year return periods. These return periods were computed by using the techniques in Johnson and Watson (1999), which involve fitting annual maximum winds to a Weibull distribution.

Other storm parameters, such as the radius of maximum wind, appear to be reasonable as well; however, this could partly be due to the use of the

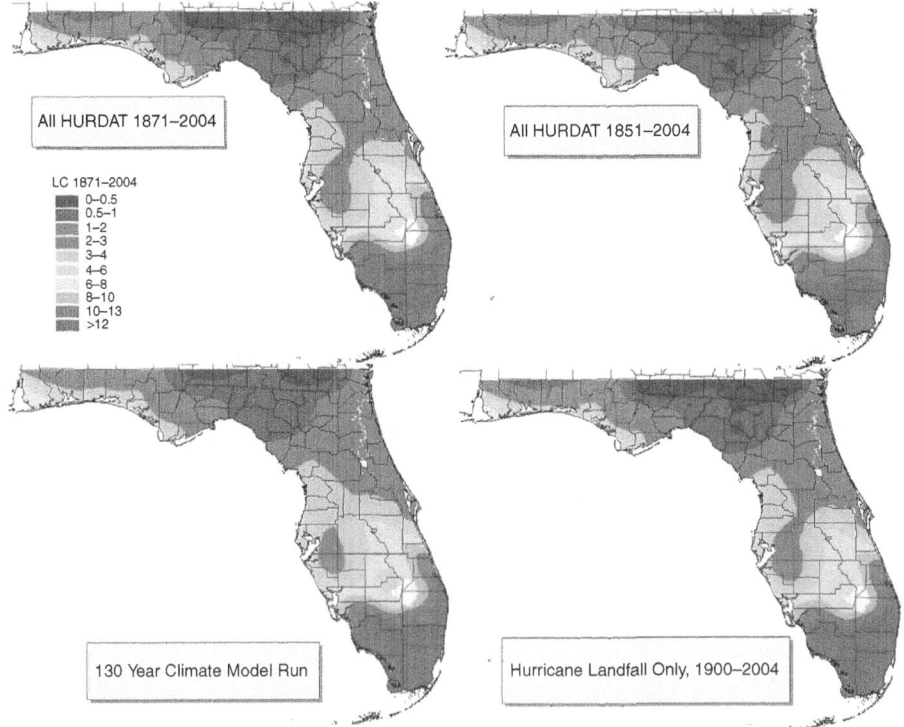

Figure 11.5. Loss costs from four different historical storm sets, including the storms generated from the climate model. The "All HURDAT" runs include all storms during the specified time period. The "Hurricane Landfall Only" run includes only storms that produced 64-knot winds over land. The legend shows the dollar loss per 1,000 dollars of exposure. For color version, see plate section.

AFGWC wind model to fit the wind profiles output by the mesoscale model. No extremely small storms (radius of maximum winds less than 8 nautical miles) were generated. Besides wind, only a cursory examination was made of other details of individual storms generated by the model, since the goal of the study was to generate realistic hurricane wind fields. It appears that other aspects (such as rainfall) were realistic, but no detailed analyses have been conducted.

Based on the results discussed above, we consider the CCSM/MM5-based outputs to be sufficient to justify using the results of the runs in loss modeling. The synthetic HURDAT file created from the GCM/MM5 outputs was subjected to the 324 combinations of models as described by Watson and Johnson (2004). Figure 11.5 shows the loss costs per US$1,000 of exposure for the climate model-based run, as well as for three different historical datasets. The loss cost contour maps are remarkably consistent: the global climate results could pass for a conventional model run based on an alternate catalog of storms as its basis. Thus, at least at this level of granularity, the global climate results would appear to be in the realm of those produced by classical modeling approaches. Much additional work remains, including validation in other basins and more detailed analyses of the structures of the storms produced by the nested grid.

11.4 Summary

Hurricane loss models that generate loss costs for insurance regulator review follow the well-worn strategy of using specific meteorological (wind field and friction), vulnerability (damage function), and actuarial (converting damage to insured loss) components. Probabilistic forecasts are predicated upon historical hurricane databases to determine appropriate input probability distributions for storm characteristics, including frequency. An alternative approach has been presented in this chapter that uses a global climate model as the primary driver of hurricane generation and evolution. The historical hurricane record serves as a reality check on the implementation of the climate model strategy. It has been demonstrated that tropical cyclone-like vortices can be generated that provide realistic hurricane characteristics, such as maximum wind speed, radius of maximum winds, forward speed, and far field pressure. The approach relies on a fully coupled ocean–atmosphere climate model with moving nested grids to monitor the generation and evolution of evolving tropical cyclone-like vortices. The strategy of generating ATCF-like files (which then produce a HURDAT file surrogate) takes full advantage of the loss cost machinery developed in the classical hurricane loss model context, particularly in conjunction with the ensemble median loss costs.

The approach was shown to produce realistic individual events and basin-wide frequency distributions. The distribution of wind speeds is reasonable. Of greatest import is that loss costs from the hurricane peril can be generated and their values also appear reasonable. Given the climate model structure, it would be natural to extend this work by running the time frame further out in time under various greenhouse gas emission assumptions.

Acknowledgments

The authors are grateful to Rick Murnane and Henry Diaz for inviting us to the Bermuda symposium and for their patience and encouragement to produce this chapter for the collected volume.

References

Brand, S., Rabe, K., and Laevastu, T. (1977). Parameterization characteristics of a wind-wave tropical cyclone model for the western North Pacific Ocean. *Journal of Physical Oceanography*, **7**, 739–46.

Camargo, S., Barnston, A., and Zebiak, S. (2005). A statistical assessment of tropical cyclone activity in atmospheric general circulation models. *Tellus*, **57A**, 589–604.

Clark, K. M. (1986). A formal approach to catastrophe risk assessment and management. *Proceedings of the Casualty Actuarial Society*, Vol. LXXIII, No. 140, November.

Clark, K. M. (1997). Current and potential impact of hurricane variability on the insurance industry. In *Hurricanes, Climate and Socioeconomic Impacts*, ed. H. F. Diaz and R. S. Pulwarty. New York: Springer, pp. 273–83.

Emanuel, K. (2005a). Increasing destructiveness of tropical cyclones over the past 30 years. *Nature*, **436**, 686–8.

Emanuel, K. (2005b). Emanuel replies. *Nature*, **438**, E13.

Friedman, D. G. (1984). Natural hazard risk assessment for an insurance program. *The Geneva Papers on Risk and Insurance*, **9**(30).

Grell, G. A., Dudhia, J., and Stauffer, D. R. (1993). *A Description of the Fifth-Generation Penn State/NCAR Mesoscale Model (MM5)*. NCAR Technical Note, NCAR/TN-398+ STR.

Ho, F. P., Su, J. C., Hanevich, K. L., Smith, R. J., and Richards, F. P. (1987). *Hurricane Climatology for the Atlantic and Gulf Coasts of the United States*. NOAA Technical Report NWS 38, Washington, D.C.: Federal Emergency Management Agency.

Iman, R. L., Johnson, M. E., and Watson, C., Jr. (2005a). Sensitivity analysis for computer model projections of hurricane losses. *Risk Analysis*, **25**, 1277–97.

Iman, R. L., Johnson, M. E., and Watson, C., Jr. (2005b). Uncertainty analysis for computer model projections of hurricane losses. *Risk Analysis*, **25**, 1299–312.

Iman, R. L., Johnson, M. E., and Watson, C., Jr. (2006). Statistical aspects of forecasting and planning for hurricanes. *The American Statistician*, **60**, 105–21.

Jarvinen, B. R., Newman, C., and Davis, M. (1984). *A Tropical Cyclone Data Tape for the North Atlantic Basin, 1886–1983: Contents, Limitations, and Uses*. NOAA Technical Memorandum NWS NHC 22, Coral Gables, Florida.

Johnson, M. E., and Watson, C., Jr. (1999). Hurricane return period estimation. *Proceedings of the Tenth Symposium on Global Change Studies*, Dallas, Texas, pp. 478–9.

Landsea, C. (2005). Hurricanes and global warming. Brief communications arising in *Nature*, **438**, E11–E13, with a reply from K. Emanuel.

Oouchi, K., Yoshimura, J., Yoshimura, H., *et al.* (2006). Tropical cyclone climatology in a global-warming climate as simulated in a 20 km-mesh global atmospheric model: frequency and wind intensity analyses. *Journal of the Meteorological Society of Japan*, **84**(2), 259–76.

Pielke, R. A., Jr. (2005). Are there trends in hurricane destruction? Brief communications arising in *Nature*, **438**, E12, with a reply from K. Emanuel.

Powell, M. D., Soukup, G., Cocke, S., *et al.* (2005). State of Florida Hurricane Loss Projection Model: atmospheric science component. *Journal of Wind Engineering and Industrial Aerodynamics*, **93**, 651–74.

Sampson, C. R., and Schrader, A. J. (2000). The Automated Tropical Cyclone Forecasting System (Version 3.2). *Bulletin of the American Meteorological Society*, **81**, 1231–40.

Schwerdt, R., Ho, F., and Watkins, R. (1979). *Meteorological Criteria for Standard Project Hurricane and Probable Maximum Hurricane Wind Fields, Gulf and East Coasts of the United States.* NOAA Technical Report NWS 23, Silver Spring, Maryland: National Weather Service.

Tenerelli, J. E., and Chen, S. S. (2001). High-resolution simulation of Hurricane Floyd (1999) using MM5 with a vortex following mesh refinement. *Preprints, 14th Conference on Numerical Weather Prediction*, July 30–August 2, 2001, Ft. Lauderdale, Florida, American Meteorological Society, J54–J56.

Vickery, P. J., Skerlj, P. F., and Twisdale, L. A. (2000). Simulation of hurricane risk in the U.S. using Empirical Track Model. *Journal of Structural Engineering*, **126**, 1222–37.

Watson, C., Jr., and Johnson, M. (2004). Hurricane loss estimation models: opportunities for improving the state of the art. *Bulletin of the American Meteorological Society*, **85**, 1713–26.

Watson, C., Jr., and Johnson, M. (2006). *Assessment of Computer Generated Loss Costs in Florida.* Report to the Florida Commission on Hurricane Loss Projection Methodology.

Watson, C., Jr, Johnson, M., and Simons, M. (2004). Insurance rate filings and hurricane loss estimation models. *Journal of Insurance Research*, **22**, 39–64.

Wu, G., and Lau, N. C. (1992). A GCM simulation of the relationship between tropical storm formation and ENSO. *Monthly Weather Review*, **120**, 958–77.

12

An exploration of trends in normalized weather-related catastrophe losses

STUART MILLER, ROBERT MUIR-WOOD,
AND AUGUSTE BOISSONNADE

Condensed summary

In order to evaluate potential trends in global natural catastrophe losses, it is important to compensate for changes in asset values and exposures over time. We create a Global Normalized Catastrophe Catalogue covering weather-related catastrophe losses in the principal developed (Australia, Canada, Europe, Japan, South Korea, United States) and developing (Caribbean, Central America, China, India, the Philippines) regions of the world. We survey losses from 1950 through 2005, although data availability means that for many regions the record is incomplete for the period before the 1970s even for the largest events. After 1970, when the global record becomes more comprehensive, we find evidence of an annual upward trend for normalized losses of 2% per year. Conclusions are heavily weighted by US losses, and their removal eliminates any statistically significant trend. Large events, such as Hurricane Katrina and China flood losses in the 1990s, also exert a strong impact on trend results. In addition, once national losses are further normalized relative to per capita wealth, the significance of the post-1970 global trend disappears. We find insufficient evidence to claim a statistical relationship between global temperature increase and normalized catastrophe losses.

12.1 Introduction

Economic losses attributed to natural disasters have increased from US\$75.5 billion in the 1960s to US\$659.9 billion in the 1990s (United Nations Development Programme [UNDP], 2004), for an annual growth rate of approximately 8%. Private sector data also show rising insured losses over a similar period (Munich Re, 2001; Swiss Re, 2005). Both reinsurers and some climate scientists have argued that these increases demonstrate a link between anthropogenically induced global warming and catastrophe losses

Climate Extremes and Society, ed. H. F. Diaz and R. J. Murnane. Published by Cambridge University Press. © Cambridge University Press 2008.

(Intergovernmental Panel on Climate Change [IPCC], 2001). However, failing to adjust for time-variant economic factors yields loss amounts that are not directly comparable and a pronounced upward trend through time that can be attributed to purely economic factors.

To allow for a comparison of losses over time, previous studies have adjusted past catastrophe losses to account for changes in monetary value in the form of inflation. However, in most countries, far larger changes have resulted from variations in human factors such as wealth and the numbers and values of properties located in the paths of the catastrophes (Van der Vink et al., 1998; Changnon et al., 2001). A full normalization of losses, which has been undertaken for the US hurricane (Pielke and Landsea, 1998; Collins and Lowe, 2001; National Oceanic and Atmospheric Administration [NOAA], 2005) and flood (Pielke et al., 2002), also includes the effects of changes in wealth and population to express losses in constant dollars. These previous national US assessments, as well as those for normalized Cuban hurricane losses (Pielke et al., 2003), have failed to show an upward trend in losses over time, but this was for the period before the remarkable US hurricane losses of 2004 and 2005.

In order to assess global trends over time, we compiled a database of normalized economic losses attributed to weather-related catastrophes from 1950 through 2005 from a large and representative sample of geographic regions. Regions were selected that had a reasonable centralization of catastrophe loss information as well as a broad range of peril types: tropical cyclone, extratropical cyclone (windstorm), thunderstorm, hailstorm, tornado, wildfire, and flood. The surveyed regions also span high- and low-latitude areas. Although global in scope, this study does not cover all regions. For example, we did not include losses from Africa or South America – first, because these continents are more affected by persistent climatological catastrophes (in particular, drought) than sudden-onset weather-related catastrophes. Also, the core economic loss data, in particular for much of Africa, were simply unavailable. However, the surveyed area included the large portion of the world's asset exposure and most of its population.

12.2 Data

We compiled economic loss data from international agencies, national databases, insurance trade associations, and reinsurers, as well as from RMS internal figures. Where possible, we tried to locate at least one government source with an official loss estimate. In some of the developing regions, this was not always possible and we deferred to private sector estimates or those provided by international bodies. We actively sought multiple loss estimates for events, and in cases where one event had different loss estimates we used the consensus mean estimate (or the median estimate if more appropriate). In cases where the quality of insured loss data exceeds that of economic losses, we estimated economic losses based upon insurance coverage ratios for the affected region and hazard type from contemporary insurance penetration rates. Data sources used for the normalization calculations are listed in Table 12.1.

Table 12.1. *Data sources used for normalization calculations*

AXCO Insurance Information Services (subscription). www.axcoinfo.com/.
Benson, C., and Clay, Edward J. (2001). Dominica: natural disasters and economic development in a small island state. Disaster Risk Management Working Paper Series No. 2. Washington, D.C.: World Bank.
Central Water Commission, Government of India. http://cwc.nic.in/.
Dartmouth Flood Observatory. www.dartmouth.edu/~floods/index.html.
EM-DAT: The OFDA/CRED International Disaster Database. www.em-dat.net/.
Emergency Management Australia (EMA). www.ema.gov.au/ema/emaDisasters.nsf.
Etkin, David, and Brun, Soren Erik (2001). Canada's hail climatology: 1977–1993. Institute for Catastrophe Loss Reduction Paper Series, No. 14.
Flood Damage in the United States. www.flooddamagedata.org.
General Insurance Association of Japan (GIAJ). www.sonpo.or.jp/e/index.html.
Guangzhou Institute of Geography (2002). *Atlas of Major Disasters and Society Responding to Them in China*. Guangzhou, China: Guangdong Science and Technology Press.
Insurance Bureau of Canada (IBC). Facts of the insurance industry (2004). www.ibc.ca/pdffiles/publications/brochures/consumer/FACTS_E04.pdf.
Insurance Disaster Response Organization (IDRO). www.idro.com.au/disaster_list/default.asp.
International Monetary Fund. International Financial Statistics. CD-ROM.
International Monetary Fund (1998). Letter of intent from Saint Kitts & Nevis. www.imf.org/external/np/loi/121098.htm.
ISO Property Claim Services (subscription). www.iso.com/products/2800/prod2801.html.
Public Safety and Emergency Preparedness Canada. Canadian Disaster Database. ww3.psepc-sppcc.gc.ca/disaster/default.asp.
Shi, P., ed. (2003). *Atlas of Natural Disaster System in China*. Beijing, China: Science Press.
Swiss Reinsurance Company. Annually 1998–2006. Sigma. *Natural Catastrophes and Man-Made Disasters*. Zürich: Swiss Reinsurance Company, Economic Research & Consulting.
United Nations National Accounts Statistics Database. (2006). http://unstats.un.org/unsd/snaama/Introduction.asp.
United Nations World Population Prospects: The 2004 Revision Population Database. http://esa.un.org/unpp/.
United States National Hurricane Center (NHC). www.nhc.noaa.gov/.

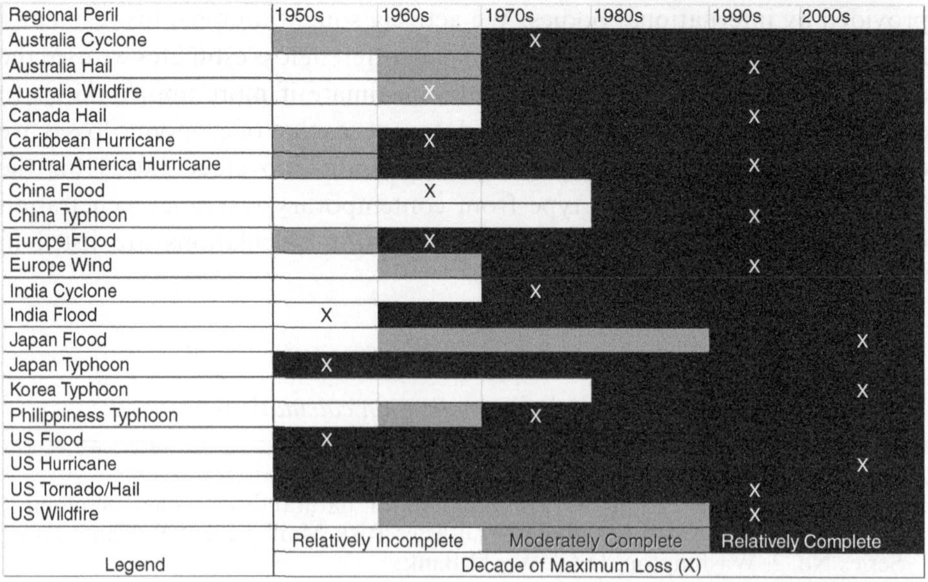

Regional Peril	1950s	1960s	1970s	1980s	1990s	2000s
Australia Cyclone			X			
Australia Hail					X	
Australia Wildfire		X				
Canada Hail					X	
Caribbean Hurricane		X				
Central America Hurricane					X	
China Flood		X				
China Typhoon					X	
Europe Flood		X				
Europe Wind					X	
India Cyclone			X			
India Flood	X					
Japan Flood						X
Japan Typhoon	X					
Korea Typhoon						X
Philippiness Typhoon			X			
US Flood	X					
US Hurricane						X
US Tornado/Hail					X	
US Wildfire					X	

Legend	Relatively Incomplete	Moderately Complete	Relatively Complete
	Decade of Maximum Loss (X)		

Figure 12.1. Post-1950 data quality for economic losses attributed to weather-related catastrophes by geographic region and catastrophe type.

Data coverage included the largest, best-known catastrophes as well as many smaller and midsize losses in developing regions. Of Swiss Re's list of the 40 costliest insured catastrophe events (Swiss Re, 2006), 36 of these were due to weather-related catastrophes that were included in the study. Data quality varies by region. Datasets from a number of territories are clearly incomplete for the period through the 1950s and 1960s; see Figure 12.1, which presents relative data completeness by decade. For this reason, any assessment of global trends prior to the 1970s has to omit a number of important contributory regions. This figure also illustrates the decade of maximum loss to provide a rough introduction to the pattern of high-loss events over time.

As an example, data for the India cyclone become less reliable for the period before 1965 and likely underreport the extent of losses during this period. Hail loss data for Canada are relatively sparse for the period before 1969. For a small proportion of countries, including China during the later stages of the Cultural Revolution, even for the 1970s the data are incomplete: China typhoon loss data underreport loss estimates prior to 1985; Korea typhoon data are unavailable for the period before 1978. However, despite these challenges, we believe it is possible to develop a global perspective of normalized losses from the 1970s and a more limited "developed world perspective" for the period from 1950.

One of the most challenging countries to which to apply the normalization methodology was Cuba, as a result of issues with converting official losses from Cuban pesos given the discrepancy between the official and black market exchange rates. There were also problems with finding reliable Cuban wealth data (Pielke *et al.*, 2003). Nevertheless, we applied a consistent normalization methodology, recognizing that results may be imprecise. Economic data were taken from the United Nations' (UN) National Accounts Statistics Database and from International Monetary Fund (IMF) statistics. Population data were taken from the UN World Population Prospects: The 2004 Revision Population Database. We used global temperature data from the Climatic Research Unit (CRU) at the University of East Anglia.

12.3 Methodology

We normalized losses to 2005 US dollars (USD) by adjusting for changes in wealth (gross domestic product [GDP] per capita in USD), inflation, and population. This methodology is consistent with that used by Pielke and Landsea (1998) and is given below:

$$NL2005 = Ly * (W2005/Wy) * (I2005)/Iy) * (P2005/Py), \qquad (12.1)$$

where normalized losses in 2005 USD (NL2005) equal the product of losses in year y and the change ratios in wealth (W), inflation (I), and population (P). Where GDP per capita is expressed in nominal terms, we omit the inflation multiplier.

12.4 Caveats

There are five issues that merit discussion before we proceed to the results.

- The term "economic loss" defies precise definition and is likely to have become broader over time. Today's estimates include direct damages such as physical damage to infrastructure, crops, housing, etc., and indirect damages such as loss of revenue, unemployment, and market destabilization (United Nations, 1992). For example, Indonesia's losses from the 2004 tsunami include an estimated US$1.53 billion for initial reduction in economic activity (Overseas Development Institute [ODI], 2005). As there is no systematic way to standardize loss estimates over time, we proceed with the caveat that recent loss estimates may report a more comprehensive and therefore higher economic loss.
- The reporting of economic loss estimates tends to improve with the size of the event (Downton and Pielke, 2005) and over time. Recent losses are almost

everywhere better recorded due to improvements in communications, literacy, news coverage, and insurance penetration. Failing to account for the summation of small to midsize event losses below a certain monetary threshold (e.g., US$1 million) will certainly affect aggregated loss estimates for most countries in earlier decades; however, data on smaller events are both harder to come by and less reliable than those for larger events. The aggregate impact on loss trends for a given country or region is also minimal, which is why the focus here has been on the largest losses.

- The method of normalization employed here assumes a constant vulnerability through time. For wind and hail, vulnerability reflects the susceptibility of buildings to direct damage, while for floods and wildfires it is the degree to which communities have been protected from risk (with flood defenses and firebreaks). The bias of assuming constant vulnerability is strongest where substantial adaptation (mitigation) has occurred, as for normalizing 1950s and 1960s storm surge losses in northern Europe, 1950s and 1960s storm surge and river flood events in Japan, or 1970s wind loss events in Australia. However, for most perils and regions, such as the US hurricane, real reductions in vulnerability have been modest. The impact of adaptation on normalized losses is considered further in Section 12.7. Vulnerability may also change for reasons unrelated to mitigation. Changes in settlement patterns, the development of new structures, and increasing population (e.g., Florida) will also impact a given region's vulnerability.

- The normalization methodology employed uses national statistics to compute the multipliers. Previous US normalizations used state- and county-level data to normalize losses. With the benefit of county-level resolution in the United States, we can see that the population growth rate for certain coastal, hazard-prone regions such as Florida is understated by using the national average. However, we consider the large-scale migration to hazardous coastal areas seen in the United States to be the exception. In the developing countries we surveyed, industrialization has led to migration to urban areas, which generally have lower risk profiles than rural areas. In other countries, there has been a greater balance between urban and coastal migration patterns. For example, high-growth coastal regions such as Queensland, Australia, have a far lower ratio of population growth relative to the national average than Florida.

- For several reasons, US losses exert a strong influence on trend conclusions. Losses are better recorded in the United States than in developing regions, which generates more data points. In addition, the United States has the largest economy and one of the highest GDP per capita levels in the world. There is consequently a higher dollar value of assets at risk in the United States than in most other countries. The US Atlantic basin and Gulf Coast contain a large stretch of coastline that is annually vulnerable to hurricane losses. While the global normalized loss results derived with the methodology above and subsequent trend analysis are technically correct, we emphasize that they are heavily influenced by the large weight of US losses on global totals.

12.5 Normalization results

On a regional basis, the greatest concentration of losses occurs in the United States (35%), followed by China and Europe with 18% and 16% of the total, respectively (Figure 12.2). That regions with higher asset values at-risk should make up the largest shares of the total is not surprising. While normalizing losses to constant dollars allows for a meaningful comparison of losses over time within a region, it may be insufficient to compare losses over time between regions. By normalizing all countries to US GDP per capita, we approximate a homogeneous distribution of wealth and control for the large impact of US normalized losses on the global total. If we remove differences that arise purely due to different levels of economic development, the distribution of losses changes noticeably (Figure 12.3). India (52%) and China (38%) account for 90% of the total, and normalized US losses drop from 38% to 3%.

When normalized losses are disaggregated by hazard type, flood is the costliest peril surveyed and accounts for 56% of normalized losses. Normalized tropical cyclone losses also constitute a sizable share (38%). Wind losses account for 4%, and other hazard types account for negligible shares of the total. Figure 12.4 shows the trend in aggregate normalized losses and the trend by hazard type. The volatile pattern of losses in the earlier

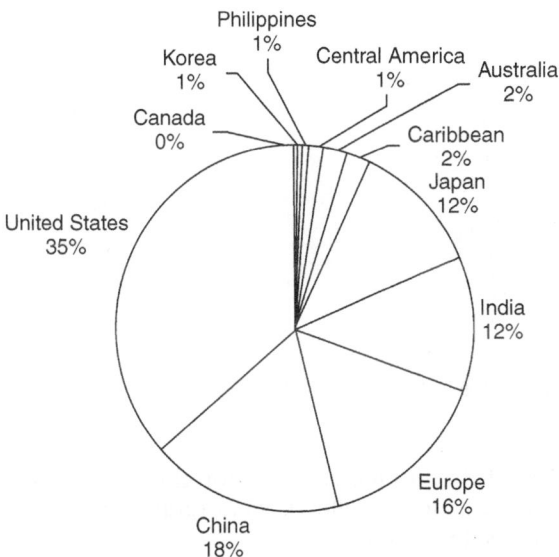

Figure 12.2. Regional distribution of aggregate economic losses from 1950 through 2005, normalized to 2005 USD.

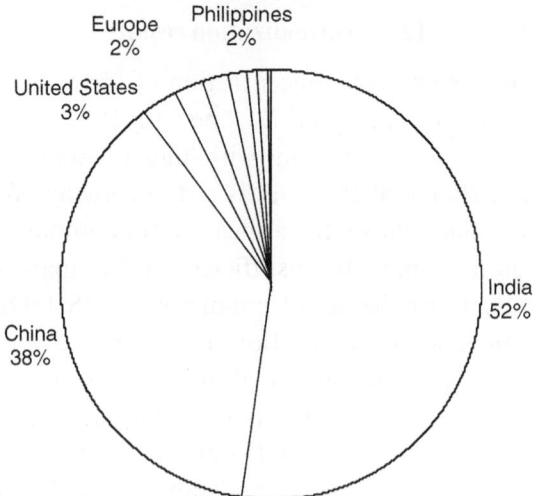

Figure 12.3. Regional distribution of aggregate economic losses from 1950 through 2005, normalized to 2005 USD and adjusted by GDP per capita. Regions not shown constitute less than 2% of the total.

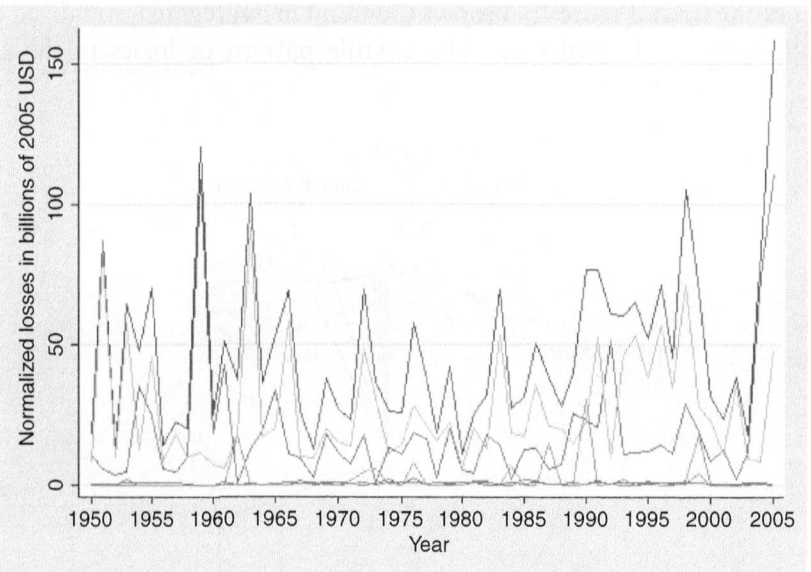

Figure 12.4. Trends in normalized losses for global totals (blue), hurricane (brown), hail (green), flood (yellow), wind (gray), and wildfire (red). For color version, see plate section.

decades is principally because the flood data are incomplete for both China and Japan for the period prior to the 1980s. It is notable that even with the incomplete data from the first two decades, the period of the 1970s and 1980s appears to have had lower levels of catastrophe losses than has been seen before or since.

12.6 Trend analysis

To test for a trend in normalized losses over time, we perform a linear regression of normalized economic losses in a given year on the year. The model is given below in Equation (12.2):

$$NL_y = \alpha + \beta_1 YEAR_y + \varepsilon_y. \qquad (12.2)$$

Normalized losses (NL) in year y are determined by the loss year (YEAR) y, where ε is the error term. If time is a significant determinant of loss level, we would expect the year to be statistically significant. The coefficient sign will indicate the direction of the trend. We fit the regression twice using global normalized loss estimates as well as hazard type and regional subsets, first with data for 1950–2005 and then with data for 1970–2005 (Table 12.A1 in the Appendix). Owing to the large impact of Katrina, 2005 losses are nearly four standard deviations from the mean and exert an upward pull on the overall trend. Overall, US hurricane losses from 2004 and 2005 as well as China flood losses exert a strong influence on the trend. To separate out the impact of these events on the overall results, we ran the regression separately with the respective losses removed (Table 12.A1). The log of normalized losses by year is shown in Figure 12.5.

When it is analyzed over the full survey period (1950–2005), the year is not statistically significant for global normalized losses. However, it is significant with a positive coefficient for normalized losses for specific regions, such as Canada at 10%, Korea at 5%, and China at 1% (in each of which the earlier record is known to be incomplete). The coefficient is negative (but not significant) for Australia, Europe, India, Japan, and the Philippines. Conclusions post-1950 are difficult to make owing to the lack of data. With the information available we do not find evidence of an upward global loss trend.

For the more complete 1970–2005 survey period, the year is significant, with a positive coefficient for (i.e., increase in) global losses at 1% with an r^2 value of 0.20, China at 1% (although again the early part of the record is likely to be incomplete), global tropical cyclones at 5%, and Caribbean losses at 10%. When Katrina losses are removed the global trend is significant at 5%. When US hurricane losses from 2004 and 2005 or China flood losses are removed, the

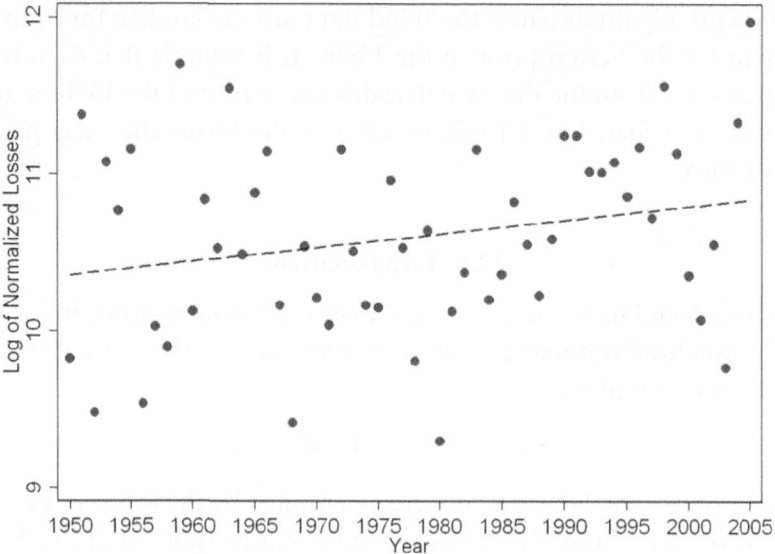

Figure 12.5. Logarithmic plot of global normalized losses from 1950 through 2005.

trend is no longer significant. There is a decreasing trend in normalized losses for the Philippines at 5%, and India at 1% (all located around the eastern Indian Ocean).

12.7 Discussion

12.7.1 Disaster loss trends

Before we consider the implications of these findings, we should first explore potential reasons for trends within the dataset. As was already noted, our methodology does not normalize for changes in the vulnerability of buildings, nor does our regression control for improved mitigation, such as reducing flood risk. However, there are several clear regional examples of declining loss trends since 1950 that merit comment. In Europe and Japan extensive investments in coastal flood defenses, in particular during the 1960s, have been well documented; the actual losses from events such as Typhoon Vera or the 1953 and 1962 North Sea storm surges would consequently be significantly reduced below the normalized values if they recurred today. For flood in Europe (Figure 12.6), the top three loss years all occurred by 1966 and recent flood years have reached less than half the value of the high-loss years in the first 20 years of the record.

While the record of river flood defense improvements in Europe is more mixed than for coastal defenses against storm surges, in Japan the dense

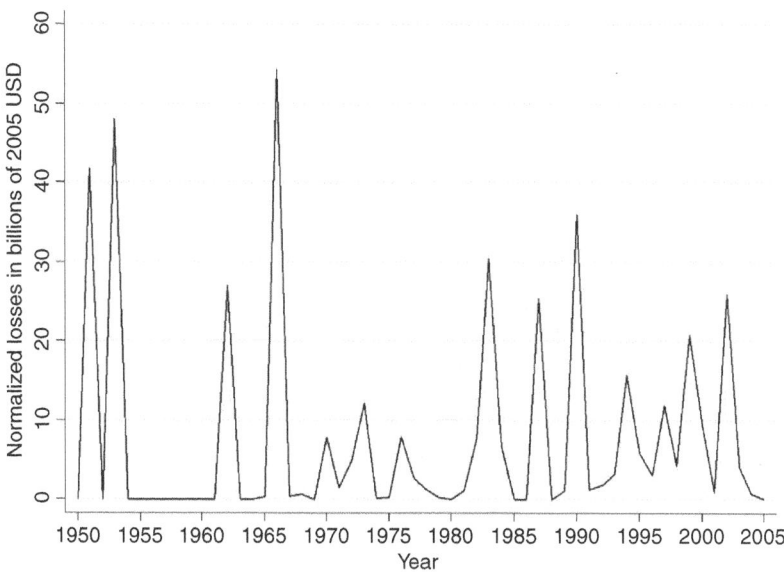

Figure 12.6. Normalized European flood losses from 1950 through 2005.

concentration of urban populations in low-lying areas has required major investments in riverine flood protection schemes throughout the periods of rapid economic growth. The impact of these investments can be seen in the dramatic reduction in the numbers of houses flooded in typhoons from 1950 through 2005 (Figure 12.7). Over the same period, improvements in building quality in Japan and the move away from traditional light wooden houses has also caused significant reductions in the numbers of properties severely damaged or lost in typhoons.

Data from Australia also show a downturn in normalized cyclone damages (Figure 12.8), with the losses from Tropical Cyclone Tracy in 1974 being a record year. However, while there have been significant improvements in building quality since that time, the principal explanation for the trend of declining losses also reflects the absence of any major cyclones hitting highly populated areas.

The reductions in flood losses that can be shown in specific regions are consistent with the overall trend analysis of global flood losses, which shows a decrease in the incidence of high annual totals. However, this has been replaced by more frequent and moderate annual losses, which range between US$20 billion and US$40 billion. The 1990s totals for flood losses are driven by very large flood losses in China, where rapid economic development began 40 years after it did in Japan, and successful flood mitigation schemes have been a feature only since the 1990s.

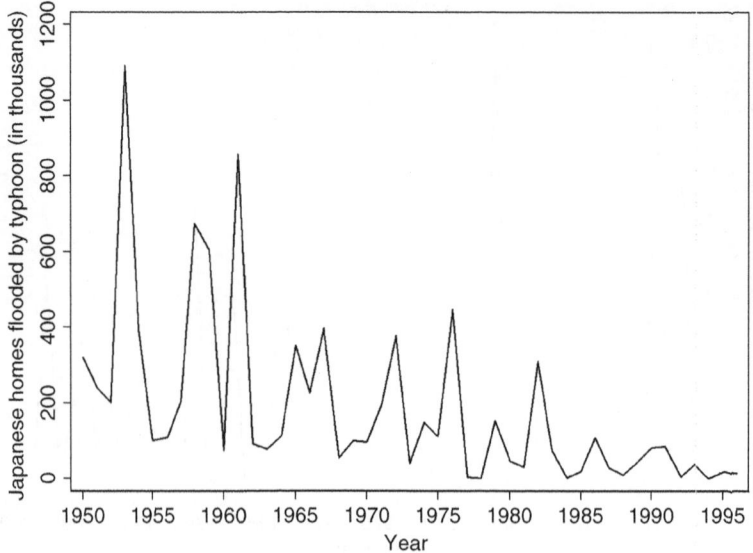

Figure 12.7. Annual estimates of houses flooded due to typhoons in Japan post-1950. Source: RMS estimates.

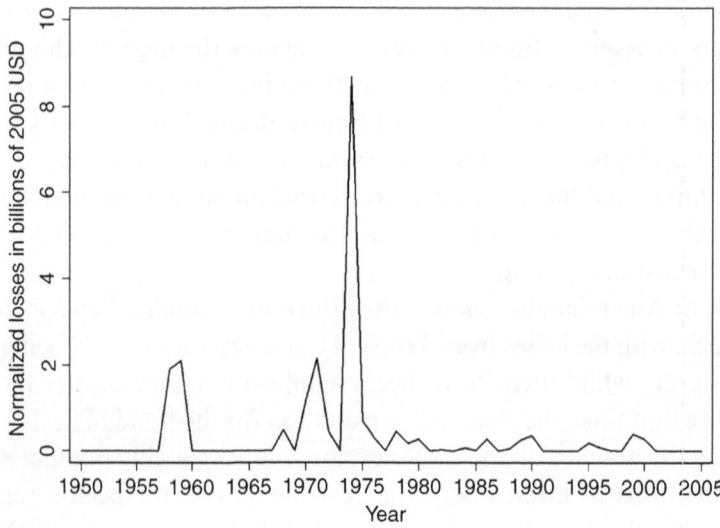

Figure 12.8. Normalized Australian cyclone losses from 1950 through 2005.

The global tropical cyclone loss trend exhibits a lower constant base level compared with floods (Figure 12.4), but it is marked by years with in excess of US$100 billion in normalized losses. A commonality to large flood and hurricane events is the ability of a single event such as Katrina or Vera to dominate losses in a given year and impact trend results over the entire survey period as well. For example, Katrina losses exceed US$100 billion and account for 8% of all normalized tropical cyclone losses, while Typhoon Vera in Japan in 1959 also exceeded US$100 billion of normalized losses. Both these storms generated more than 50% of their economic losses from flooding. The last major program of flood defense improvement in New Orleans was initiated after Hurricane Betsy in 1965. The 2005 flooding of New Orleans could be considered a failure of the 1960s levels of investment in flood mitigation for keeping pace with a rise in hurricane activity. While improved flood mitigation can help explain some part of the reduction of catastrophic flood losses since the 1950s, other causes must be sought in explaining the upward trend in global losses seen since the 1970s.

12.7.2 Climate change

The results of our trend analysis reveal an annual increase in normalized losses of 2%. Isolating the amount of this increase attributable to climate change is difficult, for the various reasons cited earlier. Without fully controlling for other factors that could affect the trend in losses, we can not draw any firm conclusions about the role of climate change in loss trends. In addition, any conclusions about a relationship between global warming and disaster losses are complicated by the sensitivity of statistical results to a few high-loss data points, the short historical loss record, and the limitations of the normalization methodology.

Instead, as a simple exercise we modeled the relationship between the annual global temperature anomaly and annual normalized losses. As a basic test, if there is an underlying link between climate change and normalized losses, we would expect the years with the highest (lowest) temperature variations to also have the highest (lowest) losses.

This relationship modeled is given below:

$$\text{NL}_y = \alpha + \beta_1 \text{TEMP}_y + \varepsilon_y, \tag{12.3}$$

where the total normalized losses in year y (NL_y) are given by the annual global temperature anomaly (TEMP) in year y, which is measured in degrees Celsius relative to the mean temperature from 1960 through 1991. Before we proceed to the results of this exercise, we emphasize that analysis of climatological trends over a span of only 50 years is insufficient to provide definitive conclusions or predictive analysis. The weight of outliers is significant, and one or two data points can exert a strong influence on the trend. Nevertheless, we present the results of this exercise for informative purposes, since the data are

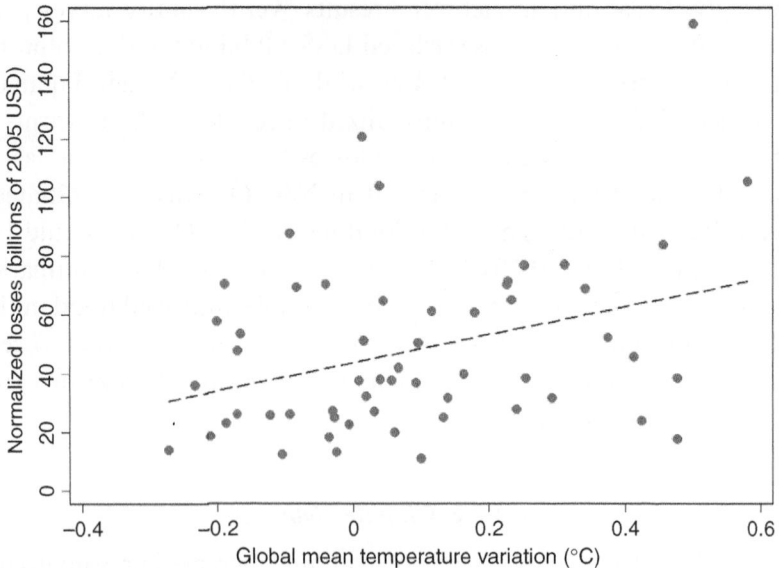

Figure 12.9. Scatter plot of normalized losses by the global temperature anomaly from 1950 through 2005.

largely unavailable for a longer period. Figure 12.9 plots the relationship between temperature variation and normalized losses post-1950.

   We caution that our model does not capture the possibility that there are no underlying factors that are common to years of high (low) temperature variation that would cause us to falsely attribute the trend in disaster losses to climatic reasons. Results suggest that the temperature anomaly is highly significant (at 1%) for normalized losses ($r^2 = 0.22$) irrespective of the survey period (Table 12.A2 in the Appendix). Results for Australia, the Philippines, and India are again significant (at the same levels as in the first model) with a negative coefficient. The rise is equivalent to an increase in normalized cata-strophe losses of US$4.8 billion (post-1950) and US$6.6 billion (post-1970) for each 0.1 °C rise in global temperatures. For details, please refer to Table 12.1A in the Appendix. However, these results are highly dependent upon recent US hurricane losses during 2004 and 2005. When the regression is rerun without these losses, the results are no longer statistically significant (Table 12.2).

12.7.3 Trend sensitivity

To test the impact of various losses on the trends, we performed four simple tests to explore the sensitivity of the results. We repeated the statistical tests in Equations (12.2) and (12.3) in order to isolate the impact of 2004 and 2005 US

Table 12.2. *Summation of trend sensitivity to high-impact events and the dominance of US losses*[a]

	Effect on significance			
	Normalized loss trend		Temperature	
Modification	Post-1950	Post-1970	Post-1950	Post-1970
Complete dataset	not significant	significant at 1%	significant at 1%	significant at 1%
Remove 2004/ 2005 US hurricane losses	not significant	not significant	not significant	not significant
Remove Katrina losses	not significant	significant at 5%	significant at 10%	significant at 5%
Remove China flood losses	not significant	significant at 10%	not significant	not significant
Renormalized wealth	significant at 5%	not significant	significant at 5%	not significant

[a] Statistical significance disappears for a post-1970 upward trend in normalized losses with the removal of 2004/2005 US hurricane losses and also disappears once we adjust for regional wealth inequalities. The relationship weakens to 5% and 10% with the removal of Katrina or China flood losses, respectively. The statistically significant relationship post-1970 (at 1%) between global temperature and disaster losses disappears once we remove 2004/2005 US hurricane losses or China flood losses, or renormalize wealth. The significance weakens to 5% with the removal of Katrina losses. This result suggests that any conclusion of a relationship between global temperature and normalized disaster losses is highly dependent upon large loss events, particularly in the United States.

hurricane losses, Katrina losses, China flood losses, and regional wealth differences on our results. In the first three instances, we simply reran the regressions with the relevant losses excluded. Since record years for hazard losses in a developing region would not have exerted such a strong pull on trend significance, we renormalized each region's normalized losses by multiplying them by the ratio of US GDP per capita to regional GDP per capita in order to approximate a homogeneous distribution of wealth (Tables 12.A3 and 12.A4 in the Appendix). When US hurricane losses from 2004 and 2005 are removed, our results are no longer significant. When losses are renormalized, the results are no longer significant for the post-1970 period. The other modifications weaken our significance findings, but do not eliminate them. Table 12.2 summarizes the impact of these modifications.

12.8 Conclusions

The original purpose of this study was to test the many statements that had been made, and charts that had been plotted, appearing to show a significant increase in global weather-related catastrophe losses over time attributed to a rise in global temperatures and anthropogenic greenhouse gas emissions. We identified several factors that must be considered in interpreting the results: (i) the variance in the definition of economic loss, (ii) improvements in loss reporting over time, (iii) changing vulnerability over time, (iv) national level statistics to adjust loss amounts that affect only a specific national region, and (v) the large weight of US losses in accounting for "global" normalized losses.

Before normalization, the annual rise in losses was about 8%. After normalization, these normalized losses did show a more modest underlying rising trend of 2% per year (from an average of US$36.4 billion in the 1970s to an average of US$64.5 billion from 1996 through 2005: a rise of almost 80%). Therefore, the large portion of the rising loss trend is explained by increases in values and exposure as well as by an increasing comprehensiveness of reporting global losses through time. For specific regions – in particular, India, Australia, and the Philippines – over this same period, there is evidence for a decline in normalized losses.

In sum, we found limited statistical evidence of an upward trend in normalized losses from 1970 through 2005 and insufficient evidence to claim a firm link between global warming and disaster losses. Our findings are highly sensitive to recent US hurricane losses, large China flood losses, and inter-regional wealth differences. When these factors are accounted for, evidence for an upward trend and the relationship between losses and temperature weakens or disappears entirely.

Finally, it appears that just as hurricane activity and intensity, correlated with a rise in equatorial Atlantic sea surface temperatures (SSTs), have shown the strongest evidence for an increase since the 1970s (Emanuel, 2005), it is hurricanes in wealthy regions that are today the principal driver of the evidence for an upward trend in global catastrophe losses.

References

Changnon, S., Pielke, R., Sylves, R., and Pulwarty, R. (2001). Human factors explain the increased losses from weather and climate extremes. *Bulletin of the American Meteorological Society*, **81**, 437–42.

Climatic Research Unit (CRU) (2006). University of East Anglia, Norwich. www.cru.uea.ac.uk/cru/data/temperature/#sciref.

Collins, D. J., and Lowe, S. P. (2001). A macro validation dataset for U.S. hurricane models. www.casact.org/pubs/forum/01wforum/01wf217.pdf.

Downton, M. W., and Pielke, R. A., Jr. (2005). How accurate are disaster loss data? The case of U.S. flood damage. *Natural Hazards*, **35**, 211–28.

Emanuel, K. A. (2005). Increasing destructiveness of tropical cyclones over the past 30 years. *Nature*, **486**, 686–8.

Intergovernmental Panel on Climate Change (IPCC). (2001). IPCC Third Assessment Report: Climate change 2001. www.ipcc.ch/pub/reports.htm.

Munich Re Group. (2001–5). *Annual Review: Natural Catastrophes*. Munich: WKD Offsetdruck GmbH.

National Oceanic and Atmospheric Administration (NOAA) (2005). The deadliest, costliest, and most intense United States tropical cyclones from 1851 to 2004 (and other frequently requested hurricane facts). www.tpc.ncep.noaa.gov/pdf/NWS-TPC-4.pdf.

Overseas Development Institute (ODI) (2005). ODI briefing paper: Aftershocks: natural disaster risk and economic development policy. www.odi.org.uk/publications/briefing/bp_disasters_nov05.pdf.

Pielke, R. A., Jr., and Landsea, C. (1998). Normalized hurricane damage in the United States: 1925–95. *Weather and Forecasting*, September 1998, 621–31.

Pielke, Jr., R. A., Downton, M. W., and Barnard Miller, J. Z. (2002). *Flood Damage in the United States, 1926–2000: A Reanalysis of National Weather Service Estimates*. Boulder, CO: University Corporation for Atmospheric Research (UCAR).

Pielke, R. A., Jr., Ribeira, J., Landsea, C., Fernández, M. L., and Klein, R. (2003). Hurricane vulnerability in Latin America and the Caribbean: normalized damage and loss potentials. *Natural Hazards Review*, August, 101–14.

Swiss Reinsurance Company. (2005, 2006). *Sigma. Natural Catastrophes and Man-Made Disasters*. Zürich: Swiss Reinsurance Company, Economic Research & Consulting.

United Nations (1992). *International Agreed Glossary of Basic Terms Related to Disaster Management* Geneva: United Nations, Department of Humanitarian Affairs (DHA), International Decade for Natural Disaster Reduction (IDNDR).

United Nations Development Programme (UNDP) (2004). Reducing disaster risk: a challenge for development. www.undp.org/bcpr/disred/rdr.htm.

Van der Vink, G. *et al.* (1998). Why the United States is becoming more vulnerable to natural disasters. *Eos, Transactions, American Geophysical Union*, **79**(44), 533–7.

APPENDIX: REGRESSION RESULTS

Table 12.A1. *Ordinary least squares (OLS) regression of normalized weather-related losses on year*

Dependent variable: normalized catastrophe losses. [a]

	Time period					
	1950–2005			1970–2005		
Survey group	Slope	Intercept	r^2	Slope	Intercept	r^2
Global losses	379.26	− 702,039	0.04	1251.08[***]	−2,437,796[***]	0.20
	(241.9)	(478,378)	—	(423.45)	(841,613)	—
Global losses	19.38	6,155.41	0.000,2	395.44	−742,591	0.04
(2004/2005	(211.95)	(419,142)	—	(350.74)	(697,107)	—
US hurricane						
removed)						
Global losses	153.69	−258,129	0.01	710.54[**]	−1,366,805[**]	0.11
(Katrina	(211.94)	(419,134)	—	(340.89)	(677,520)	—
removed)						
Global losses	99.46	−156,213	0.004	698.75[*]	−1,349,145	0.08
(China flood	(225.81)	(446,556)	—	(413.58)	(821,998)	—
removed)						
Peril						
Flood	124.01	−218,293	0.002	412.58	−792,552	0.02
	(169.37)	(334,949)	—	(263.38)	(523,477)	—
Hurricane	204.89	−386,844	0.008	837.1[**]	−1,645,767	0.16
	(181.7)	(359,324)	—	(311.51)	(619,139)	—
Hail	10.34	−20,101	0.05	−1.43	3342	0.003
	(6.35)	(12,559)	—	(14.8)	(29,408)	—
Wildfire	3.45	−6,267	0.04	0.41	−217.23	0.000,2
	(2.36)	(4,668)	—	(4.51)	(8,963)	—
Wind	46.38	−89,604	0.02	5.47	−8148	0.001
	(45.57)	(90,125)	—	(101.23)	(201,194)	—
Region						
Australia	−3.76	8,333	0.001	−50.52	101,391[*]	0.08
	(14.84)	(29,351)	—	(30.21)	(60,053)	—
Canada+	6.87[*]	−13,448[*]	0.05	7.72	−15,136	0.02
	(4.01)	(7,923)	—	(9.73)	(19,339)	—
Caribbean	31.09	−60,324	0.04	66.44[*]	−130,739[*]	0.09
	(19.83)	(39,208)	—	(35.3)	(70,167)	—
Central	3.51	−6,286	0.001	25.42	−49,953	0.03
America	(13.59)	(26,877)	—	(26.57)	(52,814)	—
China +	371.24[***]	−725,012[***]	0.12	660.15[***]	−1,300,388[***]	0.19
	(135.3)	(267,570)	—	(234.16)	(465,393)	—
Europe	−18.6	44,461	0.001	128.86	−248,958	0.02
	(109.19)	(215,936)	—	(152.46)	(303,021)	—

Table 12.A1. *(cont.)*

Survey group	1950–2005			1970–2005		
	Slope	Intercept	r^2	Slope	Intercept	r^2
India +	−24.24	53,722	0.01	−285.33[***]	573,616[***]	0.25
	(44.5)	(88,001)	—	(84.64)	(168,220)	—
Japan	−174.12	350,072	0.04	101.13	−198,148	0.06
	(120.1)	(237,504)	—	(68.98)	(137,096)	—
Korea +	16.63[**]	−32,617[**]	0.08	25.04	−49,366	0.05
	(7.89)	(15,612)	—	(19.21)	(38,175)	—
Philippines	−0.95	2,216	0.0007	−21.89[**]	43,914[**]	0.13
	(4.94)	(9,762)	—	(9.92)	(19,718)	—
United States	243.2	−463,054	0.03	652.71	−1,278,026	0.07
	(183.53)	(362,946)	—	(408.88)	(812,658)	—
$n =$	56	—	—	36	—	—

[a] Regression results presented with coefficient on top and standard error in parentheses
+, dataset incomplete and contains several zero values
[*]significant at 10%
[**]significant at 5%
[***]significant at 1%.

Table 12.A2. *OLS regression of global temperature anomaly on normalized losses* Dependent variable: normalized catastrophe losses.[a]

Survey group	Time period					
	1950–2005			1970–2005		
	Slope	Intercept	r^2	Slope	Intercept	r^2
Global losses	47,805.46[***] (17,769.29)	43,662.99[***] (4,077.56)	0.12 —	66,032.47[***] (21,557.72)	36,865.44[***] (5,835.75)	0.22 —
Global losses (2004/2005 U.S. hurricane removed)	17,287.39 (16,044.36)	42,930.31[***] (3,681.74)	0.02 —	26,203.59 (17,765.87)	38,629.04 (4,809.29)	0.06 —
Global losses (Katrina removed)	28,149.76[*] (15,836.4)	43,280.37[***] (36,34.02)	0.06 —	40,074.7[**] (17,251.5)	38,195.14[***] (4,670.11)	0.14 —
Global losses (China flood removed)	19,131.69 (17,109.21)	38,764.23[***] (3,926.09)	0.02 —	34,277.87 (21,291.59)	33,455.24[***] (5,763.71)	0.07 —
Peril						
Flood	22,379.18[*] (12,660.64)	24,926.94[***] (2,905.27)	0.05 —	28,727.42[**] (13,093.69)	22,287.77[***] (3,544.51)	0.12 —
Hurricane	21,560.86 (12,753.8)	16,392.98[***] (31,56.12)	0.04 —	35,886.66[**] (16,483.66)	11,514.39[**] (44,62.19)	0.12 —
Hail	143.67 (497.29)	340.87[***] (114.11)	0.002 —	−627.42 (751.46)	605.90[***] (203.42)	0.02 —
Wildfire	65.18 (183.89)	549.52[***] (42.20)	0.002 —	−117.75 (230.48)	628.60[***] (62.39)	0.01 —
Wind	3,875.94 (3,480.08)	1,764.97[**] (798.58)	0.003 —	1,708.32 (5,184.73)	2,419.29[*] (1,403.53)	0.003 —
Region						
Australia	−1,500.7 (1,117.65)	1,025.89[***] (256.47)	0.03 —	−3,511.97[**] (1,495.71)	1,609.05[***] (404.89)	0.14 —
Canada+	780.33[**] (296.28)	71.24 (67.99)	0.11 —	882.35[*] (480.46)	48.28 (130.06)	0.09 —
Caribbean	3,620.62[**] (1,470.62)	849.06[**] (337.47)	0.10 —	4,508.85[**] (1,738.78)	497.17 (470.70)	0.17 —
Central America	952.74 (1,032.29)	952.74 (1,032.29)	0.02 —	2,210.25[*] (1,328.3)	171.08 (359.58)	0.08 —
China+	31,417.38[***] (10,188.32)	6,294.01[***] (2,337.94)	0.15 —	30,792.82[**] (12,252.42)	6,135.57[**] (3,316.77)	0.16 —
Europe	7,723.14 (8,289.47)	6,995.16[***] (1,902.20)	0.02 —	12,094.17 (7,625.59)	4,982.11[**] (2,064.28)	0.07 —
India+	−5,531.46 (3,329.68)	62,72.84 (764.07)	0.05 —	−13,748.17[***] (4,426.33)	8,987.72[***] (1,198.22)	0.22 —

Table 12.A2. *(cont.)*

Survey group	1950–2005 Slope	Intercept	r^2	1970–2005 Slope	Intercept	r^2
				Time period		
Japan	−6,053.15	6,294.55[***]	0.008	2,861.27	2,341.79[**]	0.02
	(9,328.73)	(2,140.69)	—	(3,615.41)	(978.70)	—
Korea+	1,393.26[**]	135.84	0.09	1,468.48	141.61	0.06
	(598.98)	(137.45)	—	(977.72)	(264.67)	—
Philippines	−298.97	−298.97	0.01	−886.61[*]	567.21[***]	0.08
	(375.57)	(375.57)	—	(522.46)	(141.43)	—
United States	17,549.67	16,376.69[***]	0.03	28,339.80	14,142.37[**]	0.05
	(14,065.9)	(3,227.74)	—	(21,196.26)	(5,737.91)	—
$n =$	56	—	—	36	—	—

[a] Regression results presented with value on top and standard error in parentheses

+, dataset incomplete and contains several zero values

[*]significant at 10%

[**]significant at 5%

[***]significant at 1%.

Table 12.A3. *OLS regression of wealth-adjusted normalized weather-related losses (in billions USD) on year*

Dependent variable: wealth-adjusted normalized weather-related catastrophe losses.[a]

Survey group	1950–2005			1970–2005		
	Slope	Intercept	r^2	Slope	Intercept	r^2
Global losses	9.22**	−17,567**	0.08	3.67	−6,535	0.01
	(4.30)	(8,507)	—	(7.92)	(15,740)	—
Peril						
Flood	4.88	−9,135	0.03	4.08	−7,550	0.01
	(3.77)	(7,447)	—	(6.36)	(12,638.7)	—
Hurricane	4.11**	−7,976**	0.09	−1.34	2,879	0.003
	(1.75)	(3,462)	—	(4.11)	(8,165)	—
Hail	0.008	−15.70	0.03	−0.000,3	0.92	0.000,1
	(0.006)	(12.27)	—	(0.015)	(28.85)	—
Wildfire	0.005	−9.30	0.01	0.01	−26.08	0.02
	(0.009)	(17.58)	—	(0.01)	(28.73)	—
Wind	0.09	−168.19	0.02	0.01	−15.29	0.000,1
	(0.09)	(169.17)	—	(0.19)	(377.65)	—
Region						
Australia	−4.36	9,648.72	0.001	−58.50	117,394.5*	0.08
	(17.19)	(33,984.61)	—	(34.98)	(69,531.92)	—
Canada+	0.01*	−16.53*	0.05	0.01	−18.6	0.02
	(0.005)	(9.74)	—	(0.01)	(23.77)	—
Caribbean	0.27	−523.54	0.05	0.54*	−1,065.68*	0.09
	(0.16)	(319.33)	—	(0.29)	(571.95)	—
Central America	0.03	−47.88	0.001	0.19	−380.47	0.03
	(0.10)	(204.72)	—	(0.20)	(402.26)	—
China+	10.28**	−20,081**	0.11	19.12***	−37,679	0.18
	(3.97)	(7,842)	—	(6.91)	(13,735)	—
Europe	−0.03	83.45	0.001	0.24	−467.30	0.02
	(0.20)	(405.32)	—	(0.29)	(568.78)	—
India+	−1.47	3,252.23	0.01	−17.27***	34,725.67***	0.25
	(2.69)	(5,327.43)	—	(5.12)	(10,183.7)	—
Japan	−0.20	394.86	0.04	0.11	−223.50	0.06
	(0.14)	(267.89)	—	(0.08)	(154.64)	—
Korea+	44.69**	−87,676**	0.08	67.32	−132,700	0.05
	(21.22)	(41,966)	—	(51.63)	(102,619)	—
Philippines	0.04**	−87.68**	0.08	0.07	−132.70	0.05
	(0.02)	(41.97)	—	(0.05)	(102.62)	—
United States	243.2	−463,054	0.03	652.71	−1,278,026	0.07
	(183.53)	(362,946)	—	(408.88)	(812,658)	—
$n =$	56	—	—	36	—	—

[a] Regression results presented with coefficient on top and standard error in parentheses

+, dataset incomplete and contains several zero values

*significant at 10%

**significant at 5%

***significant at 1%.

Table 12.A4. *OLS regression of global temperature anomaly on wealth-adjusted normalized weather-related losses (in billions USD).*

Dependent variable: wealth-adjusted normalized weather-related catastrophe losses.[a]

	Time period					
	1950–2005			1970–2005		
Survey group	Slope	Intercept	r^2	Slope	Intercept	r^2
Global losses	704.50[**]	594.13[***]	0.08	315.70	711.28[***]	0.02
	(329.14)	(75.53)	—	(403.93)	(109.34)	—
Peril						
Flood	487.79[*]	462.41[***]	0.05	350.23	490.90[***]	0.03
	(284.90)	(65.38)	—	(322.62)	(87.34)	—
Hurricane	191.57	140.48[***]	0.03	−75.94	234.91[***]	0.004
	(138.17)	(31.71)	—	(210.66)	(57.03)	—
Hail	0.008	0.27[**]	0.001	−0.62	0.49[**]	0.02
	(0.48)	(0.11)	—	(0.74)	(0.20)	—
Wildfire	0.57	0.41	0.01	0.94	0.27	0.05
	(0.68)	(0.16)	—	(0.73)	(0.20)	—
Wind	7.28	3.31[**]	0.02	3.21	4.54[*]	0.003
	(6.53)	(1.50)	—	(9.73)	(2.63)	—
Region						
Australia	−1.74	1.19[***]	0.03	−4.07[**]	1.86[***]	0.14
	(1.29)	(0.30)	—	(1.73)	(0.47)	—
Canada+	0.96[**]	0.09	0.11	1.08[*]	0.06	0.09
	(0.36)	(0.08)	—	(0.59)	(0.16)	—
Caribbean	30.20[**]	6.70[**]	0.11	36.75[**]	4.05	0.14
	(11.98)	(2.75)	—	(14.17)	(3.84)	—
Central America	7.26	4.28[**]	0.02	16.83	1.30	0.08
	(7.86)	(1.80)	—	(10.12)	(2.74)	—
China+	969.97[***]	164.79[**]	0.17	1,035.22[***]	133.0	0.20
	(293.41)	(67.33)	—	(349.92)	(94.73)	—
Europe	14.50	13.13[***]	0.02	22.70	9.35[**]	0.07
	(15.56)	(3.57)	—	(14.31)	(3.87)	—
India+	−334.86	379.75[***]	0.05	−832.29[***]	544.10[***]	0.22
	(201.57)	(46.26)	—	(267.96)	(72.54)	—
Japan	—	—	—	—	—	—
Korea+	3,745.21[**]	365.16	0.09	3,947.37	380.66	0.06
	(1,610.10)	(369.47)	—	(2,628.19)	(711.46)	—
Philippines	3.75[**]	0.37	0.09	3.95	0.38	0.06
	(1.61)	(0.37)	—	(2.63)	(0.71)	—
United States	17.55	16,376.69[***]	0.03	28,339.80	14,142.37[**]	0.05
	(14.07)	(3,227.74)	—	(21,196.26)	(5,737.91)	—
$n=$	56	—	—	36	—	—

[a] Regression results presented with value on top and standard error in parentheses
+, dataset incomplete and contains several zero values
[*] significant at 10%
[**] significant at 5%
[***] significant at 1%.

13

An overview of the impact of climate change on the insurance industry

ANDREW DLUGOLECKI

Condensed summary

This chapter focuses on the types of extreme weather and climate events that are important to property insurers, and it considers evidence on how those risks have been changing and how they might change in the future with climate change. Some evidence is drawn from the United Kingdom, which has wide insurance coverage for weather perils and much familiarity with international catastrophe insurance. The author believes there is good ground to argue that climate change is already affecting the risks, although it is not the only factor that has caused change. Further, the risks of extremes may be changing very rapidly.

13.1 Introduction

The insurance industry is affected by climate change in a number of ways, although the effects may vary greatly between jurisdictions, owing to industry structure and practice as well as to climatic and geographical differences. Insurers are already encountering aggravated claims for insured property damage in extreme events, particularly floods, storms, and droughts. As yet, the main causes of the trend of rising property claims in recent decades are socioeconomic rather than climatic; however, the contribution of climatic change could rise quickly, owing to the strongly nonlinear relationship between climatic variables and damage, and the fact that a small shift in mean conditions can create a large change in extremes. The pace of change regarding weather extremes is *fast*. The underlying rate of change in the risk can be in the range 2% to 4% per year for extreme, but "insurable," events (ABI, 2004b), and much higher for very rare events (Climate Change Impacts Review Group [CCIRG], 1991, 1996). This means that in key areas like pricing and reinsurance, underwriting strategies are likely not reflecting the real risk of extremes. Already, vulnerable regions are evident on coasts and particularly

Climate Extremes and Society, ed. H. F. Diaz and R. J. Murnane. Published by Cambridge University Press. © Cambridge University Press 2008.

on deltas. Likewise, some sectors, such as residential property and agricultural buildings and crops, could be strongly affected by greatly increased risks.

A key source of evidence for the effects of climate change is the UK insurance industry, because of its wide insurance coverage and good statistical database, as well as its familiarity with overseas markets, through corporate subsidiaries and also reinsurance of external markets. Therefore, in the next sections I examine the vulnerabilities of UK insurers to climate change. The 2005 hurricane season was exceptional, creating substantial losses for British insurers and reinsurers, so that is discussed. I will also look at global trends using data from Munich Re. The subsequent sections consider projections of future losses, including those due to European storms, which are a prime cause of losses to the United Kingdom. The chapter concludes by reviewing other aspects of climate change, and finally, the future role of insurers in that issue.

13.2 Recent UK weather trends

The United Kingdom has the longest series of scientific weather records in the world (starting in 1659 for monthly temperature and in 1766 for precipitation – see the Hadley Centre website, www.metoffice.gov.uk/research/hadleycentre/obsdata/cet.html). This database gives us the ability to calculate long-term weather patterns with confidence and to assess the frequency of rare events. For insurers, it is these rare extremes that matter, as those extremes cause the largest amounts of property damage.

I first consider changes in the UK temperature and precipitation records. Any month in which the monthly temperature or rainfall fell in the top or bottom 10% of the frequency distribution, or decile, of the historical range for that month will have experienced some rather extreme days, so this decile threshold can be used to identify the months with extreme weather (hot, cold, wet, and dry). Since we are using the decile level to pick the months, on average there ought to be 12 such decile months in every decade (10% of 120 months) for each hot, cold, wet, or dry extreme. I use the data for the period before 1960 to define the climatological distributions.

The pattern of extreme months in recent decades in the United Kingdom is shown in Table 13.1 (bold for high values, italic for low values, defined as values outside the range 7 to 17, which correspond to the 5% and 95% confidence levels versus chance occurrences). For hot months, the 1960s were just below average, with only 10 hot months. Since then, the frequency has soared, and it has been running at nearly three times the expected rate since 1989, in fact. This is the highest level observed since records began. Cold months have been well below the expected level, and have now disappeared. The change to

Table 13.1. *Recent UK weather: extreme months since 1960[a]*

Number per decade (12 expected)	1960s	1970s	1980s	1990s	2000s (prorated to December 2006)
Hot	10	17	**18**	34	**36**
Cold	*5*	7	8	*3*	*0*
Wet	14	11	**19**	15	**20**
Dry	10	15	10	15	*4*

[a] See also Figure 2.1. The figures for the 2000s decade are estimated by scaling the observed occurrence of extreme months to December 2006 by the factor 120/84 to allow for the remaining months of the decade.

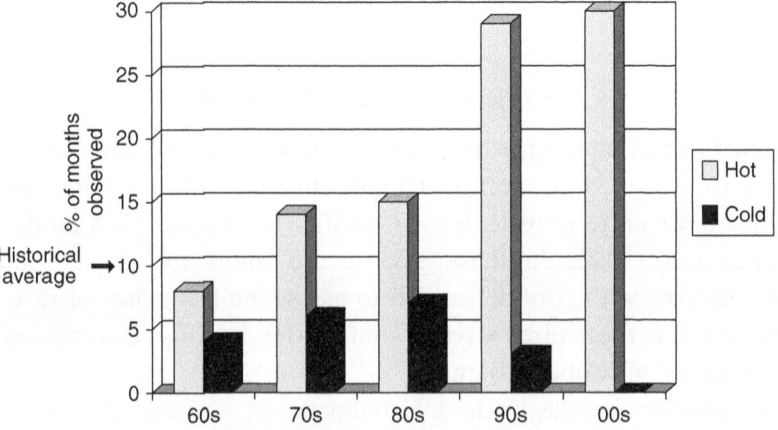

Figure 13.1. Recent trends in hot and cold months in the United Kingdom. Historical proportion of months per decade in hot (upper decile) and cold (lower decile) categories is 10%.

warmer winters is associated with more storms (the number of winter storms crossing the United Kingdom has doubled over the past 50 years [Hadley Centre, 2003]), more rain, fewer frosts, and faster thaws (CCIRG, 1991, 1996). For temperatures after 1990, these values would be seen by chance with a rate of 1 in a million for hot months and 1 in 4,000 times for cold months, respectively, confirming how very abnormal this weather is by historical standards (i.e., under the assumption of stationarity). Figure 13.1 displays the temperature data in an alternative format, such as might be used by underwriters, who use the concepts of occurrence probability, or its reciprocal, the return period of an event of a specified magnitude.

Wet months now run at double the customary rate (Table 13.1). This is particularly a winter phenomenon, important because it gives rise to floods, as

Table 13.2. *The new seasonality of precipitation, 1966–2005: frequency of decile months[a]*

	Q1	Q2	Q3	Q4	Total
Wet	20	18	8	19	65
Dry	7	14	17	4	42

[a] Twelve decile months are expected for each quarter over the 40-year period. Data from 1766 onward are used to calculate the decile values.

happened in 2000 and 2002. The total number of dry months has not shown any significant secular shift, although there has been a fall in the current decade. However, what matters from an insurance point of view is a succession of below-average months, rather than isolated very dry months. A prolonged dry period in summer leads to shrinkage of clay soil, with consequent damage to the foundations of buildings and claims for subsidence (Palutikof *et al.*, 1997; ABI, 1999).

The seasonality of precipitation is changing as well (Table 13.2). There are more dry months and fewer wet months in calendar quarter 3 (Q3; i.e., July, August, and September). At the same time, there are more wet months and fewer dry ones in quarter 1 (Q1; January–March) and quarter 4 (Q4; October–December). Quarter 2 (Q2; April–June) has seen a change to more variable precipitation, with an excess of both wet and dry months. Using a χ^2 test with three degrees of freedom, the seasonality of wet months in the period 1966–2005 is significantly different from historical patterns at the 99% level, and for wet months the change is significant at the 98% level of confidence.

If a dry winter (Q4 and Q1) precedes a dry summer, this creates a drought, which results in subsidence of clay soil and damage to buildings constructed in such soils. The concept of prolonged rainfall deficit can be quantified by using a drought index of precipitation for 18-month periods culminating in September of a year. There has been a shift in recent decades in the behavior of this drought index. For the period 1766–1965, the upper decile threshold is 1,551 mm, and the lower decile is 1,151 mm. In the past 40 years, there have been five occurrences at the high end, which is not statistically significant, although one of them (2001) was a new record and was associated with high numbers of flood claims in 2000. At the other extreme, there have been six occurrences of droughts since 1965, of which two were new lows (1976 and 1996), while three occurred in succession (1989, 1990, 1991). The sixth one occurred in 1997, adding to the effect of the 1996 low value. This result is suggestive of a change (see Figure 13.2).

A. Dlugolecki

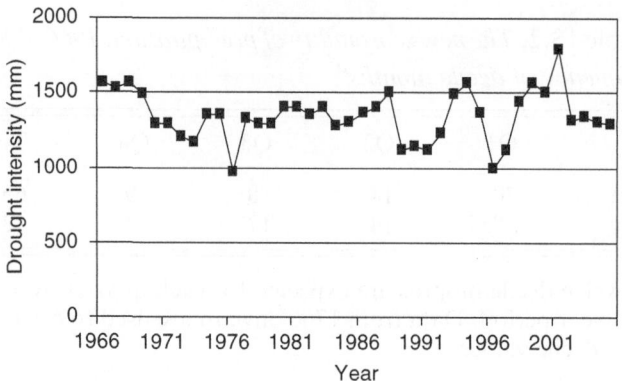

Figure 13.2. A drought index for the United Kingdom, 1966–2005. Drought intensity measured in 18-month accumulated precipitation to September of year (in millimeters).

13.3 UK property insurance experience

The UK insurance market offers a good basis for examining the effects of changing weather patterns, because the standard insurance policies provide very wide coverage for a range of weather hazards, including all types of flooding, like storm surges and sewer backups. Also, because the cover is "bundled" with non-weather perils such as fire and liability, the market penetration is high, unlike for the US system for flood insurance. Claims statistics gathered by the Association of British Insurers (ABI) cover about 95% of the domestic UK insurance market for property, but extend back to only 1988 on a comprehensive basis. Because drought is a slow-onset hazard, the subsidence claims relating to that hazard are examined separately after the "weather" claims from all types of storms and floods have been considered.

Climate impacts develop unevenly when they are looked at on an annual basis, because of catastrophes (see Figure 13.3). Experience shows that for the United Kingdom, storms with a wind speed in excess of 90 mph over urban areas produce catastrophic damage. This damage is compounded if there is a series of storms, because structures and trees weakened by the earlier storms may fail (Swiss Re, 2000; Munich Re, 2002). The massive spike in UK insured losses in 1990 was due to a 4-week period of storms commencing on Burns' Day, January 25. The fact that figures for individual years have been lower since then does not mean the risk is diminishing, because of the random occurrence pattern. There was another great storm in 1987, a near-miss in 1993, and another three near-misses in 1999; 2007 will almost certainly show an increase over 2006 due to European Storm Kyrill, which hit the United Kingdom in January 2007. This means that claims data at the national level are

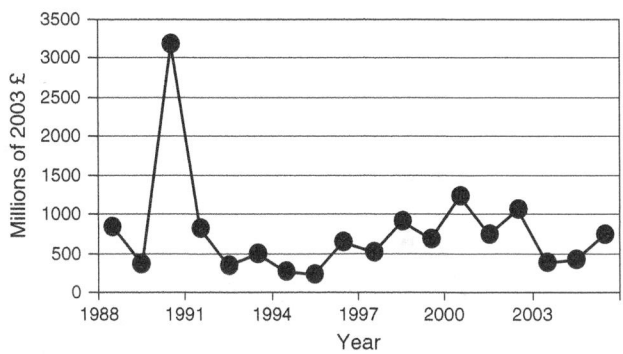

Figure 13.3. The ABI member companies' weather claims, 1988–2005 (excluding subsidence).

of limited value in identifying whole distribution trends in economic losses due to climate. Even at a global level, the question of whether there is a climate signal in economic loss data is contentious. Most recently, Muir-Wood *et al.* (2006) have reported a 2% per annum climate change trend in such data, but there is still debate (Pielke, 2006). However, if one discards the peak value of 1990, then a crude estimation of the underlying trend excluding disasters is possible. The sum of the costs from 1997 to 2004 is 48% higher than the costs from 1988 to 1996, excluding 1990. The annual average between the two periods rises from £500 million to £740 million, implying an annual rate of increase of 5.5% in non-disaster claims in constant monetary values. It is unlikely that this increase is mainly due to population growth and increases in real wealth in a relatively mature economy like that of the United Kingdom. It also seems inherently unlikely that there would be no change in weather claims patterns at a time when the weather itself is changing; i.e., the data appear to reveal a climate change signal, but we cannot estimate the scale of it precisely.

Subsidence claims respond strongly to drought because clay soils shrink as they dry. The drought index constructed in Section 13.2 is correlated (correlation coefficient $R = -0.52$) with losses from damage to structural foundations (see Figure 13.4). All of the precipitation lows have seen large numbers of subsidence claims. Equivalent claims data for the 1976 drought are not available, but that event was so severe that it resulted in amendments to the construction standards for building foundations, and prompted successful court actions by residents of damaged houses against neighbors whose trees had damaged the foundations by root invasion due to aggressive water-searching. There are of course other factors involved with subsidence claims, such as the state of the housing market. The relationship given here is only approximate, because most claims arise in the clay belt of mid- and southeastern England,

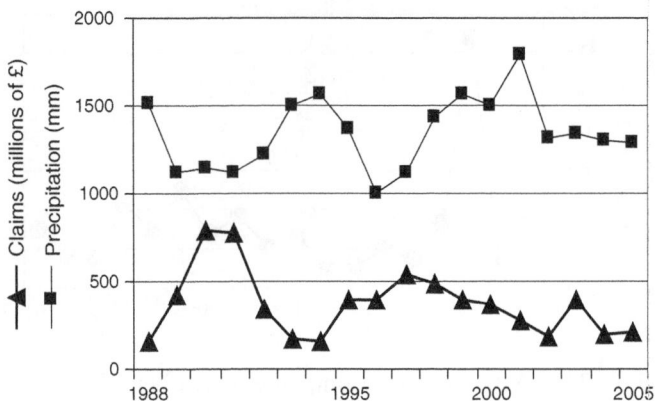

Figure 13.4. The UK household buildings subsidence damage and drought, 1988–2005. Subsidence damage measured in 2003 constant million pounds. Drought intensity measured in 18-month accumulated precipitation to September of year (in millimeters). Data supplied by ABI.

whereas the data for Figure 13.4 relate to all the United Kingdom for claims, and to all England and Wales for precipitation. If anything, the data suggest a downward trend in subsidence claims. This is not surprising, given the trend to wetter winters seen in Table 13.2, which militates against prolonged droughts.

13.4 Weather risk trends in the United Kingdom

The UK insurance industry has been willing to underwrite the broadest range of weather perils compared with other insurance markets. However, the prospective increase in UK weather damage may make such risks unacceptable for insurers. Two approaches indicate that the risk is already rising quickly, so that this could soon become a significant issue for the industry and its customers. The first approach is based on projections by the Foresight Programme (Foresight, 2004) in their recent work on flooding, while the second looks at the issue with an illustrative example for a reinsurer. It is important to remember that without reinsurance many risks would not be insured at all. This fact became evident when reinsurers withdrew from the insurance market for terrorism risk in 1993. The UK government had to step in with the creation of Pool Re to provide a backstop for the primary insurers.

13.4.1 The Foresight Programme's view of flooding in the 2080s

Table 13.3 shows the basic estimates of UK flood risk now and in the 2080s, as provided by Foresight (2004). The costs in the 2080s are based on four

Table 13.3. *Annual expected flood damage in the United Kingdom, for the present day and four socioeconomic scenarios of the 2080s*

		Socioeconomic scenarios of the 2080s			
Basis	Present day	World markets	National enterprise	Local stewardship	Global sustainability
Annual expected damage (billion pounds)	1.3	28.4	20.7	2.2	6.7

Source: Foresight (2004; Part II, Tables 9.1 and 9.2).

scenarios, or storylines, of how society might develop, based on how far societies will converge in behavior, and on the balance they reach between economic and environmental priorities. The four scenarios are:

- World Markets – a world of rapid economic growth, population that peaks in mid-century then declines, and more efficient technologies.
- National Enterprise – a very heterogeneous world with increasing population and economic growth more fragmented and slower.
- Local Stewardship – a convergent world with the same population as in World Markets but with a rapid shift to a service and information economy, and "clean" technologies.
- Global Sustainability – a world in which the emphasis is on local development, with intermediate population and economic growth.

It should be emphasized that these scenarios are noninterventionist; i.e., they assume that no action is taken on account of climate change in flood defense or planning policy. Thus they give a true view of how the underlying risk will behave. Of course, society will be able to take precautionary measures (at a cost) to control the future damage, but the inherent risk will still exist.

The increase in flood damage for Local Stewardship is £0.9 billion, for a 69% increase over the 80-year period, but this scenario seems almost utopian in that it requires a reversal of current trends in consumption and globalization, and the estimated damage lies well away from those for the other three projections. It has been ignored for the purposes of these calculations. The increase in flood damage between now and the Global Sustainability scenario is an increase of 415%. If the underlying process is uniformly continuous, reflecting the nature of the predictive curves for temperature and sea level rise in reports of the Intergovernmental Panel on Climate Change (IPCC), this implies an annual change of 2.1%, compounded for 80 years. The increase in damage between now and the World Markets scenario is an increase of over

20 times the current amount of damages. To generate this would require an annual increase of just over 3.9%. The calculation for the National Enterprise scenario is an annual increase of just over 3.5%.

Thus on the basis of the Foresight Programme view of future flood risk, the risk of flood damage is already increasing by between 2.1% and 3.9% per annum, or in round terms a range of 2% to 4% per year. In their analysis, they include a number of other factors, but climate change is a very substantial cause. They do not present separate contributions for each cause, but if we allow half of the increase for non-climatic factors, then the "pure" flood risk escalates at about 1% to 2% per annum. It is well understood in the insurance industry that when extreme events occur, the costs often increase far beyond the normal scale, owing to the compounding effect on the capacity of the repair and recovery resources and processes. This increase in cost is often termed "demand surge." This was seen after Hurricane Katrina and after the European storms in 1999. For that reason, this author believes that it is appropriate to retain the risk escalation factor at 2% to 4% per year.

13.4.2 Illustrative reinsurance example

Reinsurers tend to work on very focused risk portfolios, because they are dealing with risks that are unacceptable to primary insurers. The example below illustrates how such risks could escalate in the future and become unacceptable, even to reinsurers.

Suppose that in 2000 the reinsurer assessed there were three possible event outcomes: normality, an extreme event (defined as a 1-in-100-year probability), and a catastrophe (defined as having a 1,000-year return period). Swiss Re's current estimates for inland flood and windstorm in the United Kingdom are shown below (Table 13.4). If the reinsurer ignores the possibility of other events (i.e., slightly smaller than the 100-year flood, or between the 100- and 1,000-year flood, or greater than the 1,000-year event), then it would assess the annual expected claim cost, or risk premium due to inland flood, as

Table 13.4. *Estimated UK event costs*

Event type	Return period	UK inland flood	UK storm
Normal	1 year	£0.02 bn	£0.12 bn
Extreme	100 years	£1.7 bn	£10 bn
Catastrophe	1,000 years	£3.9 bn	£24 bn

Source: Swiss Re (personal communication).

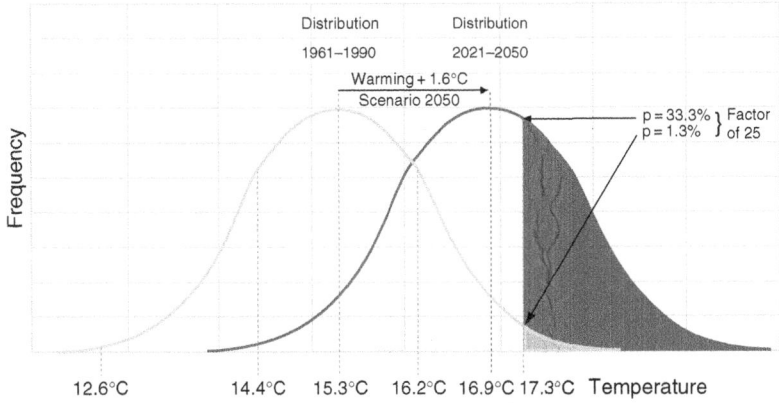

Figure 13.5. Change in distribution of UK summer temperatures: effect on extremes. (Source: CCIRG, 1991.)

£20.9 million (1% of £1,700 million, and 0.1% of £3,900 million). Of course, a single reinsurer would not be presented with the risk for the entire United Kingdom, but the same logic applies for fractions of the UK risk.

However, the balance of these events will be disturbed by climate change. Judging what the change will be is problematic, because the current generation of climate models is not well suited to dealing with extreme weather events like storms, which are, in climate-science terms, rather small-scale even though their impacts are large. However, for heat waves and intense rainfall floods, there is a growing body of evidence indicating that the frequency of today's extremes will be much higher in the future, perhaps 4 times greater (Easterling *et al.*, 2000; Huntingford *et al.*, 2003; Meehl and Tebaldi, 2004). Furthermore, this escalation factor rises rapidly as one raises the definition of "extreme", owing to the shape of the tail in statistical distributions (CCIRG, 1991, 1996; see also Figures 13.5 and 13.6).

These results might lead the reinsurer to assess that the return periods for the events in Table 13.4 will in 2050 shift to 25 years for current "extremes," and because of the escalation factor, to 100 years for "catastrophes." The scale of these changes in return period is consistent with the UK Climate Impacts Programme (UKCIP) suggestions for future flood risk (Hulme *et al.*, 2002). On that basis, the annual risk level rises to £107 million (4% of £1,700 million, and 1% of £3,900 million), an increase of 412% over a 45-year period, or roughly 3% per year – which is in the middle of the 2%–4% range based on the Foresight projections. This of course does not allow for changes in population, changing values or types of property, or other relevant factors like flood defenses – it is simply a reassessment of the raw climatic risk.

A. Dlugolecki

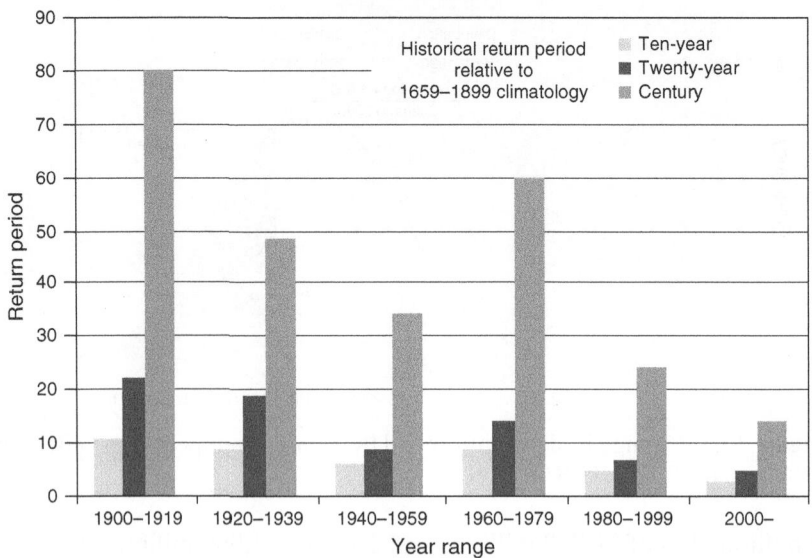

Figure 13.6. The trend to more extreme conditions: UK temperatures.

13.4.3 Potential range of increase in risk premium

The approaches discussed above support a range of 2% to 4% as an annual increase in risk premium, based on consideration of the potential weather hazard and potential changes in exposure. However, experience has shown that risk increases for other reasons also, such as vulnerability of materials and human behavior. There are no reliable measurements of these factors, but it is significant that the cost of natural disasters had been rising rapidly for several decades (see Section 13.6) before the influence of climate change began to appear. Clearly, therefore, this factor is significantly greater than zero. We can conclude that a conservative estimate of the underlying increase in weather risk premium taking account of changing climate and other factors is 2% to 4% per annum.

13.4.4 Implications for underwriting

An attitude that comes up repeatedly among underwriters is that recent extreme events are simply a feature of normal climatic variability. This conclusion is reinforced by the view that since property insurance policies are annually renewable, it would be easy to extricate oneself from a deteriorating risk situation before permanent damage was done to the balance sheet. But we are in a dynamic situation, which produces very rapid change in the likelihood of extreme events. The expected change in temperature distribution for UK summers due to climate change, as estimated by the Climate Change Impacts Review Group using the output of multiple general circulation models (GCMs), is shown in Figure 13.5. A hot summer, like that of 1995, with an average temperature of 17.3 °C, was likely to occur about once in 75 years, or 1.3% of the time on the climatic pattern of the period 1961–90, when the average was 15.3 °C. For the period 2021–50, centered on 2036, the average temperature will be 16.9 °C. On that basis, the chance of exceeding a temperature like in 1995 will be 33.3%; i.e., 1 year in 3! This implies a change in return period of 25 times, and the rate of change will be faster still for less frequent events. Clearly, this process is already under way, and it is reasonable to suppose that the rate of change is constant, because that is broadly the pattern of evolution over time for average temperature that is revealed by GCMs. One might object that temperature is not a critical variable for insurers; in fact, recent evidence on hurricanes in the Atlantic shows that rapid increases in intensity have occurred there and that sea surface temperature (SST) plays a key role (Knutson and Tuleya, 2004; Webster *et al.*, 2005; Elsner, 2006; Emanuel, 2006; Trenberth and Shea, 2006). An increase in wildfires is another cause of concern to insurers with increasing temperature (Allianz and the World Wildlife Fund [WWF], 2006; Westerling *et al.*, 2006).

The difference in time between the midpoint of the time spans from which the two distributions in Figure 13.5 are derived is 60 years. To produce a 25-fold change within 60 years implies an annual rate of change of about 5.5% per year. Such a rapid change, if ignored, soon accumulates into a significant error. In 5 years, it means that return period calculations for the event would be 30% too low. It is difficult to detect such shifts in distributions with statistical confidence. Nevertheless, such shifts suggest that one can expect "surprises" to start occurring: there are many potential low-probability events, so some of them do start to occur "too often." (Of course, the reverse happens at the other end of the distribution, where events do not happen as often as they should.) The implied figure of 5.5% per year supports the position that a conservative estimate is that the risk of weather events in the United Kingdom is changing at 2%–4% per year.

However, because we are dealing with a multidimensional, nonlinear system, the shifts can compound across more than one factor to produce very unexpected costs – for example, inland and marine flooding at the same time such as for Hurricane Katrina. Other factors that could raise costs are pressure to make *ex gratia* payments, repair price inflation due to scarcity of supplies, close repetition of events (e.g., Storms Lothar and Martin in 1999 [Munich Re, 2002], the 1990 European storms [Munich Re, 1990], or the 2004 hurricane and typhoon season [Munich Re, 2005]), and denial of access or failure of utilities, as in New Orleans in 2005 (Risk Management Solutions [RMS], 2005; Towers Perrin, 2005; Munich Re, 2006a). More frequent events increase the chance of a weather event coinciding with an uncorrelated event; e.g. an earthquake or economic catastrophe. Such a coincidence of events almost happened in the United Kingdom, when the 87 J storm arrived without warning before dawn on Saturday, October 17, 1987, just as a global stock market crash was starting. Fortunately, the fact that it was a weekend allowed emergency communications to be installed so that the London financial markets could operate on Monday, October 19.

The tendency towards more frequent extremes is already evident in the case of UK temperature. The advantage of using the UK data is that we can use actual observations rather than inferring what the frequencies were from a probability distribution. Monthly temperatures that had a return period of 10 years before 1900, now happen every 3.4 years, or 3 times as frequently (Figure 13.6). However, monthly values that happened once a century, now recur every 14 years: they are 7 times as frequent as they were previously. The graph also shows that the shortening of return periods is generally progressive, with a reversal during the period 1960–79, when global temperature increase decelerated and even reversed in some regions.

The implications of these findings for insurers and reinsurers are many.

- Risk assessment will be wrong, and prices consequently will also be wrong. If historical data are used to set prices, the error might be in the region of 20% for "typical" weather risks, and much more for catastrophic risks.
- The aggregate exposure to loss will be too large for the available insurance capital.
- Reinsurance plans may be inaccurate, and reinsurers themselves may be taking on too much risk.
- Estimates of the risk of insolvency through disasters will be incorrect; e.g., the creditworthiness of insurers will be assessed too generously.

13.5 2005 in perspective

The hurricane season of 2005 may cost insurers US$60 billion, more than double the amount for any previous year. The true cost is much greater – possibly US$450 billion including uninsured losses and property blight, and the global repercussions of higher oil prices and lost economic production (Kemfert, personal communication) – not to mention the over 1,300 dead and thousands of people traumatized. Post-Katrina issues include coverage disputes with the State of Mississippi, the cost and quality of the new levees, land zoning, business interruption, and long-term relief payments. The effects of the hurricanes linger on locally. New Orleans may never recover (as happened to Galveston in 1900), and the defenses were weaker in 2006 than in 2005.

Before Katrina, analyses of the historical record suggested that tropical cyclones were becoming more intense and frequent (Emanuel, 2005; Webster *et al.*, 2005). The records broken in the 2005 Atlantic hurricane season seem consistent with this observation: most storms (27), most hurricanes (13), most strong US landfalling hurricanes (4), costliest in total (US$200 billion +), single most expensive (Katrina), and strongest ever (Wilma). The Atlantic season also had three of the six strongest hurricanes ever recorded (Wilma, Rita, Katrina) and continued a very active phase (most storms, hurricanes, and strong hurricanes ever in a 2-year or 3-year period). However, there are problems with the observational record, and other authors suggest that the trends in the observational record are an artifact or that it is premature to ascribe them to climate change (Landsea, 2005; Pielke, 2005a; Pielke *et al.*, 2005; Trenberth, 2005; Klotzbach, 2006; Knight *et al.*, 2006; Michaels *et al.*, 2006; Zhang and Delworth, 2006). Nevertheless, evidence about a positive relationship continues to accumulate (e.g., Elsner, 2006; Hoyos *et al.*, 2006).

In terms of relevance to insurers, anthropogenic climate change will make tropical storms more intense (i.e., stronger winds, heavier rain, and higher sea-surges (Knutson and Tuleya, 2004; Emanuel, 2006; Munich Re, 2006a). In the Atlantic this pattern will possibly be reinforced by natural cyclicity (the Atlantic Multidecadal Oscillation, or AMO) over the next two decades (Goldenberg *et al.*, 2001; Chelliah and Bell, 2004; see also Mayfield, 2005). This author is skeptical of the argument of cycles (e.g., Goldenberg *et al.*, 2001) for two reasons. First, the data required to estimate a cycle length are roughly ten times the cycle, and we do not have such a long time series for hurricanes. An attempt has been made to sidestep this problem with a simulation over 1,400 years (Knight *et al.*, 2006); the model generates a cyclical AMO pattern, which is correlated with various climatic phenomena. However, this result must remain conjectural without prehistoric analogue observational data. Second, there is a tendency to view climate change simplistically as a mono-tonic increase in temperatures. In fact, we know that in the third quarter of the twentieth century a cooling effect occurred, due mainly to the emission of large amounts of industrial pollutants that suppressed warming on a regional scale. In addition, Mann and Emanuel (2006) showed how a combination of surface temperatures in the Northern Hemisphere and regional cooling associated with anthropogenic aerosols are highly correlated with the North Atlantic sea surface temperatures.

On the other hand, it could be argued that we have seen a series of large losses before. The insurance market reacted strongly to a series of great storms

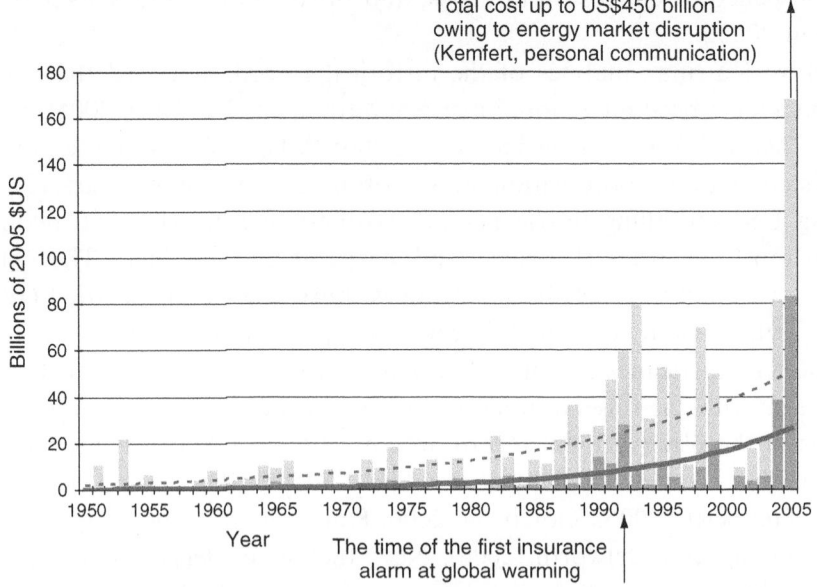

Figure 13.7. Adjusted cost of great weather-related disasters, 1950–2005. Dark color is insured cost; light color is uninsured. Trend line of insured costs is shown by a solid dark line; trend line of total costs is shown by a broken dark line. Supernormal cost of Katrina and Rita is shown by black arrow. (Source: Adapted from Munich Re, 2006b.)

between 1987 and 1992 (see Figure 13.7). However, the magnitude and number of events this time appears to be on a new level.

Munich Re has a more conservative view than Kemfert (personal communication) regarding the economic costs of Katrina and Rita, but it believes there is a watershed regarding insurable costs following the 2004 and 2005 hurricane seasons. An illustration of this change (Figure 13.8) indicates that Munich Re believes that the loss curves have shifted by almost a factor of 2, for a number of reasons (Munich Re, 2006a). These views are shared by many expert commentators (e.g., RMS, 2005; Towers Perrin, 2005).

This author believes that once damage reaches a certain level, the effects escalate dramatically, owing first to their longevity: the recovery is more extended, with consequent vulnerability and greater indirect costs like provision of temporary accommodations. The second reason cited is their greater geographic extent, through effects like migration and global market disruption. This effect might be termed "supersaturation": a change of state when the usual conditions in which a system exists are breached (Figure 13.8). We saw this happen in 2005 with Hurricane Katrina. Some economists have dismissed the likely costs of climate impacts as acceptable, around 0.2% or less of the

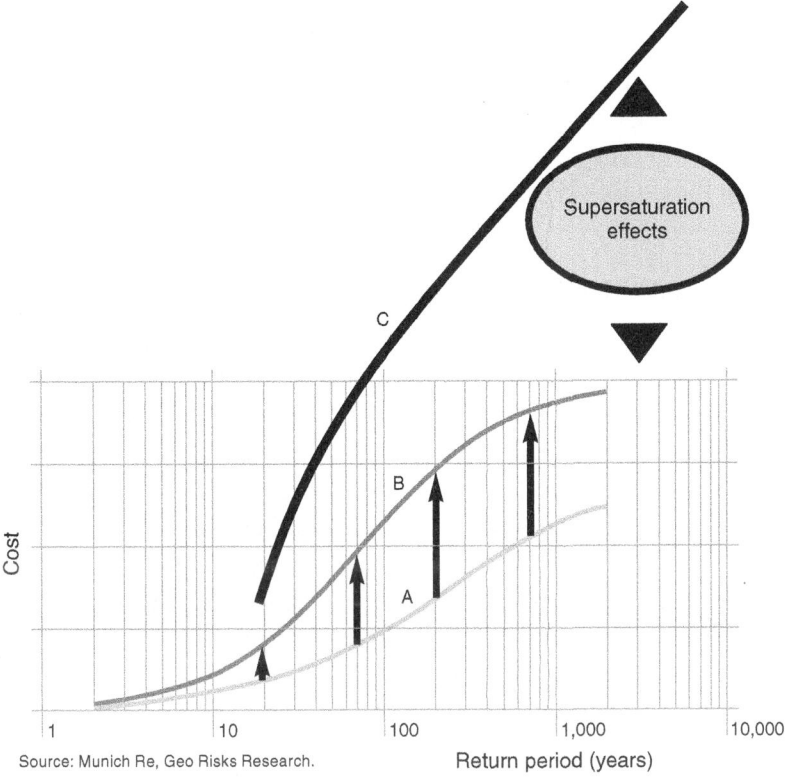

Cost

1   10   100   1,000   10,000

Return period (years)

Figure 13.8. Shift in damage curves after 2005 hurricane season. (Adapted from Munich Re, 2006a.)

global gross domestic product (GDP) by mid-century (Nordhaus and Boyer, 1999; Mendelsohn *et al.*, 2000; Tol, 2002), although Stern (2006) suggests that these rates are improbably low. In fact, Katrina may have cost 2% of the US GDP in 2005 already (Kemfert, personal communication). For New Orleans it appears to be a turning point; tens of thousands of the population may never return, and new defenses may cost more than US$32 billion (Schwartz, 2005).

In super-catastrophes like Katrina, losses become nonlinear, i.e., the scale of the event itself causes losses to increase further, through a variety of processes (see, e.g., RMS, 2005). Indeed, costs might have been even greater if Katrina had not weakened before landfall, or if Rita and Wilma had not veered away from New Orleans. After Hugo in 1989, roughly a dozen insurers were financially overwhelmed. Generally, these were smaller regional insurers that had not purchased sufficient reinsurance. After Andrew, nine insurers failed. Even some very large carriers were stressed to the limit. There have been no reports of insurance company failures as a result of Katrina (Towers Perrin, 2005), though ironically two airlines have filed for bankruptcy due to high fuel prices associated with the loss of production and refining facilities (RMS, 2005). This outcome reflects the fact that the insurance industry is much more alert to the possibility of catastrophic losses; for example, Lloyd's of London now has a specified range of catastrophe scenarios that its underwriters must use to stress-test their viability, and credit rating agencies follow the same approach in assessing the quality of insurance equities for investors.

13.6 Munich Re estimates of climate-related losses

Munich Re has been compiling statistics on natural disasters for many years because they illustrate the need for risk management. Their definition of what are "major natural catastrophes" follows the criteria laid down by the United Nations: the affected region's ability to help itself is distinctly overtaxed, interregional or international assistance is necessary, thousands are killed, hundreds of thousands are made homeless, and there are substantial economic losses and/or considerable insured losses. In this section we exclude those disasters like earthquakes and volcanoes, which are not weather related. In fact, we shall focus on the number of incidents, rather than on insurance claims, because the latter are a rather variable proportion of the total losses. Even the total losses depend on the vulnerability and wealth of the areas impacted, so they too are rather variable.

Table 13.5 shows great weather-related disasters for the period 1950–2005. The numbers rose until the 1990s, but now appear to have reversed somewhat. There has been a global increase in the cost of weather disasters, paralleling the UK picture. Of course, there are many reasons for this trend, including increased volumes of assets located in more hazardous areas. The insured losses show a very strong upward trend, which reflects the fact that as economies develop, the penetration of insurance also grows.

The number of flood events has fallen back, after very high levels in the 1980s and 1990s, but the costs have risen enormously, both in pure economic terms and in insured value, reflecting the incidence of flooding in wealthier

Table 13.5. *Great weather-related disasters, 1950–2005*

	1950s	1960s	1970s	1980s	1990s	1996–2005
Number	15	16	29	44	74	44
Economic loss (billion dollars)	47	63	89	142	477	480
Insured loss (billion dollars)	2	7	14	26	110	175

Source: Munich Re; monetary data in constant 2005 values.

Table 13.6. *Great flood disasters, 1950–2005*

	1950s	1960s	1970s	1980s	1990s	1996–2005
Number	6	6	8	18	26	12
Economic loss (billion dollars)	34	24	22	31	254	127
Insured loss (billion dollars)	—	—	1	2	9	8

Source: Munich Re; monetary data in constant 2005 values.

Table 13.7. *Great windstorm disasters, 1950–2005*

	1950s	1960s	1970s	1980s	1990s	1996–2005
Number	8	10	19	21	42	28
Economic loss (billion dollars)	12	39	56	59	201	328
Insured loss (billion dollars)	1	6	13	23	90	161

Source: Munich Re; monetary data in constant 2005 values.

countries like the United Kingdom and Germany (Table 13.6). The peak in the 1990s reflects the fact that some major events happened in the United States, where there are very large exposures; e.g., the 1993 Mississippi floods. The insured losses were relatively low because insurers (apart from the United Kingdom) do not generally cover this risk.

The number of windstorm events has trended strongly upward, with a peak in the 1990s (Table 13.7). The costs have risen very sharply, mainly due to hurricanes and typhoons.

13.6.1 Parallels with the United Kingdom

The increase in global economic losses from weather recorded by Munich Re is broadly consistent with the ABI data for the United Kingdom only. (It is better to take the economic cost, rather than the insured cost, for the global losses, because practice varies so widely in the use of insurance as a compensation vehicle.) Between the 1980s and the 10-year period 1996–2005, the cost in constant currency for global economic losses rose from US$142 billion to US$480 billion, an annual rate of increase of 8% over 16 years. During the same period in the United Kingdom, the annual increase was about 5.5% in the constant-value cost of "weather-related" claims; i.e., excluding subsidence. Given the relatively less dynamic nature of the United Kingdom's economy, and the absence of any major UK storms since 1990, this figure is consistent with the global trend. It is interesting, too, that the global figures exhibited a strong uplift in flood damage over that period, as has happened in the United Kingdom. The Munich Re statistics are corroborated by Swiss Re (see, e.g., Swiss Re, 2006a). This observation suggests that the UK experience is part of a general pattern.

Powerful non-climatic factors are driving up the potential for disasters, and magnifying the costs of them when they occur. Section 13.7 considers this multi-factor trend to see whether a climatic element can be distinguished.

13.7 Loss trends and projections

Numerous papers have argued that there is really no perceptible climatic factor in the upward trend in losses shown by, for example, Munich Re – that it is just due to socioeconomic factors, not to increases in climatic extremes (Pielke and Landsea, 1998; International Ad Hoc Detection and Attribution Group [IAHDAG], 2005; Pielke, 2005a,b). Recently, RMS has been examining global loss trends. Preliminary results seem to show that climatic factors are important. Losses in the 1990s were much worse than in previous decades, and this pattern has resumed again after a short lull in the early 2000s. This condition is not conclusive, since various factors cloud the issue, such as the effectiveness of disaster preparations and the wealth of the affected regions (due to its wealth, the United States has a disproportionate effect on the total losses), but it does mean that one of the standard objections to using loss statistics as an indicator of climate change is weakening (Muir-Wood *et al.*, 2006).

Studies of the potential net effect on the United States of climate change have produced estimates of benefits rather than losses (Mendelsohn *et al.*, 2000; Tol, 2002). This outcome seems rather optimistic now – the 2005 hurricane season may have cost 2% of the US GDP, with international repercussions beyond that; e.g., high winter fuel costs in the United Kingdom (Barclays Private Bank, 2005; Skrebowski, 2005; Lee, 2006). Recent works (Epstein and Mills, 2005; Mills *et al.*, 2005) in the United States have not provided robust figures for future loss potential, but have merely indicated that on the basis of current trends, the losses demand a coherent response from government and the private sector. Choi and Fisher (2003) provide an example linking climate change to losses and find that after correcting for wealth, population, and location, a 1% increase in precipitation results in about a 2.8% increase in US catastrophic losses.

Results for three out of four climate scenarios for the United Kingdom imply an annual increase of 2% to 4% in the cost of flood damage, which will have a large impact on the medium- and long-term planning of infrastructure

(Section 13.4.1 and Table 13.3). A preliminary study by this author for the Association of British Insurers indicated that future climate-related insurance claims in the United Kingdom might be 2–3 times higher than current levels by 2050, assuming no change in government policy on climate change adaptation (ABI, 2004b). One of the main uncertainties in this calculation is the future frequency and severity of extreme climate events, because climate models do not yet provide a consistent estimation of future storm tracks and intensity. This is a key weakness: in the United Kingdom, the cost of a 1,000-year extreme climate event is roughly 2.5 times larger than the cost of a 100-year event. In Germany, insurance claims increase as the cube of maximum wind speed, or even a power relation of the fourth or fifth degree, because of collateral damage from debris (Munich Re, 2002). A recent study of UK and German storm losses estimated an increase in insured losses by 2080 (without adaptation measures) of 37% across four GCMs, and up to 80% in one case (Leckebusch *et al.*, 2007).

The Association of British Insurers (2005) estimated the increased insurance costs of hurricanes (United States), typhoons (Japan), and winter storms (Europe) due to climate change as around two-thirds by the 2080s, keeping other factors constant, to a new annual average of US$27 billion compared to US$16.5 billion today. Losses during extreme seasons would be worse, around 75% higher than currently due to the nonlinear damage curve as wind speeds increase. These calculations may perhaps be on the low side, especially for the United States, as they came before some of the more recent papers on observed increases in hurricane intensity (see discussion in Section 13.5 above). The ABI (2005) also made a cursory estimate of future European flood costs, but that figure should be disregarded as it simply assumed European floods would parallel the UK trend.

The impact of more intense tropical cyclones as a result of climate change could have a significant impact on the insurance industry. Under high-emissions scenarios (where carbon dioxide levels double), insurers' capital requirements could increase by over 90% for US hurricanes, and by about 80% for Japanese typhoons (ABI, 2005). In total, an additional US$76 billion could be needed to bridge the gap between extreme and average losses resulting from tropical cyclones in the United States and Japan. Higher capital costs combined with greater annual losses from windstorms alone could result in premium increases of about 60% in these markets. These loss estimates do not include likely increases in society's exposure to extreme storms, due to growing, wealthier populations and increasing assets at risk. For example, if Hurricane Andrew had hit Florida in 2002 rather than in 1992, the losses would have been double, due to increased coastal development and rising asset values.

Expected losses could be significantly reduced through improved building codes and better regional planning. Strong and properly enforced building codes have been shown to prevent and reduce losses from windstorms. If all properties in south Florida were built to meet the strongest local building code requirements, damages from a repeat of Hurricane Andrew would fall by nearly 45%. If design codes for buildings in the southeastern United Kingdom were upgraded by at least 10%, increases in climate-induced damage costs from windstorms could be reduced substantially (ABI, 2003). In the United Kingdom, taking account of climate change in flood management policies, including controlling development in floodplains and increasing investment in flood defenses, could limit the rising costs of flood damage to a possible fourfold increase (to US$9.7 billion or £5.3 billion) rather than 10- to 20-fold by the 2080s (Foresight, 2004). The ABI is campaigning actively on this, owing to concern that government expenditure on coastal defenses is declining in real terms. The ABI points out that by 2080, the annual probability of a storm surge of 1.45 m will rise from 2% at present to 50% (Hulme et al., 2002) and that many investments in flood defense can yield a benefit/cost return of 7 times (ABI, 2006).

13.8 European storms

This author reported a striking correlation between winter temperatures and the occurrence and strength of great European storms, using the Central England Temperature (CET, as measured by the Hadley Centre) as a proxy for temperature, and Lamb's (Lamb, 1991) record of great storms (Chartered Insurance Institute [CII], 1994). The data in Table 13.8 show that storms are twice as frequent in warm months as in medium months, and 150% more common than

Table 13.8. *Great European storms and winter temperatures, 1690–1989[a]*

Type of winter month	Storm frequency	Storm intensity
Warm	15%	2,568
Medium	7%	2,544
Cool	6%	1,075

[a] Temperatures from CET series at Hadley Centre; storms data from Lamb (1991). Storm intensity is Lamb's Storm Severity Index; i.e., the product of (cube of maximum wind speed in knots, hours duration of damaging winds, and area of damage, in hundred thousand square kilometers). Warm and cool months are defined as being in the top or bottom decile of monthly average temperatures observed during the period 1690–1989 (CII, 1994).

in cool months. The strength of the storms is also much less in cool months than in warm or medium months. This hypothesized correlation has met several objections. First, some meteorologists dismiss the correlation because it relies on a subjective catalog rather than on instrumental observations of storm strength. However, it is currently the only long time series, and the findings seem quite robust. A second objection is that CET winter months are warm when the atmospheric flow is westerly, which is conducive to storms, rather than that warm temperatures cause storms; i.e., the relationship may be correlation rather than causation. Lamb did not notice the correlation between warm winters and storm strength, but did note that in 1990 (which fell after his record formally ended) the United Kingdom experienced great storms and a very warm winter. Since then, there was the great 1993 storm – probably the deepest European depression since records began, which destroyed the tanker *Braer* off Shetland – and the storms of 1999 and 2005. As the current chapter is being concluded, the winter of 2006–07 is also very mild, and has been characterized by stormy conditions in Europe. Therefore, it seems that these circumstances may be worthy of further investigation as an indicator of future conditions.

There is uncertainty about the future climatology of winter storms in Northwest Europe (Woolf and Coll, 2006). However, "under anthropogenic climate change the number of extreme storms could increase" (Leckebusch *et al.*, 2007). The locations of future increases in intensity and frequency are dependent on the GCM used, but the levels of change are material (up to $+15\%$ and $+200\%$ for intensity and frequency, respectively). A recent development is that stakeholders like the insurance industry are combining the results from climate models (GCMs and regional climate models [RCMs]) with those of proprietary catastrophe loss models to estimate future losses from European storms under climate change. Swiss Re (2006b) found an increase in losses of 44% for western Europe between 1975 and 2085, ranging from 16% to 68%, based on results of several climate models run by the Swiss Federal Institute of Technology; these results showed a notable bias towards more severe events, and a systematic range in the impacts between countries (from 115% in Denmark and Germany, down to 35% in the United Kingdom. The World Wildlife Fund (2006) interpreted the results from the peer-reviewed findings of the European Union-funded Modelling the Impacts of Climate Extremes (MICE) and Prediction of Regional scenarios and Uncertainties for Defining EuropeaN Climate change risks and Effects (PRUDENCE) projects with respect to storm severity and frequency for various European countries (e.g., Leckebusch *et al.*, 2006). The worst prognosis is for the United Kingdom, with an increase of up to 16% in wind speed and 25% in frequency by the 2090s; these results imply that the impacts will rise by up to 95%, allowing for the cube effect of wind speeds. Northern France and the Netherlands also face substantial increases. (This makes no allowance for adaptation actions, nor for increased exposure.) The ABI (2005) found a lower increase of wind-related insured losses from extreme European storms, of at least 5% by 2080, but this figure did not reflect more recent research on storm severity. A further source of concern is that European storms arrive in clusters, as happened in 1990 and 1999 (Mailier *et al.*, 2005), which are more costly to cope with than well-spaced events. This is systemic, but is not easy to predict, as it appears to be driven by five underlying features, one of which is the North Atlantic Oscillation (NAO).

13.9 Other issues

In this section, I briefly consider the effect of climate change impacts on the remaining classes of property/casualty insurance – i.e., other than property, and also how mitigation policy will affect insurers' underwriting and investment activities.

13.9.1 Other property/casualty classes

Other property/casualty classes may be affected by shifts in extremes – e.g., business interruption, automobile, and travel – but this author does not believe that the industry faces the prospect of a wave of climate change-related liability claims.

Automobile

In the United States, 16% of automobile accidents are attributed to adverse weather conditions, as are one-third of the accidents in Canada. Vehicles also sustain insurance losses during natural disasters, amounting to US$3.4 billion between 1996 and 2000 and averaging 10% of all disaster-related property losses. In some events, such as hailstorms, damage to automobiles can exceed 50% of total catastrophe losses (Mills *et al.*, 2005). In Hurricane Katrina, one estimate reckoned that about half a million vehicles had been damaged, mainly by floodwater, with about half of them insured. The insured costs alone fell in the band US$1.2 billion to US$2.3 billion, or between 3% and 4% of the total insured losses (Towers Perrin, 2005). While this is normally a major loss, it was so dwarfed by the concurrent buildings damage that another estimate ignored automobile entirely (RMS, 2005).

In less severe climates, such as in the United Kingdom, automobile claims are correlated with meteorological conditions, with dry/warm weather seeing fewer accidents reported, and cold/wet being the opposite. The exception to this pattern is that in very severe winters, motorcycle claims diminish owing to their drivers' greater awareness of the dangerous conditions. However, the distribution of accident types changes, with many more minor ones in icy conditions, so historically the United Kingdom is not exposed to major catastrophes in this class.

Agriculture

Another obvious impact area is agriculture, particularly crops, but the private market avoids this area, so it is hard to comment on trends.

Health/life

The effects of changes in extremes could extend into insurance industry products for life and health and pension coverage, albeit probably less strongly than in property/casualty, since privately insured people are wealthier than average and generally have better health and access to medical care. The latest estimate of excess deaths in the European heat wave of 2003 is about 50,000 (Kosatsky, 2005), but these deaths had a minimal effect on insurance markets. Similarly, the 1,300 deaths in Katrina had little effect on US life insurers. In the United States, there has been a major program of research into climate change and human health: the Climate Change and Human Health, National Health Assessment Group (NHAG), at Johns Hopkins University. Five areas were examined: heat stress, extreme events, air pollution, water-borne/food-borne disease, and vector-borne/rodent-borne disease. Among the findings were that heavy precipitation is strongly linked to outbreaks of water-borne disease; 58% of outbreaks were preceded by a rainfall event in the top decile, and 68% of outbreaks by events in the top two deciles. Water contamination was also linked to extreme precipitation, but with a greater lag effect (NHAG, 2001).

Developing countries

Insurance products have very low penetration in less developed countries, where the impacts of climate change are expected to be most acute, due to the greater vulnerability of those regions to extreme events. As an example, consider Asia. Glaciers in the Himalayas are receding faster than in any other part of the world, and many may disappear by the year 2035, with catastrophic results for rivers in India, China, and other countries in this part of the world (Hasnain, 2002). Six mega-cities in Asian deltaic regions will have populations exceeding 10 million by 2010. These deltas are shrinking due to sediment starvation; e.g., the Changjiang sediment discharge will fall by 50% after construction of the Three Gorges Dam. For a 1 m rise in sea level, half a million square hectares of Red River delta and up to 2 million square hectares of Mekong River delta are projected to be flooded. The deltas are also usually economically more developed. The GDP of the three metropolitan areas located in the Zhujiang delta, Changjiang delta, and Huang He delta will represent 80% of China's total GDP in 2050 (Sit and Cai, 2004). The current rate of sea level rise in coastal areas of Asia is reported to be between 1 mm and 3 mm per year (IPCC, 2007), marginally greater than the global average, and it is accelerating gradually. Clearly, refugees from such regions would disrupt neighboring regions also, besides placing a burden on global society (IPCC, 2007).

13.9.2 *Mitigation policy*

Insurers will also be touched by climate change through government mitigation policies. Policy changes will alter the economics of energy consumption, and these effects will have subsequent impacts on energy technologies and investment returns in a wide range of industries (beside being underwriters, insurers are also major investors; Dlugolecki *et al.*, 1995; CII, 2001; Vellinga and Mills, 2001; United Nations Environment Programme Finance Initiative [UNEPFI], 2002; Dlugolecki and Loster, 2003; Dlugolecki and Lafeld, 2005; Mills *et al.*, 2005). Insurers' willingness to support the development of new energy technologies can be crucial. Without insurance, project developers and manufacturers cannot raise capital on standard terms. For their part, insurers find that the risks of untried technologies can be very difficult to assess, with a consequent increase in uncertainty (and cost) for underwriters and claims adjusters also (Dahlstrom *et al.*, 2003). These factors are exacerbated in a rapidly developing market because there is little time to learn from mistakes. The investment arm of the insurance industry is much less aware of climate change than are underwriters, and it tends to view environmental issues as less relevant to business decisions (ABI, 2001; Mansley and Dlugolecki, 2001; Dlugolecki and Mansley, 2005; ABI, 2004a). The activation of the Kyoto Protocol and the start of the EU Greenhouse Gas Emission Trading Scheme have brought home the reality that greenhouse gas emissions do have a cost, which will inexorably rise. Even so, the convoluted chain of responsibility in managing large funds has made them slow to adopt strategies on climate change in terms of equity portfolio management and property development (Dlugolecki and Silver, 2006).

13.9.3 *The right null hypothesis?*

In the context of already evident rapid change in the climate, and the established high sensitivity of society to extreme weather, the question must be asked, whether it is right or prudent to take as the null hypothesis that damage is *not* increasing due to climate change rather than a Bayesian approach that damage should be increasing already. In the former case, it may be some time before "significant" deviations from past behavior can be discerned, which could lead to delays in taking precautionary action. In the latter case, recent observations are surely "consistent with" a new trend toward higher weather-related damage caused by climate change. In the author's opinion, scientists have a particular responsibility in advising lay stakeholders *not* to convey the impression that because the null hypothesis of no change cannot be defeated, therefore there is no change. It is surely better to counsel that severe change is possible, and may arrive quite soon. In effect, this approach is adopting the precautionary principle as advocated by Stern (2006). It is better to take mitigation measures that may not be necessary, than not to take them and find out later that they were necessary. The cost of being "cautiously wrong" is much lower than the cost of being "optimistically wrong."

13.10 The future role of the insurance industry

Previous sections have shown that climate change will generate an increasing impact of extreme events, and that probably this process has already started to happen. In one sense, this is good news for insurers; more disasters mean a greater need for insurance. However, this idea is simplistic. The scale of impacts is hard to assess accurately, and that uncertainty deters underwriters and, behind them, investors in insurance and reinsurance companies. This is evidenced by the progressive decline in reinsurers' credit ratings – from a financial strength of AA- in 2002 to one of A in 2006 (Fitch Ratings, 2006) – and the rating agencies' increased emphasis on catastrophic events (Fitch Ratings, 2005). Also, insurance works by pooling risks, often from diverse locations, to reduce the variability of the total risk through diversification. Climate change may act to increase risks simultaneously across regions. One estimate is that before the year 2040, economic losses from climatic disasters might reach US$1 trillion (in 2005 values) in a single year (UNEPFI, 2006). Stern (2006) observes that additional capital is available to underwrite risks, through traditional insurance and innovative products like catastrophe bonds and weather derivatives, but ultimately the risk/reward balance may not be attractive, and insurers and investors could withdraw. The industry stands at a crossroads, therefore: it can become actively involved in adaptation and mitigation, or it can take a passive stance while safeguarding its assets from harm (Mills, 2005). In the United Kingdom, the industry has made it clear that it sees climate change as an important issue, in which it can play a strong role

(ABI, 2004b). There are signs that this decision may be made elsewhere as well (Hewitt, 2003; Dlugolecki and Lafeld, 2005; Mills *et al.*, 2005; Chief Risk Officers [CRO] Forum, 2006; Hanson *et al.*, 2006; Lloyd's, 2006; UNEPFI, 2006), with major companies issuing position statements and promoting action in public–private partnerships. However, the industry is much less concentrated than, say the oil industry or automobile manufacturers, and regulatory systems and practices vary considerably, even in individual countries, so there are considerable obstacles facing the development of a coherent approach.

Acknowledgments

The author acknowledges the support received from the Association of British Insurers during the research for much of this work, and also the benefit of the literature review provided by the IPCC process.

References

Allianz and WWF (2006). *Climate Change and Insurance: An Agenda for Action in the United States*. Munich and Washington: Allianz and WWF.

Association of British Insurers (ABI) (1999). *Subsidence – A Global Perspective* (author R. Radevsky). London: ABI.

Association of British Insurers (ABI) (2001). *Investing in Social Responsibility* (author R. Cowe). London: ABI.

Association of British Insurers (ABI) (2003). *The Vulnerability of U.K. Property to Windstorm Damage*. London: ABI.

Association of British Insurers (ABI) (2004a). *Risk Returns and Responsibility* (author R. Cowe). London: ABI.

Association of British Insurers (ABI) (2004b). *A Changing Climate for Insurance* (author A. Dlugolecki). London: ABI.

Association of British Insurers (ABI) (2005). *Financial Risks of Climate Change*. London: ABI.

Association of British Insurers (ABI) (2006). *Coastal Flood Risk: Thinking for Tomorrow, Acting for Today*. London: ABI.

Barclays Private Bank (2005). *Special Focus: Energy Markets after Katrina and Rita – and How to Invest*. London: Barclays.

Chartered Insurance Institute (CII) (1994). *The Impact of Changing Weather Patterns on Property Insurance* (ed. A. Dlugolecki). London: CII.

Chartered Insurance Institute (CII) (2001). *Climate Change and Insurance* (ed. A. Dlugolecki). London: CII.

Chelliah, M., and Bell, G. (2004). Tropical multidecadal and interannual climate variability in the NCEP-NCAR reanalysis. *Journal of Climate*, **17**, 1777–803.

Choi, O., and Fisher, A. (2003). The impacts of socioeconomic development and climate change on severe weather catastrophe losses: Mid-Atlantic Region (MAR) and the U.S. *Climatic Change*, **58**, 149–70.

Chief Risk Officers (CRO) Forum (2006). Position Paper on Climate Change & Tropical Cyclones in the North Atlantic, Caribbean and Gulf of Mexico, Chief Risk Officers Forum, Emerging Risk Initiative. www.cronetworks.org/ Cro%20Products/CRObriefing.pdf.

Climate Change Impacts Review Group (CCIRG) (1991). *The Potential Effects of Climate Change in the United Kingdom* (ed. M. Parry). London: Her Majesty's Stationery Office.

Climate Change Impacts Review Group (CCIRG) (1996). *Review of the Potential Effect of Climate Change in the United Kingdom* (ed. M. Parry). London: Her Majesty's Stationery Office.

Dahlstrom, K., J. Skea and Stahel, W. (2003). Innovation, insurability and sustainable development: sharing risk management between insurers and the state. *Geneva Papers on Risk and Insurance*, **28**(3), 394–412.

Dlugolecki, A., and Lafeld, S. (2005). *Climate Change – Agenda for Action: The Financial Sector's Perspective*. Munich: Allianz Group and WWF.

Dlugolecki, A., and Loster, T. (2003). Climate change and the financial services sector. *Geneva Papers on Risk and Insurance*, **28**(3), 382–93.

Dlugolecki, A., and Mansley, M. (2005). *Climate Change and Asset Management*. Technical Report #20, Tyndall Centre for Climate Change Research, Norwich, UK.

Dlugolecki, A., and Silver, N. (2006). Environmental professionalism and sustainability: too important to get wrong. In *Actuaries: A Case Study in Professional Development and Environmental Sustainability*. Greener Management International, pp. 49, 95–109.

Dlugolecki, A., Clark, K., Knecht, F., *et al.* (1995). Financial services. In *Climate Change 1995: Impacts, Adaptation, and Mitigation of Climate Change: Scientific-Technical Analysis*, ed. T. Watson, M. Zinyowera, R. Moss, and D. Dokken. Cambridge, UK: Cambridge University Press, for Intergovernmental Panel on Climate Change, Working Group II.

Easterling, D., Meehl, G., Parmesan, C., Changnon, S., Karl, T., and Mearns, L. (2000). Climate extremes: observations, modeling, and impacts. *Science*, **289**, 2068–74.

Elsner, J. (2006). Evidence in support of the climate change: Atlantic hurricane hypothesis. *Geophysical Research Letters*, **33**, L16705, doi:10.1029/ 2006GL026869.

Emanuel, K. (2005). Increasing destructiveness of tropical cyclones over the past 30 years. *Nature*, **436**, 686–8.

Emanuel, K. (2006). Environmental influences on tropical cyclone variability and trends. *Proceedings of 27th AMS Conference on Hurricanes and Tropical Meteorology*, Paper #4.2. http://ams.confex.com/ams/pdfpapers/107575.pdf.

Epstein, P., and Mills, E. (eds.) (2005). *Climate Change Futures: Ecological, Health and Economic Dimensions*. Boston, MA: The Center for Health and Global Environment, Harvard Medical School, November 2005.

Fitch Ratings (2005). *New Thinking on Catastrophic Risk and Capital Requirements*. New York: Fitch.

Fitch Ratings (2006). *Reinsurance Review and Outlook: Cycle Management – A Bumpy Ride Ahead*. New York: Fitch.

Foresight (2004). *Future Flooding*. Foresight Programme, Office of Science and Technology. www.foresight.gov.uk/Previous_Projects/ Flood_and_Coastal_Defence/Reports_and_Publications/Executive_Summary/ Executive_Summary.html.

Goldenberg, S., Landsea, C., Mestas-Nuñez, A., and Gray, W. (2001). The recent increase in Atlantic hurricane activity: causes and implications. *Science*, **293**, 474–9.

Hadley Centre (2003). *Climate Change Observations and Predictions: Research on Climate Change Science from the Hadley Centre*. Exeter, UK: The Meteorological Office.

Hanson, C., Palutikof, J., Dlugolecki, A., and Giannakopoulos, C. (2006). Bridging the gap between science and the stakeholder: the case of climate change research. *Climate Research*, **31**, 121–33.

Hasnain, S. (2002). Himalayan glaciers meltdown: impact on South Asian rivers. In *Regional Hydrology: Bridging the Gap between Research and Practice*, ed. H. van Lanen and S. Demuth. International Association of Hydrological Sciences (IAHS) Publication No. 274, pp. 417–25.

Hoyos, C. D., Agudelo, P. A., Webster, P. J., and Curry, J. A. (2006). Deconvolution of the factors contributing to the increase in global hurricane intensity. *Science*, **312**(5770), 94–7. doi:10.1126/science.1123560.

Hewitt, R. (2003). The economy of climate: consequences for underwriting. Munich Re, Munich. *Topics*, **2003**(2), 18–19.

Hulme, M., Jenkins, G. J., Lu, X., *et al.* (2002). *Climate Change Scenarios for the United Kingdom: The UKCIP02 Scientific Report*. Norwich, UK: Tyndall Centre, School of Environmental Sciences, University of East Anglia.

Huntingford, C., Jones, R., Prudhomme, C., Lamb, R., Gash, J. and Jones, D. (2003). Regional climate-model predictions of extreme rainfall for a changing climate. *Quarterly Journal of the Royal Meteorological Society*, **129**, 1607–21.

Intergovernmental Panel on Climate Change (IPCC) (2007). *Fourth Assessment Report: Working Group II, Asia*. (in press).

International Ad Hoc Detection and Attribution Group (IAHDAG) (2005). Detecting and attributing external influences on the climate system: a review of recent advances. *Journal of Climate*, **18**, 1291–314.

Klotzbach, P. J. (2006). Trends in global tropical cyclone activity over the past twenty years (1986–2005). *Geophysical Research Letters*, **33**, L10805, doi:10.1029/2006GL025881.

Knight, J., Folland, C., and Scaife, A. (2006). Climate impacts of the Atlantic Multidecadal Oscillation. *Geophysical Research Letters*, L17706, doi:10.1029/2006GL026242.

Knutson, T., and Tuleya, R. (2004). Impact of CO_2-induced warming on simulated hurricane intensity and precipitation: sensitivity to the choice of climate model and convective parameterization. *Journal of Climate*, **17**, 3477–95.

Kosatsky, T. (2005). The 2003 European heatwaves. *Eurosurveillance*, **10**(7), 148–9.

Lamb, H. (1991). *Historic Storms of the North Sea, British Isles and Northwest Europe*. Cambridge, UK: Cambridge University Press.

Landsea, C. (2005). Hurricanes and global warming. *Nature*, **438**, E11–12.

Leckebusch, G., Koffi, B., Ulbrich, U., Pinto, J., Spangehl, T., and Zacharias, S. (2006). Analysis of frequency and intensity of European winter storm events from a multi-model perspective, at synoptic and regional scales. *Climate Research*, **31**, 59–74.

Leckebusch, G., Ulbrich, U., Froelich, L., and Pinto, J. (2007). Property loss potentials for European mid-latitude storms in a changing climate. *Climate Research* (in press).

Lee, J. (2006). Fundamentals of the oil market. Presentation at Global Risk Management Seminar, February 15, 2006, Centre for Global Energy Studies, London. www.cges.co.uk.

Lloyd's (2006). *Climate Change – Adapt or Bust*. London: Lloyd's.

Mailier, P., Stephenson, D., Ferro, C., and Hodges, K. (2005). Serial clustering of extratropical cyclones. *Monthly Weather Review*, **134**, 2224–40.

Mann, M., and Emanuel, K. (2006). Atlantic hurricane trends linked to climate change. *Eos*, **87**(24), 233–44.

Mansley, M., and Dlugolecki, A. (2001). *Climate Change – a Risk Management Challenge for Institutional Investors*. London: Universities Superannuation Scheme.

Mayfield, M. (2005). Testimony of the Director, National Hurricane Center, National Oceanic and Atmospheric Administration (NOAA), to Hearing on NOAA Hurricane Forecasting before the Science Committee, House of Representatives, October 7, 2005.

Meehl, G., and Tebaldi, C. (2004). More intense, more frequent, and longer lasting heat waves in the twenty-first century. *Science*, **305**, 994–7.

Mendelsohn, R., Morrison, W., Schlesinger, M., and Adronova, N. (2000). Country-specific market impacts from climate change. *Climatic Change*, **45**, 553–69.

Michaels, P. J., Knappenberger, P. C., and Davis, R. E. (2006). Sea-surface temperatures and tropical cyclones in the Atlantic basin. *Geophysical Research Letters*, **33**, L09708, doi:10.1029/2006GL025757.

Mills, E. (2005). Insurance in a climate of change. *Science*, **309**, 1040–14.

Mills, E., Roth, R., and Lecomte, E. (2005). *Availability and Affordability of Insurance under Climate Change: A Growing Challenge for the U.S. Investor*. Boston, MA: Network on Climate Risk, CERES Inc.

Muir-Wood, R., Miller, S., and Boissonnade, A. (2006). The search for trends in a global catalogue of normalized weather-related catastrophe losses. Presented at *Climate Change and Disaster Losses Workshop: Understanding and Attributing Trends and Projections*. http://sciencepolicy.colorado.edu/sparc/research/projects/extreme_events/munich_workshop/index.html.

Munich Re (1990). *Windstorm: New Loss Dimensions of a Natural Hazard*. Munich: Munich Re.

Munich Re (2002). *Winter Storms in Europe (II)*. Munich: Munich Re.

Munich Re (2005). *Annual Review: Natural Catastrophes 2004*. Munich: Munich Re.

Munich Re (2006a). *Hurricanes: More Intense, More Frequent, More Expensive: Insurance in a Time of Changing Risks*. Munich: Munich Re.

Munich Re (2006b). *Annual Review: Natural Catastrophes 2005*. Munich: Munich Re.

National Health Assessment Group (NHAG) (2001). *Climate Change and Human Health*. Baltimore, MD: Johns Hopkins University.

Nordhaus, W. D., and Boyer, J. (1999). *Role the DICE Again: the Economics of Global Warming*. New Haven, CT: Yale University.

Palutikof, J. P., Subak, S., and Agnew, M. D. (eds.) (1997). *Economic Impacts of the Hot Summer and Unusually Warm Year of 1995*. London: Department of the Environment.

Pielke, R., Jr. (2005a). Are there trends in hurricane destruction? *Nature*, **438**, E11.

Pielke, R., Jr. (2005b). Attribution of disaster losses. *Science*, **310**, 1615.

Pielke, R., Jr. (2006). White paper prepared for workshop on climate change and disaster losses: understanding and attribution of trends and projections. Presented at *Climate Change and Disaster Losses Workshop: Understanding and Attributing Trends and Projections*. http://sciencepolicy.colorado.edu/sparc/research/projects/extreme_events/munich_workshop/index.html.

Pielke, R., Jr., and Landsea, C. (1998). Normalized hurricane damages in the United States: 1925–95. *Weather and Forecasting*, **13**, 621–31.

Pielke, R., Jr., Landsea, C., Mayfield, M., Laver, J., and Pasch, R. (2005). Hurricanes and global warming. *Bulletin of the American Meteorological Society*, **86**, 1571–5.

Risk Management Solutions (RMS) (2005). *Hurricane Katrina: Profile of a Super Cat: Lessons and Implications for Catastrophe Risk Management.* Newark, CA: Risk Management Solutions Inc.

Schwartz, J. (2005). Category 5: levees are piece of a $32 billion pie. *New York Times*, late edition, final, Section A, p. 1, November 29, 2005.

Sit, V., and Cai, J. (2004). Formation and development of China's extended metropolitan areas. *Geographic Research*, **33**(5), 531–40.

Skrebowski, C. (2005). A stunning achievement, but have we reached the yield points? Editorial, *Petroleum Review*, November 2005. Energy Institute.

Stern, N. (2006). *Stern Review on the Economics of Climate Change.* London: The Treasury. www.sternreview.org.uk.

Swiss Re (2000). *Storm over Europe: an Underestimated Risk.* Zurich: Swiss Re.

Swiss Re (2006a). *Annual Review: Natural Catastrophes 2005.* Sigma 2006(2). Zurich: Swiss Re.

Swiss Re (2006b). *The Effects of Climate Change: Storm Damage in Europe on the Rise.* Zurich: Swiss Re.

Tol, R. S. J. (2002). Estimates of the damage costs of climate change, part II. Dynamic estimates. *Environmental and Resource Economics*, **21**, 155–60.

Towers Perrin (2005). Hurricane Katrina: analysis of the impact on the insurance industry. www.towersperrin.com.

Trenberth, K. (2005). Uncertainty in hurricanes and global warming. *Science*, **308**, 1753–4.

Trenberth, K., and Shea, D. J. (2006). Atlantic hurricanes and natural variability in 2005. *Geophysical Research Letters*, **33**, L12704. doi:10.1029/2006GL026894.

United Nations Environment Programme Finance Initiative (UNEPFI) (2002). *Climate Change and the Financial Services Industry*: Geneva, Switzerland: Module 1. *Threats and Opportunities.* Module 2A. *Blueprint for Action.* Climate Change Working Group.

United Nations Environment Programme Finance Initiative (UNEPFI) (2006). *CEO Briefing: Adaptation and Vulnerability to Climate Change: The Role of the Finance Sector.* Climate Change Working Group. www.unepfi.net.

Vellinga, P., and Mills, E., et al. (2001). Insurance and other financial services, ed. A. Dlugolecki. In *Climate Change 2001: Impacts, Adaptation, and Vulnerability*, ed. J. McCarthy, O. Canziani, N. Leary, D. Dokken, and K. White. Cambridge, UK: Cambridge University Press, for Intergovernmental Panel on Climate Change Working Group II.

Webster, P., Holland, G., Curry, J., and Chang, H.-R. (2005). Changes in tropical cyclone number, duration and intensity in a warming environment. *Science*, **309**, 1844–6.

Westerling, A., Hidalgo, H., Cayan, D., and Swetnam, T. (2006). Warming and earlier spring increase in western U.S. forest wildfire activity. *Science* **313**(5789), 940. doi:10.1126/science.1128834.

Woolf, D., and Coll, J. (2006). *Impacts of Climate Change on Storms and Waves in Marine Climate Change Impacts. Annual Report Card 2006*, ed. P. J. Buckley, S. R. Dye, and J. M. Baxter. Online Summary Reports, Marine Climate Change Impacts Partnership (MCCIP). www.mccip.org.uk.

World Wildlife Fund (WWF) (2006). *Stormy Europe: The Power Sector and Extreme Weather.* Berlin: WWF International. (See also Methodology and References Annex.)

Zhang, R., and Delworth, T. (2006). Impact of Atlantic Multidecadal Oscillation on India/Sahel rainfall and Atlantic hurricanes. *Geophysical Research Letters*, **33**, L17712, doi:10.1029/2006GL026267.

14

Toward a comprehensive loss inventory of weather and climate hazards

SUSAN L. CUTTER, MELANIE GALL, AND CHRISTOPHER T. EMRICH

Condensed summary

A comprehensive national loss inventory of natural hazards is the cornerstone for effective hazard and disaster mitigation. Despite federally demanded mitigation plans (Disaster Mitigation Act of 2000, DMA 2000) that are supposed to accurately represent the risks and losses, there still is no systematic and centralized inventory of *all* hazards and their associated losses (direct, indirect, insured, uninsured, etc.) – at least not in the public domain. While a variety of agencies and groups collect hazard-related information, differences among their goals result in a patchwork of data coverage. In lieu of a central, systematic, and comprehensive events and loss inventory, the Hazards Research Laboratory at the University of South Carolina developed the Spatial Hazard Events and Losses Database for the United States (SHELDUS). This database is currently the most comprehensive inventory for the United States. However, issues that emerged during its development – such as standardizing losses, spatial coverage, and so forth – stressed the need for a national clearinghouse for loss data. It is imperative that such a clearinghouse be developed to promote standardized guidelines for loss estimation, data compilation, and metadata standards. Otherwise, the development of a national loss inventory will remain deficient and the nation will continue to lack baseline data against which trends in hazardousness and the effectiveness of mitigation efforts could be evaluated.

14.1 Introduction

There is significant uncertainty about the severity and magnitude of disaster losses in the United States. Depending on the estimate used, US flood losses from 1994 through 2004 were either US\$55 billion according to the National Weather Service (Hydrologic Information Center, 2004) or US\$24 billion

Climate Extremes and Society, ed. H. F. Diaz and R. J. Murnane. Published by Cambridge University Press. © Cambridge University Press 2008.

according to the National Climatic Data Center (Hazards Research Lab, 2006). The National Flood Insurance Program, another source of flood loss data, recorded almost US$9 billion in paid insurance claims during the same time period (Federal Emergency Management Agency [FEMA], 2006). In other words, we really do not know the totality of disaster losses in the United States, and, at this point, one guess is just as good as another. As Mileti (1999) pointed out, "Determining losses with a higher degree of accuracy is impossible because the United States has not established a systematic reporting method or a single repository for the data".

Comprehensive knowledge of incurred losses from natural hazards is essential for the survival of the insurance and reinsurance industries, and reinsurance companies maintain large databases on the financial impacts of natural hazards (Guha-Sapir and Below, 2002) precisely for that reason. However, the US Federal Emergency Management Agency and local emergency management agencies lack similar corporate loss databases. While FEMA keeps track of individual and public assistance claims and payouts, there is no single database that provides sufficient information for the historic reconstruction of losses (e.g., crop losses, casualties) caused by natural hazards in the United States. How can the effectiveness of risk reduction and mitigation policies be evaluated without a historic baseline of losses? And most important, how does FEMA prioritize the allocation of mitigation money – geographically and by hazard type – in the absence of such data?

The key to effective hazard and disaster mitigation is a comprehensive inventory of losses: how much, and where they occur. Without it, the development of sound mitigation plans that build upon a locale's hazard loss history with the intent to reduce future losses is difficult, if not impossible. Despite federally demanded mitigation plans (Disaster Mitigation Act of 2000, DMA 2000) that are theoretically based on hazard assessments derived from previous loss history or loss estimation models, it is still difficult to accurately represent the risks and losses from all hazards. There have been consistent calls for such an inventory (Mileti, 1999; National Research Council, 1999a; Cutter, 2001; Changnon, 2003), yet explanations for its absence remain speculative; they are, however, clearly associated with political will and/or agenda setting. It is clear that losses will continue to escalate and that weather-related hazards exceed hundreds of billions of dollars annually. In fact, 2005 was the costliest weather year on record, with more than 200 billion US dollars in losses, the majority attributed to windstorms (tropical storms and hurricanes) generally and Hurricane Katrina (US$125 billion) specifically (Shein, 2006).

14.2 Who needs a loss inventory?

The United States is exposed to many different types of natural hazards, and to weather and climate hazards in particular. Media-friendly hurricanes batter the coasts while floods, earthquakes, tornadoes, droughts, and other common and powerful hazards affect not only coastal areas but also the interior areas of the country. The destruction caused by Hurricane Katrina along the Gulf Coast in 2005 is just the current placeholder for the next "big hazard event." Death and destruction can occur anywhere in the United States, not just along the coastlines. In times of increasing losses from natural hazards at both a global and a national scale (McBean, 2004; Cutter and Emrich, 2005), the country should not and cannot plan for the future without some systematic accounting or a central repository of past losses.

The National Planning Scenarios, developed by the Homeland Security Council in collaboration with the US Department of Homeland Security (DHS), are designed to support mitigation planning and emergency preparedness. These scenarios "represent threats and hazards of national significance with high consequence" (Homeland Security Council, 2005). However, the National Planning Scenarios include only hurricanes and earthquakes in their list of 15 worst-case scenarios, besides the abundance of other hazards and their proven destructive powers. Hurricane Katrina, in August 2005, has surpassed the national planning hurricane scenario that anticipated 1,000 fatalities and one million evacuees. More alarming is that flood hazard events were not deemed hazardous enough to give rise to a National Planning Scenario, although they are "the major source of monetary loss in the United States from natural hazards" (Cutter, 2001, p. 86). Such a misperception of the relative impacts of natural hazards and extreme events could be avoided with a national loss database – one that facilitates policy making and mitigation planning at all levels.

The current competition between antiterrorism and hazard mitigation goals within the Department of Homeland Security heightens the need for more efficient hazard mitigation strategies. The success of FEMA relies partially on the maximization of limited financial resources in order to protect people and reduce disaster losses. Efficient mitigation targets high-risk areas through tailored actions. However, the identification of high-risk areas depends on expert knowledge and experience, since there is no procedure, mechanism, or dataset on which to base such a delineation and associated decision making.

Even worse, the effectiveness of mitigation measures is difficult to assess in the absence of baseline data on losses. This lack renders it almost impossible to perform cost–benefit analyses to assess the efficiency and usefulness of mitigation. Furthermore, cost–benefit analyses could be an essential tool in prioritizing specific hazard types and areas at risk, irrespective of the hazard de jour and policy agendas.

14.3 Data on hazard events and losses

There is a wealth of information on natural hazards in the United States. A variety of agencies and groups collect hazard-related information; however, differences in their goals result in a patchwork of data coverage. Useful sources of data currently available for analyzing historic and current hazard event and loss trends are summarized in Table 14.1. The mission agencies collecting hazard-related data have a broad range of mandates that are reflected in the type of data they compile and how they collect, analyze, archive, and disperse it to the public and private sectors. For example, during a major hazard event, each federal agency has responsibility for a *segment* of the response, impacts, or losses, as specified under the Federal Response Plan. Unfortunately, archiving much of the data is not mandatory and is not funded; therefore, valuable information is typically lost in the months or years following the event.

Some of the agencies are more interested in specific types of event information, such as strength, magnitude, intensity, and other physical characteristics (as for earthquakes and hurricanes), but do not collect data on losses. Others are more concerned with loss information (property damage, insurance claim payouts) and do not focus on specific hazard event characteristics or risk. The time frames for the listed datasets also vary substantially, limiting historic comparisons and time-series analyses. Finally, geographic scale is an issue for almost all of the hazards for which data are currently collected.

The scale of the impact is also a determinant as to whether data are collected on losses. In many instances, unless an event receives a presidential disaster declaration or is a so-called billion-dollar weather disaster (Table 14.2), valuable event and loss data are not collected. Midsize disasters are documented better than small events that, for example, cause only a few thousand dollars in hail damage. As a result, data on US hazards and their impacts are very fragmented, they are not cataloged in a unified manner, and they exhibit inconsistent coverage with regard to hazard type, magnitude, and locality.

The multitude of data sources hampers the consistency in estimating and reporting losses. Estimating losses is problematic especially when the boundaries between direct and indirect losses, insured and uninsured losses, as well as the distinction between costs and losses are blurry. This is particularly true for direct and indirect casualties from a hazardous event. The International

Table 14.1. *Selected data sources for hazard events and losses*

Hazard	Data source	Dates	Variables
Tornado[a]	Storm Prediction Center	1959–present	Date, time, latitude, longitude, deaths, injuries, damage category
Thunderstorm wind[a]	Storm Prediction Center	1959–present	Date, time, latitude, longitude, deaths, injuries, damage category
Hail[a]	Storm Prediction Center	1959–present	Date, time, latitude, longitude, deaths, injuries, damage category
Lightning[b,c]	National Climatic Data Center	1959–present	Date, time, injuries, deaths, damage, location of strike, county of strike
Meteorological events	Storm Data:[e] National Climatic Data Center; Natural Hazards Center	1959–present	Date, time, location, deaths, injuries, damage category (property and crop), monetary losses for flood, hurricane, tornado
	Extreme Weather Sourcebook[b], National Center for Atmospheric Research (NCAR)	Varies by hazard	
Hurricane[e]	National Hurricane Center; *Monthly Weather Review*	1851–present (Atlantic); 1949–present (Pacific)	Date, time, wind speed, pressure, deaths, damage
Earthquake	Epicenters:[d] Council of the National Seismic System	1970–present	Time, latitude, longitude, depth, magnitude
	Significant earthquakes:[c] National Geophysical Data Center	2150 BC–1995	Date, time, latitude, longitude, magnitude (Richter), intensity, death, damage category
	Significant earthquakes: Earthquake Research Institute	3000 BC–1994	Date, time, latitude, longitude, magnitude (moment), intensity, death, injuries, damage category
Flood	Flood Damage Report:[b] US Army Corps of Engineers	1987–1996	Damages, lives lost
	Global Register of Major Flood Events:[e] Dartmouth College	1994–present	Date, origin, deaths, cost, displaced peoples

Table 14.1. (cont.)

Hazard	Data source	Dates	Variables
	National Flood Insurance Program:[d] FEMA	1978–present	Total losses, payments, policies, amount insured
	Summary of Natural Hazard Fatalities:[e] Storm Data	1988–1995	Deaths
	Stream gauge data[d] US Geological Survey (USGS)	Varies by gauge date	Latitude, longitude, county, drainage area
Drought	National Drought Mitigation Center	1895–1995	Droughts by drought management district
Wildfire	National Fire Incident Reporting System; National Fire Data Center	New in 1999	Injuries, deaths, losses, and others undetermined
Volcano[d]	Global Volcanism Program, Smithsonian Institution	8,000 BC–Present	Latitude, longitude, elevation, type (morphology), status, last eruption
Nuclear power plants[d]	US Department of Energy	Current	Location, type, reactor count, safety check status
Hazardous materials spill[b]	US Department of Transportation Hazardous Materials Information System	1970–present	State, injuries, death, damage
Hazardous sites[d]	Environmental Protection Agency	1966–present	ID, name, latitude, longitude
Toxic chemical releases[d]	Environmental Protection Agency	1987–present	Facility information, amount of release, chemical released, media of release

Data limitations:

[a] Limited historical data time frame;

[b] limited geographical coverage;

[c] only those events with death, damage or injury;

[d] no deaths, injuries, or damages;

[e] difficulty in assigning losses to specific location.

Table 14.2. *Billion-dollar US weather disasters between 1995 and 2005*

Year	Hurricanes	Flooding	Tornadoes	Drought	Hail
2005	Wilma Rita Katrina Dennis	—	—	Midwest	—
2004	Jeanne Ivan Frances Charley	—	—	—	—
2003	Isabel	—	Midwest; Mississippi, Ohio, and Tennessee valleys	—	Southern Plains, lower Mississippi valley
2002	—	—	—	30-state drought	—
2001	Tropical Storm Allison	—	Midwest, Ohio valley	—	—
2000	—	—	—	South-central and southeastern United States	—
1999	Floyd	—	Arkansas, Tennessee, Oklahoma, Kansas	Eastern United States	—
1998	Georges Bonnie	Texas	—	Southern United States	Minnesota
1997	—	West Coast, Northern Plains, Midwest	Midwest	—	—
1996	Fran	Pacific Northwest, Mid-Atlantic, Northeast	—	Southern Plains	—
1995	Opal Marilyn	Texas, Oklahoma, Louisiana, Mississippi, California	—	—	—

Source: NOAA's National Climatic Data Center (www.ncdc.noaa.gov/oa/reports/billionz.html).

Classification of Diseases (ICD-10) makes it possible to relate a death to the initial cause, which could be a hazard such as an earthquake, fire, or flood. Thus, every death certificate should theoretically indicate whether a natural hazard was the direct or indirect cause of death. Unfortunately, many cause-of-death statements are poorly written and incomplete. Therefore, there is a tendency to either under- or over-report direct and indirect deaths and injuries from natural hazards.

Many of these issues were highlighted in the *Second Assessment* (Mileti, 1999; Cutter, 2001). Gilbert White (1994, p. 1237) stated, "Accurate and comparable data on losses of lives and property from extreme events are very difficult to assemble; standards and methods of data collection are far from uniform and consistent." While the task may seem daunting, it does not obviate the need for this national dataset. Rather, it makes the compilation of a national database an important and challenging task.

14.4 Spatial Hazard Events and Losses Database for the United States

In lieu of a central, systematic, and comprehensive events and losses inventory, the Hazards Research Lab at the University of South Carolina developed the Spatial Hazard Events and Losses Database for the United States (SHELDUS). The first release of the database, in July 2003, spanned events from 1960 through 2000. Currently, SHELDUS (www.sheldus.org) is in its fourth version, covering events through May 2005 and featuring a total of more than 400,000 records. Updates and additions to the database become available twice a year. The database includes loss information for 18 different hazard types: avalanches, coastal hazards (e.g., currents, erosion), droughts, earthquakes, floods, fog, hail, heat, hurricanes/tropical storms, landslides, lightning, severe thunderstorms, tornadoes, tsunamis/seiches, volcanic eruptions, wildfires, wind, and winter weather (e.g., blizzards, heavy snowfall). Various sources provide the input data for SHELDUS, which acts as a compilation tool for some of the many hazard event and loss datasets that reside in disparate places around the country. The Storm Data and Unusual Weather Phenomena publications by the National Climatic Data Center (NCDC) serve as a major data source, with complementary data from the National Geophysical Data Center (NGDC); the National Hurricane Center; the Storm Prediction Center, in Norman, Oklahoma; the US Geological Survey; major newspapers; and others.

Each SHELDUS record indicates incurred casualties, injuries, and property and crop losses, as well as the beginning and ending date of an event. It is important to note that SHELDUS reports these losses not necessarily by event but rather by location; i.e., it is possible to query SHELDUS by county and state but not by event (e.g., Hurricane Ivan). This means that SHELDUS geo-references a large event to the affected counties and splits the event's total losses across those counties. Ultimately, SHELDUS provides a unique opportunity to create loss and hazard profiles by US county from 1960 through May 2005 – a major component of funding requirements established by DMA 2000. SHELDUS does not include loss information on Puerto Rico, Guam, and other US territories.

14.4.1 Standardizing losses in SHELDUS

Yet another reason that a comprehensive standardized database of hazard events and losses at the national scale is needed can be seen in the way historical loss data are reported. SHELDUS reports losses by using a conservative "yardstick," meaning that all losses based on SHELDUS should be considered as minimum estimates. This conservative approach is required because the input data, which are derived from different federal agencies, report losses in diverse metrics. Some entities give loss figures in (logarithmic) categories rather than as actual dollar amounts. A good example is NCDC, which reported losses in logarithmic classes (US\$0; US \$1–5; US\$5–50; US\$50–500; US\$500–5,000; US\$5,000–50,000; US\$50,000–500,000; US\$500,000–5,000,000; US\$5,000,000–50,000,000; US\$50,000,000–500,000,000; > US\$500,000,000) until 1994/1995 and then switched to exact values. In order to avoid exaggeration, SHELDUS followed a conservative route and adopted the lower boundary of such loss classes wherever applicable.

An example of the various ways in which different agencies collect hazard event and loss data are the loss reports by the National Geophysical Data Center (NGDC). Occasionally, the center lists numerous estimates for a single earthquake event. This is due to the fact that NGDC's database collates information from various sources, similar to the procedures in SHELDUS. Where there are two or more estimates of losses, SHELDUS selects the lowest loss estimate. Again, this practice maintains the minimum or conservative loss approach.

Furthermore, adjustments to losses (e.g., for inflation) were avoided within the SHELDUS database. Loss figures in SHELDUS always represent event year dollar figures. This enables users to standardize losses to any given base year.

14.4.2 Spatial coverage

The level of detail in regard to spatial extents of losses varies significantly by data source and/or hazard type. Lightning strikes are generally reported at a local level, whereas droughts exhibit a regional and even multi-state character.

A good example of differences in data collection due to spatial extent is the "30-state drought" in 2002, which caused more than US$10 billion in reported damages (NCDC, 2006). SHELDUS does not include a single record on the 30-state drought event. Instead, the total of US$10 billion in losses was split evenly across every affected county. This means that only the summation of the losses in all affected counties results in a total of US$10 billion. Omission of affected counties causes an underestimation of the total losses of the event. On the other hand, whenever a specific hazard location was given, SHELDUS geo-referenced the event to the county level and simply included the exact location among miscellaneous information.

It is anticipated that future versions of SHELDUS will contain an event query function; i.e., the database will be searchable by event type. This change should help eliminate the risk of underestimating hazard-specific losses by omitting affected counties from the query.

14.4.3 Caveats

Particularly challenging to SHELDUS was the switch from categorical loss reporting to actual dollar loss reporting by NCDC in 1995 – again the major (but not sole) data provider for SHELDUS. To accommodate this change, SHELDUS dropped its initial threshold procedure for data entries. From 1960 through 1994, SHELDUS included only events that generated more than US$50,000 in either property or crop losses. This threshold applied to every data source used for SHELDUS during this time period.

Since 1995, SHELDUS has included every event from all its data sources that causes some sort of human or economic loss, regardless of the threshold. This methodological adaptation was necessary because NCDC not only started reporting actual dollar losses but also improved the spatial resolution of its reports. More events were reported on a county basis and fewer on a regional or state basis. Regional events generally top the US$50,000 threshold, while county events often fall well short of it.

Furthermore, certain hazard types are underreported by NCDC and consequently also by SHELDUS. The extent of loss underestimation is nebulous, however. According to well-known drought researchers (Wilhite, 2000; Svoboda et al., 2002), droughts result in the largest dollar losses among the weather-related hazards. However, in our database hurricanes and floods top the ranks, with a share of 20% each, while drought losses generate only 7% of the total losses over the study period.

Such caveats are indicative of the pitfalls of non-standardized data collection. The need for data verification by the end user applies to every dataset that is collected outside a standardized context. A national clearinghouse or consortium for hazard and disaster data could reduce such problems by consolidating and standardizing the collection and reporting system. Despite the limitations noted above, SHELDUS continues to be the most comprehensive geo-referenced database on natural hazards events and losses in the USA.

14.5 Increasing losses from weather-related hazards

Since 1960, dollar losses from natural hazards have steadily increased (Figure 14.1). Weather-related events constitute most of these losses (more than 75%), with hurricanes, floods, and severe storms (including hail and tornadoes) as the major causal agents. The year 2004 would have been the most costly year, with almost US$30 billion in losses, had there not been Hurricane Katrina in 2005. The tallies for the storm of the century stand currently at more than US$100 billion in damages with more than 1,000 fatalities (NCDC, 2006).

It is not surprising that weather-related events dominate the loss statistics. Their short return intervals and abundant occurrences outweigh geophysical events, at least over the 40-year period considered in the loss analysis. The dramatic increase in dollar losses from weather events, particularly over the past several years, seems frustrating given the investments and improvements in hazard mitigation, detection, monitoring, and warning systems.

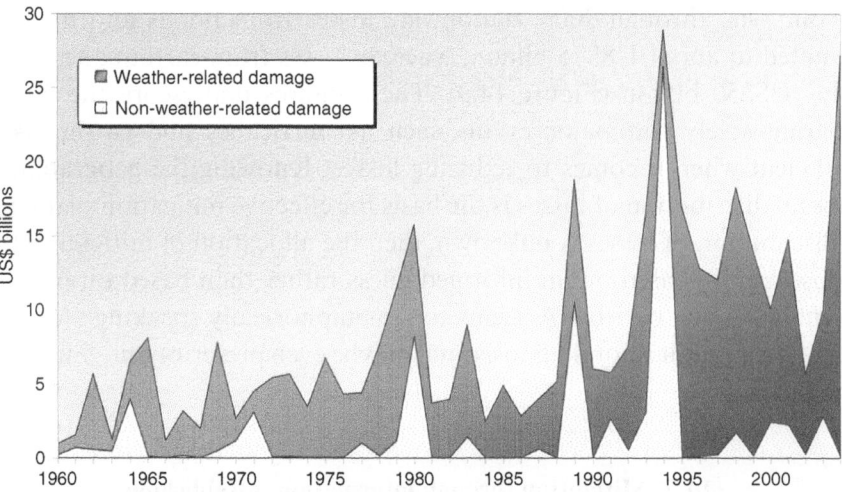

Figure 14.1. Cumulative losses from natural hazards for the period 1960 through 2004 (in billions of dollars; adjusted to 2005 US dollars), based on the Spatial Hazard Events and Losses Database for the United States (SHELDUS Version 4.2).

While scientific discussions about potentially increased severity of events are ongoing (see other chapters in this volume), there is evidence that the frequency has remained relatively constant (van der Vink, 1998; National Research Council, 1999b; Changnon et al., 2000; Changnon, 2003). While the climatic causality between losses and hazard intensity/frequency is inconclusive, the anthropogenic driving factors behind rising losses are certain. A major cause of exorbitant losses is that more people are living in hazardous places – along coastlines or in floodplains – placing more wealth, real estate, and infrastructure in exposed areas subject to damage.

14.6 Weather-battered states

As can be seen in Figure 14.2a, most losses from natural hazards occur along the US coastlines. In considering only weather-related losses, the pattern remains essentially the same, with the exception of California, where earthquake losses are dominant (Figure 14.2b). Interestingly, earthquakes and other geophysical events do not significantly alter the historic spatial pattern of losses. Instead, they simply add another loss burden to areas that also suffer from weather-related losses, mainly flooding. To reduce losses and improve preparedness, it is suggested that communities prepare for the impacts of weather-related events to the same degree that they plan for the impact from earthquakes, or at least adopt an all-hazards mitigation approach as has been suggested by the research community (Mileti, 1999).

From 1960 through 2004, nationwide losses from floods and hurricanes amounted to about US$75 billion, whereas losses from earthquakes reached "only" US$50 billion (Figure 14.3). These figures underscore the fact that preparing solely for major events such as hurricanes and earthquakes is insufficient when it comes to reducing losses. Knowing the geographic and temporal distribution of losses is the basis for effective mitigation planning. If the distribution of losses is unknown, then the allocation of mitigation funds is consequently based on an informed guess rather than based on facts. The current approach by FEMA seems to – metaphorically speaking – prescribe mitigation without prior diagnosis, and maybe even prescribe mitigation to the wrong patient.

14.7 Mitigation through information: establishing
a clearinghouse for loss data

Currently, there are two databases in the public domain that attempt to document hazard events and losses in Latin America, and those mid- to

(a)

(b)

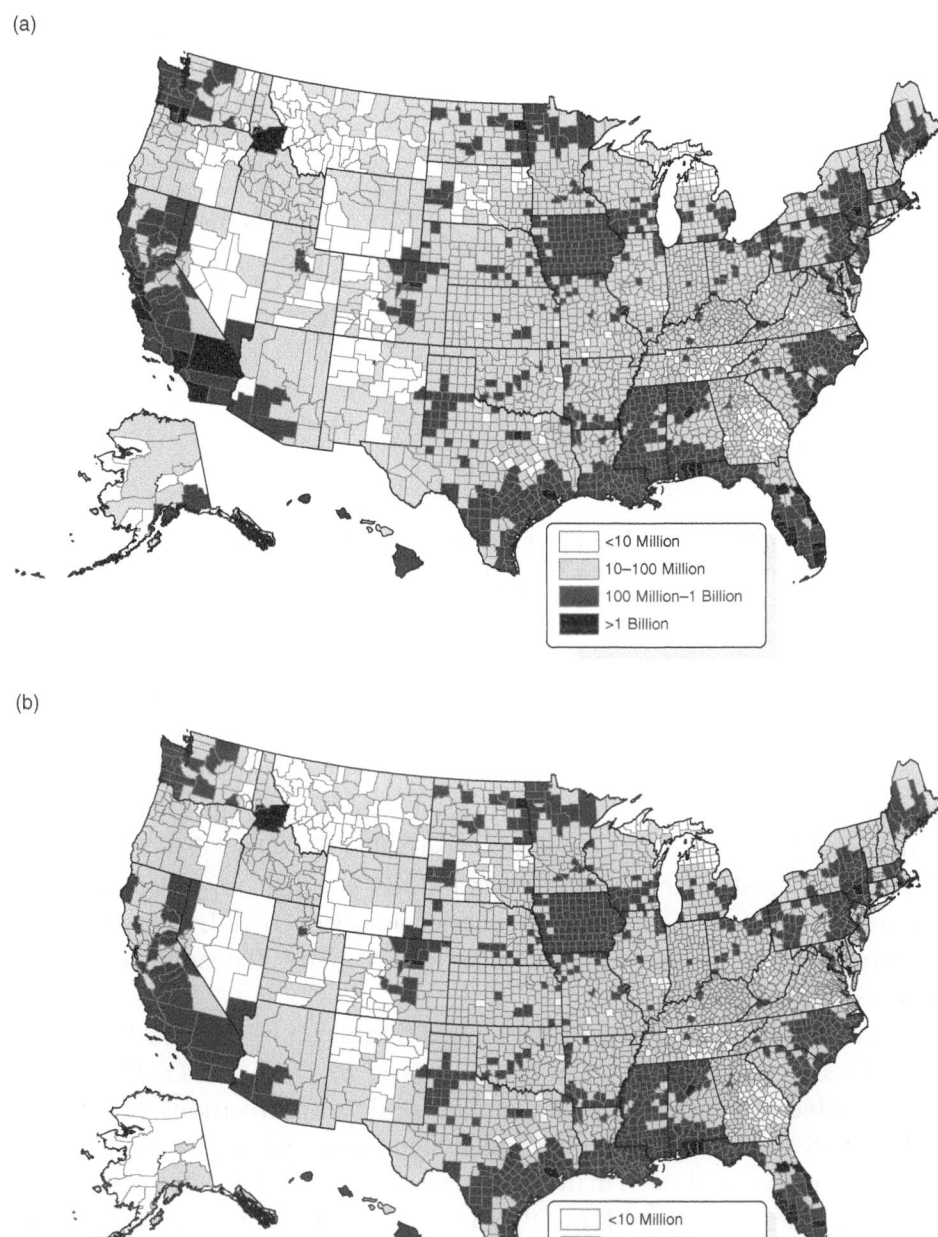

Figure 14.2. (a) Total losses from all natural hazards from 1960 through 2004; (b) total losses from only weather-related hazards from 1960 through 2004. Based on the Spatial Hazard Events and Losses Database for the United States (SHELDUS Version 4.2), and adjusted to 2005 US dollars.

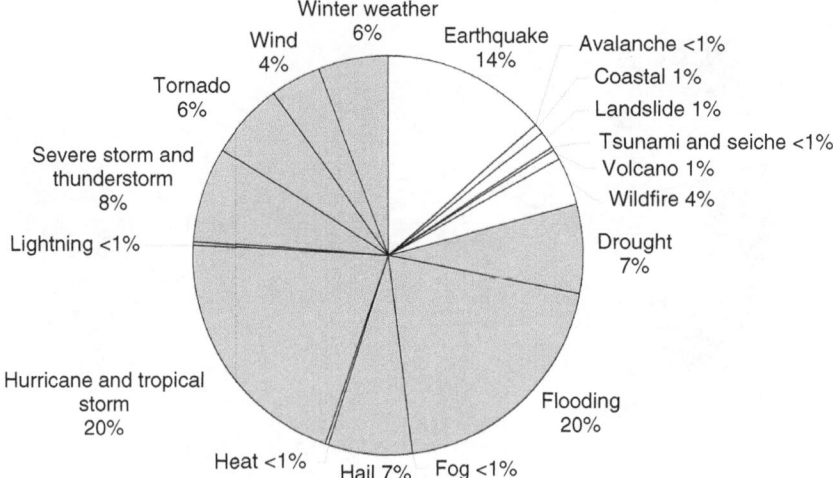

Figure 14.3. Weather-related events (highlighted in gray) caused most of the losses between 1960 and 2004, based on the Spatial Hazard Events and Losses Database for the United States (SHELDUS Version 4.2).

large-sized disasters that occur throughout the world, recorded at the country level. The Network of Social Studies in Disaster Prevention in Latin America – better known by its Spanish acronym LA RED – maintains the Latin American loss database called DesInventar. The Centre for Research on the Epidemiology of Disasters (CRED), of the University of Louvain in Brussels, is responsible for a global disaster database called EM-DAT. The CRED and the Asian Disaster Reduction Center (ADRC), which proposed the idea, recently agreed to assign a global identifier number (GLIDE) to every event to harmonize the documentation of disasters (see www.glidenumber.net) and their associated losses.

A national loss inventory for the United States can be improved based on several lessons from SHELDUS, DesInventar, EM-DAT, and GLIDE. First, it is possible to consolidate and standardize data from various origins (see SHELDUS and DesInventar). Second, documenting only major events might satisfy initial needs, though in the long run a complete and detailed coverage of all losses is desirable (see SHELDUS and EM-DAT). Third, geo-referencing the impact of hazards to subnational levels supports local policy makers in their decision making and allows any user, researcher, or stakeholder to gain a full understanding of (a) where the event occurred, (b) what losses the event produced, (c) when the event happened, and (d) what type of hazard occurred. Finally, a national US loss inventory that included specific identification numbers for each event would improve not only national, but also global, knowledge on the losses from natural hazards.

The data maintained by the National Climatic Data Center, in particular its Storm Data database, would be a very good starting point for a national inventory. However, there are no internal guidelines for loss estimation, and quality control is minimal (Downton *et al.*, 2005). As a result, the estimates between NCDC and the National Weather Service differ significantly, as was alluded to earlier in this chapter. To establish a reliable national loss inventory, the following issues must to be considered:

- Integrate and coordinate stakeholders. For instance, create a US consortium for hazard and disaster data with tasks similar to those of the Federal Geographic Data Committee (FGDC).
- Designate *one* institution as the national clearinghouse for loss data from all natural hazards, with responsibility for archiving and disseminating data through a Web-based distributed geographic information (DGI) system.
- Develop guidelines for loss estimation, data compilation, metadata standards, and other factors used in evaluating a disaster.
- Set minimum standards for loss data so that losses are comparable across hazard type, space, and time. Minimum specifications should include event type; beginning and end date of the event; geographic coordinates or political boundary identifiers (county); incurred fatalities, injuries, and property and crop losses; indirect losses; insured losses; and so forth.
- Establish partnerships with insurance and reinsurance companies to generate a comprehensive assessment of direct, indirect, insured, and uninsured losses.

Ultimately, a national loss inventory would provide the necessary baseline data against which trends in hazardousness and the effectiveness of mitigation efforts could be evaluated. According to a recent study by the Multihazard Mitigation Council (Ganderton *et al.*, 2006), mitigation pays off. The study concluded that every mitigation dollar spent by FEMA from 1993 through 2003 generated about US$4 in future benefits. Unfortunately, though, the study based its analyses on modeled losses (using the software HAZUS-MH[1]). Not surprisingly, one of the study's recommendations is to systematically measure the benefits from mitigation activities. A national hazard loss inventory would be an invaluable piece of such an accounting system.

Coordination, harmonization, and standardization efforts are essential to realizing the goal of a national US loss inventory. The currently scattered, heterogeneous, and mission-oriented documentation of losses needs to be molded into consistent loss reports that can be consolidated into a single loss

---

[1] The software Hazards US Multi-Hazard (HAZUS-MH) is available in its second release, MR2, from FEMA (www.fema.gov/plan/prevent/hazus/index.shtm).

database with nationwide all-hazards coverage. SHELDUS is a positive step in that direction.

Acknowledgments

Support for this research was provided by the National Science Foundation (Grant No. 99053252 and 0220712) and the National Consortium for the Study of Terrorism and Responses to Terrorism (START), a Center of Excellence of the US Department of Homeland Security.

References

Changnon, S. A. (2003). Measures of economic impacts of weather extremes. *Bulletin of the American Meteorological Society*, **84**(9), 1231–5.
Changnon, S. A., Pielke, R. A., Jr., Changnon, D., Sylves, R. T., and Pulwarty, R. (2000). Human factors explain the increased losses from weather and climate extremes. *Bulletin of the American Meteorological Society*, **81**(3), 437–42.
Cutter, S. L., ed. (2001). *American Hazardscapes: The Regionalization of Hazards and Disasters*. Washington, D.C.: Joseph Henry Press.
Cutter, S. L., and Emrich, C. T. (2005). Are natural hazards and disaster losses in the U.S. increasing? *Eos, Transactions of the American Geophysical Union*, **86**(41), 381, 388–9.
Downton, M. W., Barnard Miller, J. Z., and Pielke, R. A., Jr. (2005). Reanalysis of U.S. National Weather Service flood loss database. *Natural Hazards Review*, **6**(1), 13–22.
Federal Emergency Management Agency (FEMA) (2006). National Flood Insurance Program: flood insurance statistics (Internet). Last updated March 30, 2006. Cited June 3, 2006. Available at www.fema.gov/business/nfip/statistics/statscal.shtm.
Ganderton, P. T., Bourque, L. B., Dash, N., *et al.* (2006). Mitigation generates savings of four to one and enhances community resilience: MMC releases study on savings from mitigation. *Natural Hazards Observer*, **30**(4), 1–3.
Guha-Sapir, D., and Below, R. (2002). The quality and accuracy of disaster data: a comparative analysis of three global data sets (Internet). ProVention Consortium, The World Bank Group. Cited May 31, 2003. Available at www.proventionconsortium.org/themes/default/pdfs/data_quality.pdf.
Hazards Research Laboratory (2006). Spatial Hazard Events and Losses Database for the United States (SHELDUS), version 4.2 (online database). University of South Carolina. Cited June 3, 2006. Available at www.sheldus.org.
Homeland Security Council (2005). National Planning Scenarios. Draft, version 20.1. Washington D.C.: Department of Homeland Security. Available at media.washingtonpost.com/wp-srv/nation/nationalsecurity/earlywarning/NationalPlanningScenariosApril2005.pdf.
Hydrologic Information Center (2004). Flood losses: compilation of flood loss statistics (Internet). National Weather Service (NWS). Last updated January 05, 2004. Cited June 3, 2006. Available at www.weather.gov/oh/hic/flood_stats/Flood_loss_time_series.shtml.

McBean, G. (2004). Climate change and extreme weather: a basis for action. *Natural Hazards*, **31**, 177–90.

Mileti, D. S. (1999). *Disasters by Design: A Reassessment of Natural Hazards in the United States*. Washington, D.C.: Joseph Henry Press.

National Climatic Data Center (NCDC) (2006). Billion dollar U.S. weather disasters (Internet). Cited February 14, 2006. Available at lwf.ncdc.noaa.gov/oa/reports/billionz.html.

National Research Council (1999a). *The Impacts of Natural Disasters: A Framework for Loss Estimation*. Washington, D.C.: National Academies Press.

National Research Council (1999b). Mitigation emerges as major strategy for reducing losses caused by natural disasters. *Science*, **284**, 1943–7.

Shein, K. A. (2006). State of the climate in 2005. *Bulletin of the American Meteorological Society*, **87**(6), 801–5.

Svoboda, M., LeComte, D., Hayes, M., *et al.* (2002). The drought monitor. *Bulletin of the American Meteorological Society*, **83**(8), 1181–90.

van der Vink, G. (1998). Why the United States is becoming more vulnerable to natural disasters. *Eos, Transactions of the American Geophysical Union*, **79**(44), 533, 537.

White, G. F. (1994). A perspective on reducing losses from natural hazards. *Bulletin of the American Meteorological Society*, **75**(7), 1237–40.

Wilhite, D. A. (2000). Drought as natural hazard: concepts and definitions. In *Drought: A Global Assessment*, vol. 1, ed. D. A. Wilhite. London: Routledge Publishers, pp. 3–18.

15

The catastrophe modeling response to Hurricane Katrina

ROBERT MUIR-WOOD AND PATRICIA GROSSI

15.1 Introduction

Hurricane Katrina was the most destructive US natural disaster in history and the most expensive catastrophe loss ever for the global insurance industry. The occurrence of Hurricane Katrina, and the associated flooding of New Orleans, was also by far the greatest US catastrophe to have occurred since the widespread application of catastrophe models in the mid 1990s. The combination of catastrophic wind and flood losses proved to be a potent test of the underlying methodologies and procedures of catastrophe loss modeling. This chapter reviews the impact of Hurricane Katrina and the response of Risk Management Solutions (RMS) to the event, both in the immediate aftermath and then in galvanizing new research to expand the agenda for catastrophe loss modeling in order to gain a more comprehensive perspective on catastrophe risk costs.

The legacy of Hurricane Katrina has meant that the agenda of catastrophe modeling has become more complex and comprehensive in attempting to capture all facets of loss, incorporating an expanded range of nonlinearities that ramp up losses caused by the largest catastrophes. The insurance industry has also learned from the experience to question the validity and comprehensiveness of the data that are entered into the models and how the models are employed to explore the sensitivity of predicted losses.

Moreover, with evidence that hurricane frequencies and severities in the Atlantic basin are being affected by global warming (e.g., Emanuel, 2005; Mann and Emanuel, 2006), the experience of the 2005 hurricane season has highlighted one of the principal consequences of climate change for catastrophe risk. Rising sea levels and increasing hurricane intensities mean that some of the largest increases in risk can be expected in the coastal zone. Catastrophe models will continue to be the principal means by which changes

Climate Extremes and Society, ed. H. F. Diaz and R. J. Murnane. Published by Cambridge University Press. © Cambridge University Press 2008.

in hazards become communicated into adjusted metrics for pricing and managing hurricane risk.

15.2 Hurricane Katrina

15.2.1 First landfall in Florida

The National Hurricane Center (NHC) initiated advisories on Tropical Depression 12 (TD12) at 4:00 p.m. Central Daylight Time (CDT) on Tuesday, August 23, 2005, while it was located over the Bahamas. TD12 became Tropical Storm Katrina at 10:00 a.m. CDT on August 24, continuing to strengthen and becoming a category 1 hurricane just prior to its first landfall in southeastern Florida between Hallandale Beach and North Miami Beach at 6:00 p.m. CDT on August 25.

The impact of the storm was borne by south Florida residents with weary familiarity after the four hurricanes of 2004. Katrina's southern outflow generated more than 13 cm of rainfall across a large area of southeastern Florida. Localized heavy rainfall of more than 25 cm was experienced in Miami-Dade County and further south, which caused flash flooding and 11 deaths as cars were caught in flood torrents.

15.2.2 Reemergence into the Gulf of Mexico

By the next day, on August 26 at 2:00 a.m. CDT, Hurricane Katrina emerged off the southwestern coast of Florida into the Gulf of Mexico as a tropical storm (Knabb *et al.*, 2005). Because of the extremely warm waters ahead of the storm, the NHC forecast it to be a "dangerous hurricane in the northeastern Gulf of Mexico in about 3 days." By midday on August 26, Air Force reconnaissance reports indicated that it had strengthened to a category 2 hurricane with sustained winds of 160 km h^{-1} and a central pressure of 971 hPa. As it turned west and then north-northwest, the storm encountered the warmest waters, 2 °C above the average expected for that time of year. During a series of rapid intensification periods, Hurricane Katrina strengthened to a category 4 hurricane at 1:00 a.m. CDT on August 28 with 233 km h^{-1} sustained winds, and to a category 5 hurricane with 258 km h^{-1} sustained winds only 6 h later. Katrina's maximum sustained winds occurred at midday on August 28, reaching 282 km h^{-1}, and by 4 p.m. CDT, the minimum central pressure had fallen to 902 hPa. Katrina also had one of the fastest pressure drops ever recorded, from 930 to 909 hPa in 6 h at a rate of 3.5 hPa per hour.

15.2.3 Second landfall in Louisiana

Just 80 km before landfall in Louisiana, Katrina's wind field and eyewall weakened slightly. At its second landfall – on the Gulf Coast just south of Buras, Louisiana, at 6:10 a.m. CDT on August 29 (Figure 15.1) – sustained winds were $225\,\mathrm{km\,h^{-1}}$, which made it a category 4 hurricane, with a central pressure of 920 hPa.

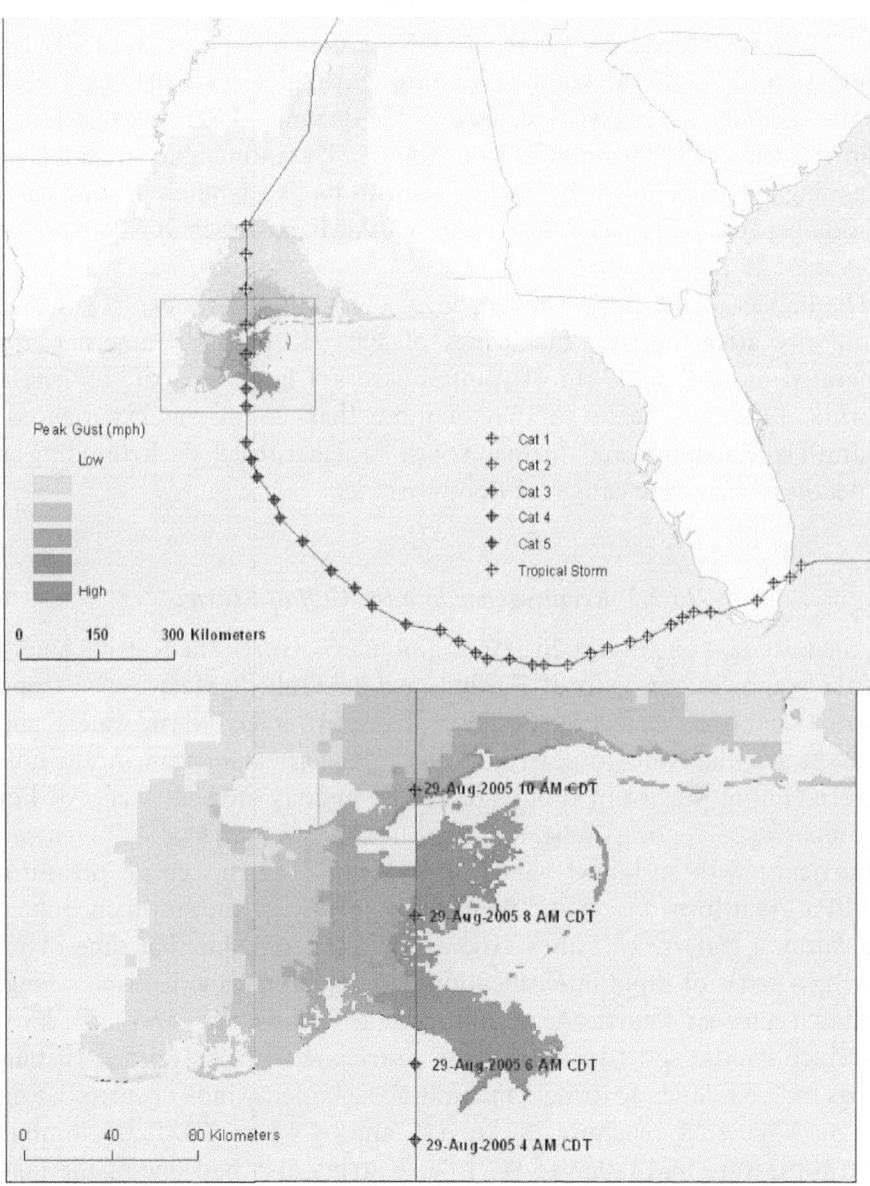

Figure 15.1. The path of Hurricane Katrina, showing first and second landfalls in Florida and Louisiana (above) and a close-up of the second landfall in Louisiana (below). (From RMS, 2005.) For color version, see plate section.

Table 15.1. *Hurricane Katrina landfall characteristics*

Landfall region	Pressure[a] (hPa)	1-minute sus-tained winds[a] (km h^{-1})	Radius to maximum winds[b] (km)	Translational speed[a] (km h^{-1})
Southeast Florida	985	129	28	10
South of Buras, Louisiana	920	225	30	24
Louisiana/ Mississippi border	927	200	52	26

[a] National Hurricane Center/Tropical Prediction Center, National Weather Service.
[b] NOAA/AOML Hurricane Research Division.

The center of the storm then crossed the Mississippi delta and the Chandeleur Sound and came onshore again near the Louisiana/Mississippi border 4 h later, at 10:00 a.m. CDT; by this time, the sustained winds had weakened to 200 km h^{-1}, making Katrina a category 3 hurricane. Significantly, the radius of the storm, R_{max}, had almost doubled: from 30 to 52 km. This radius is unusually large and puts it 1.0–1.5 standard deviations from the mean R_{max} recorded for hurricanes in the Gulf of Mexico (Table 15.1). The outer bands of the storm started to impact the Gulf Coast before Katrina made landfall. The landfall region in the Gulf received cumulative rainfall in excess of 25 cm, adding to the water damage in many locations.

15.2.4 A formidable storm

When Hurricane Katrina made landfall in Louisiana, its central pressure was the third lowest recorded at US landfall since 1900 (Knabb *et al.*, 2005). Hurricane-force winds swept across 400 km of coastline, embracing three states and reaching more than 160 km inland. The storm drove an unusually large storm surge onto the coast of southern Mississippi that, combined with major wave action, reached 7.5–8 m above sea level. Everywhere along the Mississippi coast, the surge was 1–1.5 m higher than in Hurricane Camille in 1969, a category 5 storm with a smaller radius that caused minor damage in New Orleans.

According to the correlation between intensity and surge heights assumed in the Saffir–Simpson scale, the storm surge was commensurate with that of a severe category 5 hurricane, even while the winds experienced along the coast-line at landfall were consistent with a category 3 hurricane. The extraordinary height of the storm surge was subsequently understood to have been derived from the period of nearly 24 h prior to landfall, when Hurricane Katrina had maintained a category 5 intensity. The large radius to maximum winds had also pushed a broad front of water into the shallow bathymetry of the Louisiana–Missisippi embayment.

15.2.5 The Great New Orleans Flood

As the center of the storm moved inland after its second landfall in Louisiana, easterly winds pushed a storm surge up the Mississippi River as well as into Lake Borgne and Lake Pontchartrain, two gulfs of the sea located to the north of the Mississippi River delta (see Figure 15.2). Surge generation becomes amplified in shallow water, and the surge into Lake Borgne reached water levels of more than 5.5 m, but this surge rapidly dissipated as the winds shifted around to the north and then northwest. However, in Lake Pontchartrain, high water levels persisted for longer and the counterclockwise rotation of the winds that had first driven water into the lake shifted to bring the highest water levels along the northern shore of New Orleans.

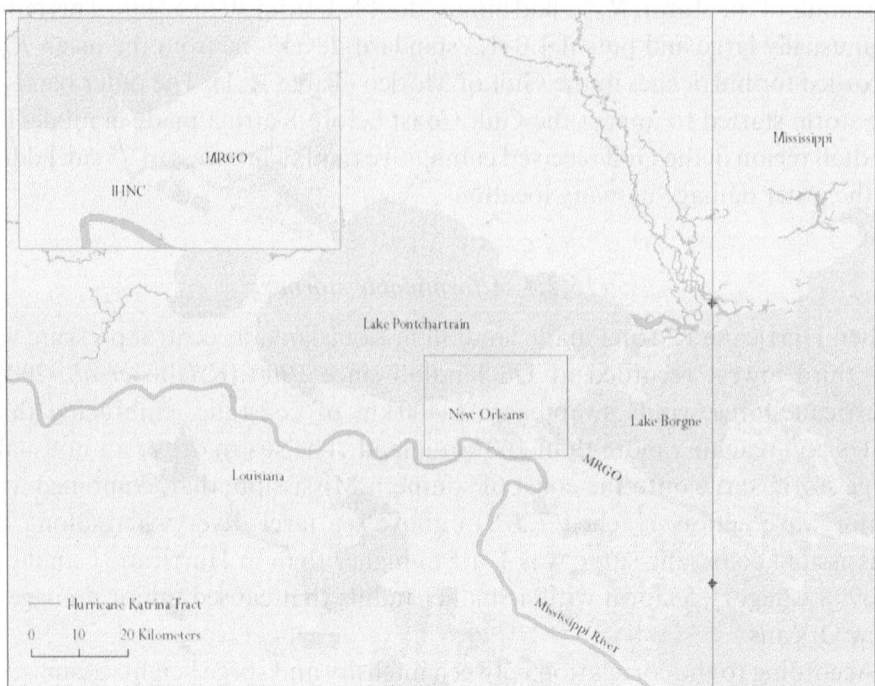

Figure 15.2. The region at landfall of Hurricane Katrina, with key locations labeled and close-up of New Orleans (in upper left). For color version, see plate section.

New Orleans was first hit by the storm surge arriving up the Mississippi River Gulf Outlet (MRGO) from Lake Borgne to the east, between 4:00 a.m. and 6:00 a.m. CDT on Monday, August 29, 2005. By 6:00 a.m. CDT, St. Bernard Parish had been submerged, with many houses being pushed off their foundations by the speed of the advancing floodwaters. At approximately 4:45 a.m. CDT, the first breaching occurred – on the western side of the Inner Harbor Navigation Canal (IHNC) at the western end of the MRGO – allowing water to spread into the area to the north of the French Quarter and contributing between 10% and 20% of the water that flowed into the downtown area. On the eastern side of the IHNC, at the western edge of the Lower Ninth Ward, water 4.6 m high spilled over the 4.3 m defenses, and at about 7:45 a.m. CDT catastrophic breaching occurred as the foundations of the levees were ripped out and the rush of water floated off buildings several blocks away. Around the same time, the surge overtopped and breached the levees at a number of locations on the northern side of the MRGO along the southern edge of the New Orleans East area.

The second phase of flooding in New Orleans came from the north via Lake Pontchartrain, with water entering along the drainage canals that had originally been dug in the late nineteenth century to allow water pumped out of the low-lying downtown areas to flow back into the lake. While water levels in these canals were insufficient to overtop the floodwalls, at three locations the foundations of the floodwalls failed (two along the London Avenue Canal and one along the 17th Street Canal), leading to breaches that rapidly expanded and became scoured to many meters below ground level. Water continued to flow into the city for more than a day before the waters of Lake Pontchartrain receded. Pumping equipment was restored or brought into the city within a week, so that most of the 946 billion liters that had flowed into the city had been removed by September 24. The city was claimed dry on October 12, 2005.[1]

Evacuation

As a result of the initial warnings about the hurricane, including forecasts predicting category 5 intensity at landfall, the majority (about 75%) of the population of New Orleans had fled. However, about 20,000 had gathered in the city's Superdome while the remainder stayed in their properties. These people included many of the old and poor without transport, as well as those in hospitals and nursing homes.

---

[1] For more details on the flooding of New Orleans, see Independent Levee Investigation Team (ILIT), 2006; RMS, 2005; and US Army Corps of Engineers (USACE), 2006.

This was the second major evacuation, after Hurricane Georges in 1998, and the third major alarm in a decade; therefore the majority of the city's population treated the experience as a "fire drill." Precious possessions had not been evacuated or elevated, and few people had appreciated that their homes might be completely ruined. For several days after the flood, those people left in New Orleans suffered increasing deprivation and lack of facilities, and there were outbreaks of civil disorder and looting. The floodwaters prevented fire crews from being able to suppress fires, and several fires burned for hours in industrial buildings and residential homes across the city. Finally, a compulsory evacuation was ordered on September 6, a week after the flood had begun, as troops went house-to-house to ensure that everyone alive was forced to leave. In all, 800 bodies were recovered in the first month even while many other citizens were reported missing.

15.3 Initial RMS response: estimation of losses

For catastrophe loss modelers, the initial challenge in the hours and days after a hurricane has made landfall is to provide users of the models with detailed information to guide them as to the magnitude of their expected losses. There was little doubt that Hurricane Katrina was going to be an unprecedented insurance loss, and therefore the need to provide information would be of the highest importance. Also, in contrast to the situation of the four hurricanes in 2004, much of the loss from this event would pass to international reinsurers.

However, the very scale of Katrina, compounded with the flooding in New Orleans, presented a number of very significant challenges in loss estimation. In particular, not all of the components of loss were represented in the RMS catastrophe model (the US Hurricane model in RiskLink® 5.0). There soon became two perspectives on loss: the "modeled loss" and the "projected loss." The modeled loss was the loss that a model user would derive for a storm with a comparable track and wind field. The projected loss was the expected loss that RMS reconstructed based on rapidly acquired additional information – in particular related to the extraordinary extent of flooding. Inevitably, the gap between these two perspectives became of concern to model users who were struggling to present their expectations regarding losses to boards of management, investors, and regulators.

15.3.1 *First landfall projected loss from wind*

The wind field from Katrina as it tracked as a category 1 hurricane across Florida on August 25 was modeled and validated from meteorological data with a maximum modeled gust wind speed of 151 km h^{-1}. Using information on industry-insured exposures, the RMS US Hurricane model generated wind losses just below US$900 million, with an expected increase in losses due to the intense rainfall and associated flooding, for total projected losses of US$1 billion to US$2 billion.

15.3.2 *Second landfall projected loss from wind and storm surge*

Wind damage

The wind field on the second landfall in Louisiana was assessed from meteorological modeling, recorded wind speeds, aerial reconnaissance, and assessments of damage by a field survey team. The population of wind speed recordings was notably patchy, as many monitoring stations had gone offline due to loss of power. The RMS field survey team was in the area as the hurricane made landfall and conducted reconnaissance across the affected region for 3 days afterwards, observing and photographing a large number of damaged facilities. At the coast where the strongest winds occurred (e.g., Bay St. Louis), damage was found to be consistent with that produced by 210–225 km h^{-1} gust wind speeds.

On the left-hand side of the hurricane track in New Orleans, observed wind damage was significantly less extensive, as winds with peak gusts of 160 km h^{-1} occurred in this region. There was some damage to roofs, windows, and non-structural elements. Previous hurricanes had demonstrated that buildings in Louisiana and Mississippi have higher vulnerability to wind damage than those in many other hurricane states, reflecting poorer construction practices and building code enforcement. In the RMS US Hurricane model, these factors were taken into consideration in estimating the vulnerability of exposures.

Even if New Orleans had not flooded, Hurricane Katrina would still have been the most destructive hurricane on record. The hurricane wind field impacted the towns of Biloxi, Gulfport, and Camp Shelby in Mississippi, with their ports, oil refineries, casino resorts, and associated commercial facilities. The hurricane-force winds (above 119 km h^{-1}) were felt across an area containing about US$500 billion of property value, while the area of potentially damaging winds (above 80 km h^{-1}) contained about US$1.5 trillion of property value.

Storm surge damage

In Mississippi, the coastal communities of Gulfport and Biloxi were severely damaged by a combination of storm surge and Katrina's strongest winds.

Along this coastline seawater flooded inland on average about 0.8 km, reaching elevations in excess of 7.5 m. Based on aerial reconnaissance of damage to structures and debris lines, along with ground-based observations, RMS survey teams categorized the inundation extent of the storm surge along the Mississippi coast into three zones:

• **Total**: an area extending inland, about 0.6 km, where the storm surge reaching overland water depths of up to 6 m caused complete destruction, pushing almost all structures off their foundations.
• **Major**: an area where large waves caused structural damage to property, and the storm surge was approximately 0.6–1.8 m deep.
• **Moderate**: areas subject to shallow flooding of up to 0.6 m with light wave action.

The observed surge inundations extended further inland than the 500-year return period limits denoted for the National Flood Insurance Program (NFIP).

Projected losses

Estimates of the total wind and storm surge losses for the second landfall – excluding New Orleans – were obtained by superimposing the wind field footprint (Figure 15.3) on the RMS proprietary exposure data and employing the best representation available of vulnerabilities, incorporating yet-to-be-released results of the analysis of 2004 claims data. In addition, a conservative

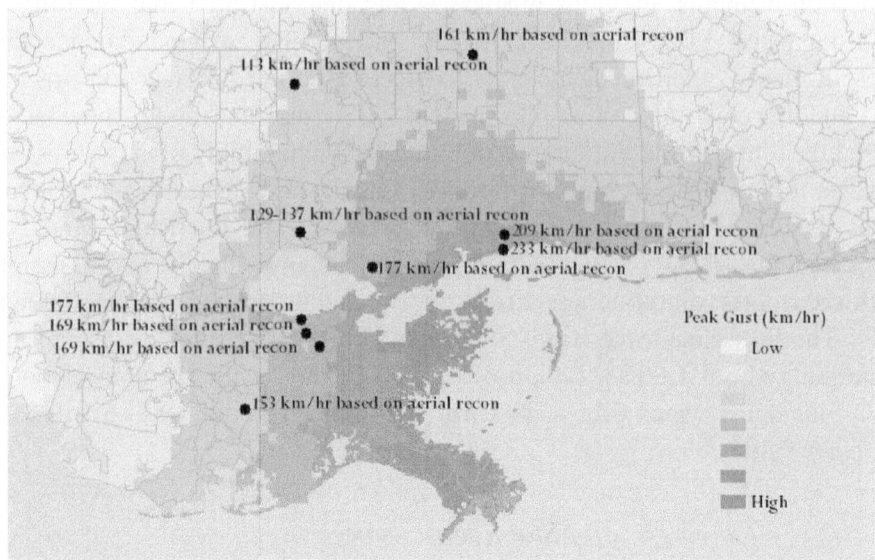

Figure 15.3. Map of the RMS-modeled wind field of the Gulf Coast landfall of Hurricane Katrina. (Based on RMS, 2005.) For color version, see plate section.

view was taken concerning expected demand surge (supplementing losses by 40%) and additional sources of loss, such as aggravated business interruption and off-premises power interruption.

The projected insured losses at second landfall derived by this process were US$20 billion to 25 billion, reflecting the overall effects of the wind and including a proportion of the costs arising from storm surge damage. It was assumed that where a building had been damaged both by water and by wind, in determining the cause of the damage the insurer would likely end up paying a greater proportion of the loss than might strictly be required under the terms of the contract. This estimate did not include the flood losses in New Orleans or the full costs of the storm surge.

15.3.3 Offshore energy projected losses

While Hurricane Katrina was at its peak intensity, it crossed the offshore oil and gas fields of the Mississippi Canyon Corridor and South Timbalier, causing damage to the platforms and pipelines throughout the area. More than US$20 billion in platform replacement value (600 platforms) was exposed to wind gusts in excess of $225 \, \text{km} \, \text{h}^{-1}$ (i.e., the wind speeds of Hurricane Ivan the previous year) and the associated extreme waves.

As of September 19, 2005, 46 shallow-water platforms had been reported destroyed, representing more than 1% of the total production of the Gulf. Other platforms had damage to varying degrees, including four drilling rigs destroyed, nine rigs extensively damaged, and six rigs drifted in the storm. Four large deepwater platforms suffered extensive damage (Figure 15.4). In addition, there were reports that damaged pipelines in the Mississippi Canyon Corridor and South Timbalier areas would take months to repair.

The initial projected loss estimate for the offshore energy industry was based on taking the locations, values, and ages of the platforms and modeling the wind and wave fields of the storm, as tested against available reports of lost and severely damaged facilities. The vulnerability functions relating repair and replacement costs to wind and wave loads had recently been refined (but again not yet implemented in the standard model) based on research conducted following Hurricane Ivan in 2004. In all, insured losses of between US$2 billion and US$5 billion were estimated to offshore energy, including US$1 billion to US$2 billion of direct damage to platforms, while the remainder would be expected to come from impacts to pipeline infrastructure, removal of wreckage, and the expected interruption of production.

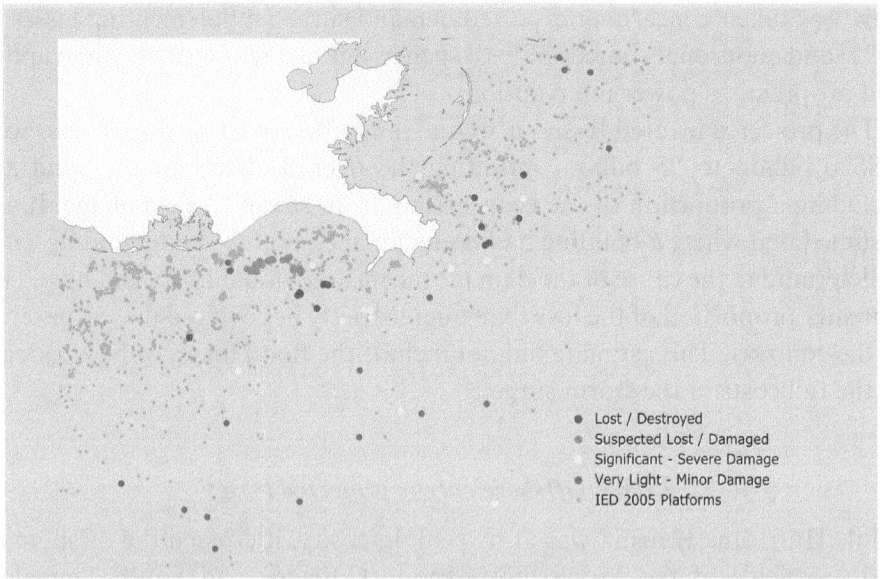

Figure 15.4. Offshore platforms affected by Hurricane Katrina: reports as of September 19, 2005. (In RMS, 2005.) For color version, see plate section.

15.3.4 New Orleans projected flood losses

The flooding in New Orleans covered an estimated 80% of the city, with 55% of the city's properties being inundated by more than 1.2 m of water, with maximum flood depths up to 6 m. During the flood, the water washed out oil tanks, sewerage systems, and two major landfill sites, along with several industrial sites, gas stations, and other locations where hazardous materials were stored. Concerns about the health hazards of the polluted floodwaters were among the principal reasons for the compulsory evacuation of the city.

The flood extent (Figure 15.5) was determined from four different sources of data: Landsat 5 imagery taken on August 31, 2005; Digital Globe imagery taken on September 3, 2005; Federal Emergency Management Agency (FEMA) flood extent maps as of August 31, 2005; and aerial reconnaissance photos taken at 1,525 m on August 30, 2005. Flood depth was determined by using high-resolution (5 m horizontal) LIDAR data, from which RMS constructed a digital terrain model with ground elevation values assigned to each cell in a 100 m × 100 m grid over the greater New Orleans area. The modeled flood depths were validated using aerial reconnaissance imagery taken on August 30, from which flood depths were assessed relative to surrounding structures, automobiles, and other distinguishable objects. A mapping of the

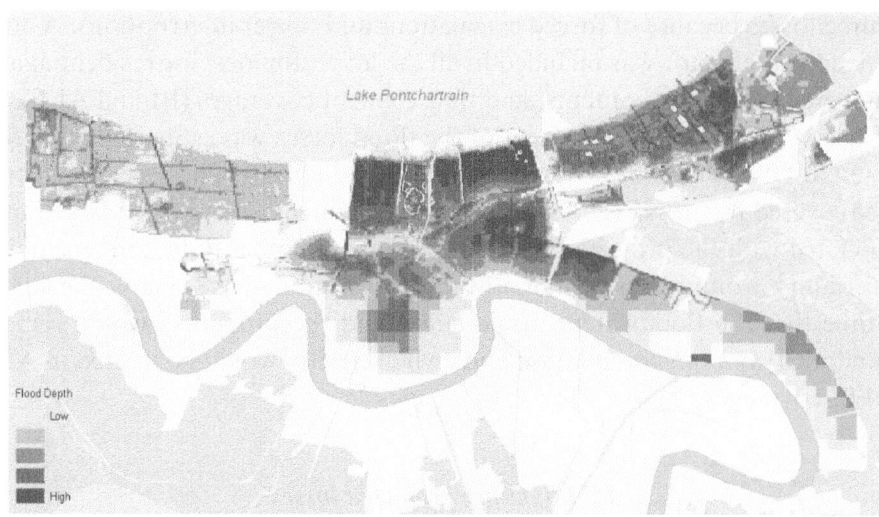

Figure 15.5. RMS-modeled flood depth in New Orleans. (Based on RMS, 2005.) For color version, see plate section.

city was also made to determine for each 100 m grid square the number and value of exposed properties based on heights and occupancy types.

A suite of damage curves for structure and contents relative to flood depths were developed based on US Army Corps of Engineers (USACE) studies. An economic loss was then calculated for both residential and commercial structures and contents, by using the flood depth in each grid square and the likely loss to the mix of buildings at that location.

Based on an analysis of National Flood Insurance Program aggregate data, US Census data, and insurance market data, an estimate of the insurance industry's likely proportion of the total economic flood loss was determined. For residential properties, it was assumed that 50% of the residential structure and contents losses would be taken up by the NFIP, while the remainder would be largely uninsured, with less than 3% of the residential loss attributable to excess NFIP residential coverages. Subsequent surveys showed that between 34,000 and 35,000 of the flooded homes carried no flood insurance, including many that were outside designated flood risk zones (Hartwig, 2006).

As for other assumptions in the loss calculation process, the conservative assumption was made that homeowners with all-risk policies, but without flood cover, would still be able to claim additional living expense (ALE) because of the compulsory evacuation of the city under civil exclusion orders. For commercial lines, it was assumed there was a high penetration of excess NFIP commercial flood cover for structure and contents in the flood-affected region and that commercial business interruption (BI) would be a large part of insured losses because of forced evacuations and power interruptions. A 40% demand surge factor was included in all the loss estimates, for residential and commercial building, contents, and time element coverages (BI and ALE).

The overall insured component of the flood losses was estimated at US$15 billion to US$25 billion. Losses of an additional US$10 billion were predicted to be covered by the NFIP. As with other components, there was a large range associated with the projected losses, reflecting a number of factors, including uncertainty around the splitting of water and wind damage claims, the length of time that the floodwaters would stand and the properties would remain abandoned, and how evacuation and civil exclusion would be treated in ALE and BI coverages.

15.3.5 Other projected losses

Additional sources of loss due to Hurricane Katrina stemmed from impacts on the power and energy industries, the cost to clean up polluted water and soils, and the increased costs due to economic demand surge.

On August 31 as the storm hit, over 30 power stations across the region were taken out of service, either to avoid damaging equipment or from failures of generating or transmission systems. Power was lost to an estimated 4.5 million residential and commercial customers. Although power was restored to most critical facilities such as hospitals within 24 h, on September 19 – three weeks after the storm – about a quarter of a million customers were still without power. These interruptions in power caused additional losses for businesses, even those undamaged by the storm itself.

In addition, six major oil spillages occurred across southeastern Louisiana. For example, an estimated 80% of the properties in St. Bernard Parish were polluted with oil in the floodwaters from a ruptured storage tank at the Murphy Oil refinery. Where oil had sunk into the earth, officials estimated that as many as 4,000 homes would have to be razed and 0.6–0.9 m of soil would have to be removed before the area could be inhabited again.

Finally, as the full impact of Katrina became known, concerns about capacity constraints for labor and materials drove price increases for reconstruction. By mid-September, Hurricane Katrina was estimated to have destroyed over 250,000 homes – an order of magnitude larger than the 28,000 destroyed by Hurricane Andrew in 1992.

Prices of steel and lumber reached record highs in 2004 following the active hurricane season, but had reduced in the months before Katrina as production had been increased to meet demand. Lumber prices jumped about 15% after Katrina but remained well below levels of the previous year, while cement shortages had been reported in more than 30 states. Labor costs were expected to be driven upwards by significant shortages of labor and upward pressure on wages well beyond the impacted region.

The effects of Katrina traveled well beyond the region of the Gulf. More than 91% of oil production and 83% of gas production was shut down in the Gulf of Mexico with the damage to rigs and refineries. The region provides a quarter of US oil and gas production. These closures led to immediate rises in energy prices, with oil peaking at nearly US$70 a barrel in the days following Katrina.

15.3.6 Initial consolidated projected loss

The RMS total expected gross industry insurance losses covers Louisiana, Mississippi, Alabama, and Florida, and the offshore oil and gas industry (Table 15.2). As was outlined above, the range in the loss estimates reflects the results of the modeling and analyses of the wind, surge, flood, and socio-economic and infrastructure impacts on insurance losses, including the reconstruction of the flooded areas and associated damage in New Orleans.

These initial loss estimates can be compared with the private insurance losses published 15 months after the event, estimated as US$40.6 billion by Property Claims Services (PCS), excluding offshore energy losses. Flood losses in New Orleans remained towards the lower end of the initial RMS estimate, although they still had the potential to rise significantly if insurers ended up losing lawsuits concerning policy wordings, which were being argued should cover the inundation losses as these were the result of the failure of the man-made flood defenses. On November 27, 2006, a federal judge rejected an attempt by insurers to dismiss the litigation to force them to pay for flood damage in the city, ruling that the language in some policies excluding coverage for water damage was "ambiguous."

Table 15.2. *Hurricane Katrina: RMS insurance industry loss estimates*[a]

Loss component	Gross industry loss
First landfall in Florida	US$1–US$2 billion
Offshore energy	US$2–US$5 billion
Wind and surge, second landfall	US$20–US$25 billion
New Orleans flooding	US$15–US$25 billion
Additional sources of loss	US$2–US$3 billion
Total estimated loss	**US$40–US$60 billion**

[a] Published September 9, 2005.

15.4 The agenda of catastrophe modeling after Hurricane Katrina

While it became possible to construct a representation of the losses from Hurricane Katrina within a few days of the event, a number of the components of this loss reconstruction represented elements not in the standard RMS US Hurricane model (RiskLink® 5.0) as it existed in summer 2005. In addition, Katrina had revealed some new elements of loss generation not recognized in the earlier generation of catastrophe models. Then there was the whole question of activity rates: Hurricane Katrina had occurred in the middle of the highest 2-year period of intense hurricane landfalling ever known.

Thus Hurricane Katrina came to reset the agenda for innovations and enhancements in catastrophe loss modeling, in a way that had not been seen since the first generation of models built in the aftermath of Hurricane Andrew and the 1994 Northridge earthquake. The remainder of this review concerns how Katrina and the hurricanes of 2004 and 2005 have driven the agenda for revisions in all the different elements of catastrophe modeling.

15.4.1 Hurricane wind fields

The RMS US Hurricane model contains a large stochastic set of hurricane events (about 18,000), capturing a significant number of events with wind fields comparable to or greater than that of Katrina. Although the combination of wind field and storm size allied with forward speed was towards the upper end of the expected extent of a potential wind field of a storm of this intensity, it was not outside the range of wind speed and track parameters. The hurricanes of 2004 and 2005 included very small hurricanes, such as Charley and Wilma (while it was close to the Yucatan), as well as very large hurricanes, such as Frances and Jeanne in Florida.

15.4.2 Hurricane activity rates

Since the first generation of models, one of the basic principles of hurricane catastrophe loss modeling had been that activity rates should be based on the average of history (over the maximum time period for which historical observations could be considered complete). For landfalling storms in the United States, this has generally been taken to be the period since 1900. However, hurricane activity in the Atlantic basin had remained persistently high (in all but El Niño years) since 1995, and then in 2004 and 2005 the high activity in the basin had broken through to US landfall, creating the highest loss years ever. Relative to the average of the low-activity period from 1970 through 1994, the overall number of hurricanes in the basin had increased by more than 60% and the number of more intense category 3–5 storms had increased by more than 120%.

Acknowledging that the long-term historical baseline is no longer the best measure of current hurricane activity means it is necessary to be explicit about the intended time horizon of the catastrophe model. Opinions were solicited among those who use and apply the results of such models to find the duration over which they sought to characterize risk. It was concluded that a seasonal (few months) perspective would be too volatile for most business planning and regulatory applications, but that a 5-year horizon envelopes all standard applications around long-term capital allocation, dealing with investors and rating agencies, as well as the issuing of catastrophe bonds. Based on this consultation, starting in May 2006, the 5-year "medium-term" horizon became the explicit time horizon for all new and upgraded RMS catastrophe models.

In order to determine the appropriate 5-year medium-term rates for US Hurricane, in October 2005, RMS called an expert meeting in Bermuda of four leading hurricane climatologists to analyze the activity rate data and develop a consensus forecast for medium-term hurricane activity in the Atlantic basin, at US landfall, and in the Caribbean (for 2006 through 2010) (Lonfat *et al.*, 2007). This procedure was repeated in October 2006 with seven climatologists and a full range of statistical and climatological models of activity rate fore-casting to develop the updated perspective for 2007 through 2011. For the United States, the medium-term perspective represents a 21% increase in the mean activity of category 1–2 hurricanes and a 36% increase in the mean activity of category 3–5 hurricane landfall rates relative to the 1900–2005 historical baseline.

15.4.3 *Storm surge*

Before 2005, the representation of storm surges in catastrophe models did not have the same priority given to it as the representation of hurricane wind fields. Based on hurricanes of the previous 30 years, storm surge had been viewed as only a minor determinant of the overall insured loss. Not only is the inland extent of storm surge flooding generally small relative to the wind field, but the greatest portion of flood losses would not be covered by standard insurance policies. After Katrina, it became a priority to improve the representation of the storm surge hazard and the associated losses. It was also no longer safe to assume that the surge at landfall would be a simple function of the wind field at landfall, but rather it could reflect the preceding strength of the storm, as Katrina had demonstrated.

15.4.4 Flooding of New Orleans

In the 2005 RMS US Hurricane model (as available in RiskLink® 5.0), extreme storm surges were modeled as extending into New Orleans (for hurricanes of category 3 and above), but it had been assumed that there would not be significant breaching of levees and that deep flooding would be prevented by the pumping capacity in the city. These widely shared assumptions had proved highly unconservative (Independent Levee Investigation Team [ILIT], 2006).

With respect to the potential for levee breaching, as was experienced along the 17th Street and London Avenue Canals in Hurricane Katrina, levee fragility relationships would in the future need to include the potential for failure and breaching to occur at surge water levels lower than the defense crest (see Post-Katrina/Interagency Performance Evaluation Taskforce [IPET], 2006).

New Orleans presents two fronts along which storm surges have the potential to flood the city. The weakest link is in the southeast of the city where the expanded Mississippi River Gulf Outlet shipping channel leads directly into the Inner Harbor Navigation Canal from the open sea of Lake Borgne. The city is also vulnerable to rarer extreme surges from Lake Pontchartrain to the north, but the defenses to the north of the city remain far stronger than those to the east. Following Hurricane Katrina, the USACE applied protective remedies by blocking off the northern ends of the three drainage canals passing south from Lake Pontchartrain, thus reducing the potential for floodwaters to enter the city from this direction. As of the end of 2006, nothing had been done to resist the arrival of surges from Lake Borgne through the MRGO.

Rather than model each section of levee independently, a simple holistic model of levee fragilities, probabilities of breaching, and the ingress of water into the city was employed that related water levels of the storm surge to the height to which floodwater is expected to rise within the city. All else being equal, breach sizes tend to reflect the extent of the inland floodplain, as the breach will continue to expand as long as water is driven to flow at high velocities through the hydraulic gradient at the breach (Muir-Wood and Bateman, 2005). The new model was used to generate results for New Orleans as part of a study of future flood risk to the city (Grossi and Muir-Wood, 2006).

15.4.5 Industry exposure data

One challenge in modeling industry losses from Hurricane Katrina (and the other hurricanes of 2004 and 2005) was the vintage of the available insurance industry exposure data. This might reflect a reconstruction of values and exposure locations that could be 18 months out of date by the time it was used to generate losses. Owing to the rapid inflation in building values and significant new construction in states such as Florida, this gave the potential to understate the value of the losses by 10%–15%. Going forward with catastrophe response activities, it was decided to provide exposure data that were as contemporary as possible, even if that meant projecting values forward to attempt to match the values expected to be in place in the next hurricane season.

15.4.6 Vulnerabilities

Throughout the late 1990s and early 2000s, insurance companies using catastrophe loss models had come to collect much more detailed information on what they were insuring. The four major hurricanes of 2004 provided the first occasion in more than a decade when very large numbers of properties had been exposed to a wide range of hurricane wind speeds (up to 225 kmh^{-1}). In the aftermath of the 2004 hurricanes, it had been possible to work with a total of 42 insurance companies underwriting risk in Florida to obtain US$14 billion of detailed claims data, along with the associated insured exposure information. The highest quality datasets were employed by a team of claims analysts to revisit all the vulnerability functions in the hurricane model, as well as to explore how new occupancies should be modeled and new catagories of vulnerability subdivision should be defined.

Inevitably, the claims data became available most rapidly for the simplest insurance classes, and vulnerability functions for manufactured homes were updated in the May 2005 RiskLink® 5.0 model. Results for commercial property lines were becoming available during summer 2005 and were included in the RiskLink® 6.0 release in May 2006.

Changes in vulnerabilities all tended to increase modeled losses, with increases being about 10% for residential lines and up to 40% for some commercial lines.

15.4.7 Uncertainty in loss modeling

The way in which contracts are defined with deductibles and limits means that insurance and reinsurance losses can be very sensitive, not only to modifications in the mean, but also to the overall distribution of modeled losses (as represented by the standard deviation – or the coefficient of variation, the value of the standard deviation divided by the mean). For example, a high-excess policy will pay out only if the loss is above a threshold, which might be 10% or 20% of the value of the insured property. Where the mean modeled loss is significantly below this amount, a 10% or 20% increase in loss can lead to a two- to threefold increase in what is predicted to fall within the contract.

15.4.8 Loss amplification

The traditional approach in catastrophe loss modeling has been to employ an engineering definition of damage as the basis from which to determine insured losses. In both the two major catastrophe losses of the 1990s – Hurricane Andrew in 1992 and the 1994 Northridge earthquake – it had been recognized that an additional factor termed "demand surge" was required to reflect all elements of the increase in losses, beyond the expectations of the predicted costs to repair the damages. In the absence of other US examples, since the 1994 Northridge earthquake, demand surge was relatively poorly understood in terms of its underlying economic drivers, and it was modeled by using factors relating the magnitude of the loss to the size of the regional economy.

The losses of all the hurricanes of 2004 and 2005 highlighted how simple models of demand surge failed to capture the full range of ways in which the insured losses can increase significantly when that damage happens within a major catastrophe: a set of phenomena labeled "loss amplification." It became possible to identify four components of loss amplification:

Economic demand surge reflects the underlying inflation in the costs of materials and labor that results when widespread demand overwhelms the supply and capacity across a region. Demand surge encompasses all those elements of the costs that are resource-constrained. The modeling of demand surge requires an understanding of the capacity that existed in the repair economy prior to the event, as well as the degree to which supplies or materials and laborers can be supplied from outside the region of damage.

Repair delay inflation occurs when a building is left unrepaired and the damage increases as a result of the delays. This reflects deterioration from humidity and high temperature, as well as damage from water ingress when the roof has been punctured. In situations where the property is located within an evacuation zone, deterioration may become a source of loss (and extended time element coverages) even when the property itself is not damaged.

Claims inflation occurs when insurers are faced with large numbers of claims and they relax claims assessment procedures, leading to rising levels of fraud and exaggeration by claimants. After the largest catastrophes, insurers may implicitly or even publically, set "no claims assessor" loss thresholds. About 1.75 million insurance claims resulted from Hurricane Katrina, and while the focus of claims assessment will be on the most costly claims, the auditing of the smaller claims is likely to have been relaxed.

Coverage expansion occurs in the largest catastrophes, when political pressure and litigation encourage insurers to extend payments for losses outside the original terms of policy coverage. Insurers may conclude that generous claims settlements are good for preserving the loyalty of their customers, or may simply reflect the onslaught of court actions to force insurers to pay residential flood losses from Katrina.

These four components of loss amplification require different classes of model to capture their behavior and influence in magnifying losses. In the RMS RiskLink® 6.0 model released in May 2006, initial loss amplification models were introduced to capture more of these inflationary factors found in the largest catastrophes. Research is continuing to make the modeling of loss amplification as detailed as the other elements of catastrophe models.

15.4.9 Super catastrophes

The term "Super Cat" was coined after Hurricane Katrina to describe the situations encountered in the largest catastrophes, when secondary consequences become significant factors in loss generation. In some situations, secondary consequences can even be considered catastrophes in their own right, a situation termed a "Cat following Cat" – as with the fire in San Francisco following the great 1906 earthquake, and the flooding of New Orleans following Hurricane Katrina. Three principal elements of the secondary consequences found in Super Cat events are containment failures, evacuation, and systemic economic impacts.

Containment failures

Containment failures could be the failure of flood defenses, allowing the flooding of a city, or the failure of systems designed to contain fire or prevent chemicals or toxic agents from escaping. Pollution was extensive in Hurricane Katrina, from the floodwaters lifting tanks and causing them to crack and spill, as well as from sewage systems and toxic chemical stores releasing hazardous chemicals to be washed through inhabited areas. Pollution cleanup greatly increases the costs of repair and lengthens the time it takes to reoccupy properties that otherwise have not had significant structural damage. Pollution has been a primary area of litigation and has caused additional liability claims for insurers covering industrial facilities.

Evacuation

The mandatory evacuation of the population of New Orleans after the flooding of the city became a source of consequential economic and physical damages. The evacuation was an inevitable response to the way in which the principal functions of the city had collapsed and there was a serious public health problem. The mandatory evacuation of urban areas can be expected in any major catastrophe that causes widespread contamination such as a toxic release or where the primary functions of a city are significantly disrupted.

While evacuation can cause additional losses through deterioration of properties, the biggest impact on insurance coverage is for time element coverages (BI and ALE) when residents and businesses are unable to return to their premises. Businesses that depend on the labor force from an evacuated area will also be affected. For example, the evacuation of key personnel in New Orleans led to the collapse of critical facilities, including the pumping systems, the fire department, emergency response personnel, and law enforcement.

Systemic economic impacts

The greater the level of destruction and disruption, the greater the likelihood that systemic economic impacts will start to become additional factors in expanding the losses. Businesses such as hotels and restaurants may delay reopening owing to a lack of customers, while builders may be unable to find laborers. The lower the proportion of the losses restituted by insurance, the longer systemic economic impacts will persist. However, economic dysfunction can also reduce the cost of repairing damages, where uncertainties and delays in permitting reconstruction mean that other forms of loss amplification have declined.

Modeling Super Cats

Before Hurricane Katrina, some "Cat following Cat" events, such as fire following earthquake (FFEQ), were already explicitly included in RMS models. However, following Katrina, work was undertaken to attempt to explore a much wider range of potential consequences of this kind. The lessons of Super Cat amplification of business interruption were first enabled in the RiskLink® 6.0 release to assess the full scale of losses of the most extreme loss events impacting principal urban concentrations across the globe.

A Super Catastrophe can increase the number of damage agents and the lines of business and asset classes damaged, adding to the losses. In essence, it "switches on" exposure to a wider range of insured lines of business. There can also be situations where the correlation of the uncertainty of loss outcomes is increased – as, for example, with different properties that share the same flood defense or among different coverages for the same property. The correlation in loss outcomes may also be much higher for certain lines of business, such as oil refineries or casinos, which are situated in similar coastal locations with high levels of surge damage.

15.4.10 Catastrophe models in practice

The mismatch between modeled losses and actual losses highlighted by Katrina was recognized to reflect not only missing elements in the models but also deficiencies in how insurers collected and coded exposure information and applied appropriate assumptions in running the models. The most notorious example of inappropriate data coding concerned casino barges in Biloxi, Mississippi, that had been modeled and underwritten as reinforced concrete land-based structures. In the face of uncertainty and missing data, underwriters were tempted to make optimistic assumptions about what was not known, so as to be able to write the risk at a competitive rate.

Chastened by the Katrina experience, the best-run insurers changed their practice to focus on obtaining accurate data on insured properties and using the models to stress test assumptions where such data were missing. Instead of being rewarded for the volume of business written, the catastrophe model could itself be used to determine the profitability of the business in relating the technical rate for the risk to the premium obtained.

Some of the innovations in the RMS catastrophe model were deliberately made to match this demand for conservative underwriting. While in previous models demand surge had to be explicitly switched on, in the new models (since RiskLink® 6.0 in 2006) the default is that loss amplification is considered part of any standard analysis.

15.5 Conclusions

Hurricane Katrina has transformed how catastrophe models are structured and parameterized as well as how they are applied within insurance risk management. The event highlighted a wide range of catastrophic nonlinearities – from the systemic effects of loss amplification to the way in which a small failure in a flood defense can lead to the large-scale abandonment of a major city.

The agenda of catastrophe modeling has also had to expand to include the potential that the terms of an insurance contract may not protect an insurer from having to pay for losses outside the defined coverage. Given that the overlap of wind and flood losses remains the subject of ongoing litigation the eventual insurance loss in the event, even after 18 months, remains uncertain.

While it was possible to develop, implement, and release necessary adjustments to catastrophe modeling assumptions within 9 months of the date that Katrina made its landfall, it will still be several years before loss amplification and Super Cat "secondary consequence" losses can be considered to be modeled to the same level of detail found in the science that underlies the reconstruction of a hurricane wind field, for example. An agenda that started out from principles of engineering, meteorology, and statistics has now had to expand to encompass economics and the behavioral sciences.

Significantly, in implementing increased activity rates in its US Hurricane model, RMS has made the first implicit link within catastrophe modeling between climate change and increased insurance costs.

For all these reasons, Hurricane Katrina can be seen to have been a transformative event – redefining much of the agenda of catastrophe loss modeling. There is the much to be learned from the most significant catastrophes. Those events that are most challenging to reconstruct the losses have the greatest impact in increasing the accuracy and comprehensiveness of the next generation of catastrophe models.

References

Emanuel, K. A. (2005). Increasing destructiveness of tropical cyclones over the past 30 years. *Nature*, **436**, 686–8.

Grossi, P., and Muir-Wood, R. (2006). Flood risk in New Orleans: implications for future flood risk and insurability. RMS Technical Report. Available at www.rms.com/Publications/.

Hartwig R. (2006). Hurricane season of 2005: impacts on US P/C Markets, 2006 and beyond. March 2006, presentation, Insurance Information Institute. Available at www.iii.org/media/presentations/katrina/. http://server.iii.org/yy_obj_data/binary/744085_1_0/katrina.pdf.

Independent Levee Investigation Team (ILIT) (2006). Investigation of the performance of the New Orleans flood protection systems in Hurricane Katrina on August 29, 2005. July 31, 2006. Available at www.ce.berkeley.edu/~new_orleans/.

Post-Katrina/Interagency Performance Evaluation Taskforce (IPET) (2006). Performance evaluation report 2 and interim status. March 2006. Available at https://ipet.wes.army.mil/.

Knabb, R. D., Rhome, J. D., and Brown, D. P. (2005). Tropical cyclone report, Hurricane Katrina, August 23–30, 2005. National Hurricane Center (NHC), December 20, 2005. Available at www.nhc.noaa.gov/pdf/TCR-AL122005_Katrina.pdf.

Lonfat, M., Boissonnade, A., and Muir-Wood, R. (2007). Atlantic basin, U.S. and Caribbean landfall activity rates over the 2006–2010 period: an insurance industry perspective. *Tellus* (in press).

Mann, M. E., and Emanuel, K. A. (2006). Atlantic hurricane trends linked to climate change. *Eos: Transactions of the American Geophysical Union*, **87**(24), 233–44.

Muir-Wood, R., and Bateman, W. (2005). Uncertainties and constraints on breaching and their implications for flood loss estimation. *Philosophical Transactions of the Royal Society of London* A, **363**(1831), 1423–30.

Risk Management Solutions (RMS) (2005). Hurricane Katrina: profile of a super cat, lessons and implications for catastrophe risk management. www.rms.com/publications/.

US Army Corps of Engineers (USACE) (2006). Draft final report of the Interagency Performance Evaluation Task Force (June 1, 2006). Available at https://IPET.wes.army.mil.

US Navy (USN) (1983). *Hurricane Havens Handbook for the North Atlantic Ocean.* Naval Research Laboratory, NAVENVPREDRSCHFAC TR 82–03 (modified August 2005). Available at www.nrlmry.navy.mil/port_studies/tr8203nc/0start.htm.

16

The Risk Prediction Initiative: a successful science–business partnership for analyzing natural hazard risk

RICHARD J. MURNANE AND ANTHONY KNAP

Condensed summary

The Risk Prediction Initiative (RPI) has been working with companies active in the catastrophe reinsurance market since 1994. The goal of the RPI is to support scientific research on topics of interest to its sponsors and to provide connections between the scientific and business communities, mainly through science–business workshops on a variety of topics. The major topics of RPI-funded research include paleotempestology, the relationship between tropical cyclone activity and climate, improvement of best-track data, and European storms. A workshop sponsored by the RPI in Bermuda in October 2005 catalyzed the compilation of this volume. This chapter provides an overview of RPI's history, its efforts at making science on natural hazard risk available and understandable to its sponsors, and suggestions for similar endeavors.

16.1 Introduction

Extremes in climate and weather have a wide range of societal consequences. Many examples are offered in chapters throughout this book. This chapter focuses on an indirect consequence of an extreme event: the formation of a unique science–business partnership, the Risk Prediction Initiative (RPI), which is part of the Bermuda Institute of Ocean Sciences (BIOS; formerly known as the Bermuda Biological Station for Research). The RPI was formed, in part, as a consequence of the unexpectedly large insured losses caused by Hurricane Andrew in 1992. Here we provide a short history of the Risk Prediction Initiative and an overview of its activities as a science–business partnership that has survived for more than 10 years and proved beneficial for both scientists and business.

An example of RPI's activities is the workshop held in Bermuda in October 2005 (Murnane and Diaz, 2006). Some of the presentations at that workshop

Climate Extremes and Society, ed. H. F. Diaz and R. J. Murnane. Published by Cambridge University Press. © Cambridge University Press 2008.

form the nucleus for this edited volume. The workshop brought together insurers, reinsurers, and climate scientists in an effort to assess the current understanding of climate extremes. The US National Oceanic and Atmospheric Administration (NOAA) co-sponsored the workshop with the RPI. Workshop speakers were asked to address (1) expected changes in extreme event frequency in response to global warming, (2) the possibility of setting upper and lower bounds for changes in extreme events, and (3) information and observations needed to improve models and statistics for extreme events. The range of extreme events covered by workshop speakers suggests some of the extremes of interest to the reinsurance community, and the questions were framed to provide answers that would help provide guidance for decisions related to changing risk.

The next three sections provide an overview of RPI's genesis, its activities and a discussion of RPI-funded research. These are followed by a discussion of RPI's experience in trying to connect the world of science and academia with the world of catastrophe reinsurance. We end with a discussion of recommendations that might be of value to other scientists and agencies interested in forming partnerships analogous to the RPI with other parts of the private sector. The presentations and discussions at the October 2005 workshop are used to illustrate key points regarding the (re)insurance industries' interest in extreme events.

16.2 Genesis of the Risk Prediction Initiative

The large losses resulting from Hurricane Andrew put many insurance companies out of business or in difficult straits, and a significant fraction of capital available for use by the insurance industry was used to pay claims. The resultant depleted pool of capital available for insurance drove up reinsurance prices and increased the potential returns for companies offering reinsurance.[1] A number of new property catastrophe reinsurers were formed in Bermuda to take advantage of this business opportunity.[2]

In general, these new companies, as well as other reinsurance and insurance companies that survived the large losses, were highly motivated to learn more about their exposure to risk from natural hazards. The RPI was created in 1994 as a result of discussions between BIOS scientists and individuals in these companies who were seeking novel approaches for understanding natural hazard risk. The goal is to support research on natural hazards and to transform the science into knowledge that sponsors can use to assess their risk

[1] Reinsurance is essentially insurance for insurance companies.
[2] This cycle of events was repeated after the large losses associated with the terrorist attacks in 2001 and after the losses during the 2005 hurricane season.

(Malmquist, 1997). The October 2005 workshop was motivated in part by the active 2004 hurricane season and was designed to meet RPI's goals, but it was received with extreme interest because it was closely preceded by Hurricane Katrina and two publications (Emanuel, 2005; Webster *et al.*, 2005) that raised the specter of greatly enhanced losses due to an increase in the intensity and number of the strongest tropical cyclones.

A number of leading companies active in the property catastrophe insurance industry have sponsored the RPI through restricted donations to BIOS, a US not-for-profit 501c(3) corporation. The companies that sponsor RPI have changed through the years in response to changes in market conditions, mergers, changes in business strategies within individual companies, and the formation of new companies.

Sponsors of RPI represent different sectors of the insurance community that are active in property catastrophe reinsurance. The 2006 sponsors represent companies that are active in the primary insurance market (State Farm Fire and Casualty Company), in catastrophe risk modeling (Risk Management Solutions), and in catastrophe reinsurance (XL Capital Re Ltd., Renaissance Reinsurance Ltd., AXIS Capital Holdings Specialty Ltd., PartnerRe, Aspen Insurance, Flagstone Re, and Arch Reinsurance Ltd.).

Companies sponsor RPI for a range of reasons. Some companies hope that RPI research results will improve the catastrophe risk models that are used to estimate probable loss distributions. Other companies build their own proprietary versions of the commercial risk models and use RPI research results for their own model-development purposes. Some companies support RPI as a way of meeting their company's charitable goals. Regardless of a company's interests, their satisfaction with RPI's activities is assessed each year when it is time for the annual renewal of a company's sponsorship of RPI.

An early issue tackled by RPI sponsors involved the identification of research topics that would be supported through RPI funding. The decision to focus on tropical cyclones was made for three major reasons. The first concerned the business relevance of meteorological hazards in general and tropical cyclones in particular. Meteorological hazards are responsible for a large portion of insured property losses, and hurricanes are the leading cause of insured loss (Table 16.1). The recent large losses from landfalling US hurricanes in 2004 and 2005 confirm the rationale supporting the decision to focus on hurricanes.

The second reason for focusing on tropical cyclones was the fact that the ocean is a major mover and reservoir of heat, and heat from the ocean powers tropical cyclones. The BIOS has the longest record of ocean measurements in the world through the Panulirus Hydrographic Stations, and BIOS has strong connections to the ocean climate community. The production of oceanographic

Table 16.1. *Distribution of top 30 insured losses by hazard for the period 1970–2005*

Hazard type	Loss (billions 2005 US dollars)	Percent of total
US hurricane	136	57
Man-made	24	10
European wind	21	9
US earthquake	19	8
Japanese typhoon	17	7
US tornado and hail	8	3
European flood	8	3
Japanese quake	3	1
Wildfire	2	1

Data from Zanetti and Schwarz (2006).

data and access to climate scientists provides BIOS with close connections to scientists with tropical-cyclone-related knowledge.

The third reason is related to the difference in our ability to forecast hurricanes and earthquakes. Earthquakes cannot be predicted in "real time," and as a result much of the federal research support for seismology is directed towards basic research. In contrast, hurricanes can be predicted in real time and a large fraction of federal support for hurricane research is aimed at improving real-time prediction. However, most insurance-related business decisions related to hurricanes are made on annual or longer timescales, so this federal support is directed towards problems on timescales that do not match those of the insurance business cycle. In addition, funding that RPI could provide to the seismological community would have trouble rising above the noise of federal support for earthquake research. In contrast, RPI funding could produce a significant signal in the hurricane research community interested in problems relevant to the insurance community. Thus, the combination of hurricane losses and potential influence within the research community resulted in RPI's decision to focus on aspects of basic hurricane research that have business relevance.

16.3 RPI activities

In many ways, the RPI acts as a facilitator of communication among a variety of entities interested in extreme events that produce large insured losses (Figure 16.1). The major hazards of interest are listed in Table 16.1. The

Table 16.2. *Examples of RPI-sponsored workshops*

Year	Workshop title
2005	Assessing, Modeling, and Monitoring the Impacts of Extreme Climate Events
2004	Fire at the Wildland–Urban Interface: What Does the Future Hold?
2003	Recent Advances in Earthquake Research and Probability Assessment
2002	Weather Extremes and Atmospheric Oscillations
2001	The Potential Development of a Unified Northwestern Pacific (NWPAC) Tropical Cyclone Best-Track Data Set
2000	Tornadoes and Hail
2000	Extreme European Windstorms and Floods
1999	Uncertainty of Damageability
1999	Hedging with Weather Derivatives: How Reliable are Seasonal Predictions of Temperature and Precipitation?
1999	European Windstorms and the North Atlantic Oscillation
1998	Transition of Tropical Cyclones to High-Latitude Storms
1997	Wind Field Dynamics of Landfalling Tropical Cyclones

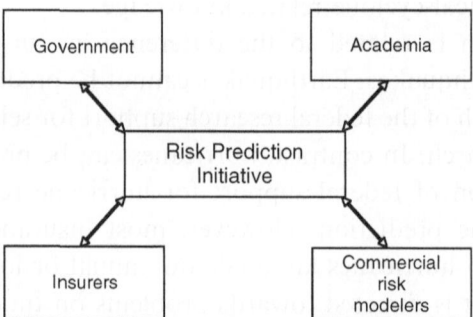

Figure 16.1. The RPI acts as an interface between different societal sectors interested in insured losses caused by extreme events.

interactions among the groups displayed in Figure 16.1 occur in two major ways: workshops that are sponsored, organized, and led by the RPI, and research that is supported by RPI funding.

The workshops occur on an approximately semiannual basis and take two major forms. One is the annual RPI Research Update workshop. This workshop features presentations on research results obtained by scientists who received RPI funding during the year. A summary of RPI-funded research is presented in the next section. The other generally involves an exploration of the state of the science on topics chosen by the sponsors for their business relevance (Table 16.2). These workshops often results in a publication such as a research agenda or a summary of a topic for the scientific community (Malmquist, 1997; Murnane, 2004; Murnane and Diaz, 2006; Murnane *et al.*, 2000, 2002), or a primer on a topic written for non-specialists (Malmquist, 1997, 1998, 1999, 2000; Murnane, 2003). The RPI also attempts to bring together scientists who would not normally cross paths so that the workshops offer the participants exposure to disciplines they would not normally experience, as well as an opportunity for scientists and people in the business community to interact.

The October 2005 workshop provides an example of the interaction between scientists and insurers. The landfalling tropical cyclone, the extreme event of greatest interest to property catastrophe insurers, was the main topic of interest. However, other relevant extremes were covered, including drought, precipitation extremes, changes in significant wave heights, and changes in temperature extremes. These extremes do not cause the largest insured losses, but they do provide a context for assessing changes in climate. In addition, the quality and quantity of temperature data make temperature-related extremes the focus of many climate studies. In contrast, precipitation, wind, and wave data tend to be of lower quality and less complete. These data limitations led to several discussions regarding the necessity for model studies of how extreme events might respond to climate change and highlighted for RPI sponsors the importance of time series observations and the homogeneity of data.

An extension of RPI's efforts in fostering science–business communication is its attempts to enhance the scientific community's awareness of the property catastrophe insurance industry's interests in a variety of topics. This is done through the sponsorship and facilitation of specific workshops as well as through other oral and written presentations.

The second major RPI activity is support of scientific research. A feature that distinguishes the RPI research program from standard National Science Foundation (NSF)-style requirements is that RPI-supported research must pass reviews for business relevance in addition to being of high scientific quality. Although NSF proposals are reviewed on the basis of scientific quality and societal relevance, the perception, if not the reality, is that the latter standard receives less emphasis during funding decisions. The requirement that funded research be scientifically strong as well as relevant to the business of RPI sponsors is an important key to RPI's continued success. The large number of peer-reviewed publications by RPI-funded scientists is a testament to the quality of RPI-funded science. The continued support by RPI sponsors documents the business relevance of RPI-funded science.

The RPI's research projects are funded through a variety of mechanisms. The most familiar process involves a request for proposals (RFP) on general or specific research projects. The proposals we receive in response to an RFP are sent out for scientific review and then are subjected to review for business relevance by RPI sponsors. An example of a project funded through this mechanism is the research on seriality in European windstorms. A second funding mechanism, generally used for small projects, is to support projects that come up through discussions with scientists. An example of this approach is the Extended Best-Track Database, whose funding was initiated through discussions at an RPI workshop. A third funding mechanism involves the RPI directly soliciting proposals from scientists for specific projects. An example of a research project initiated through this mechanism is the use of satellite data to estimate tropical cyclone wind radii.

Sponsors of RPI research belong to one to three different research groups. The names of these groups reflect the type of research they support: Benchmark Development, Forecasting, and Emerging Markets. The Benchmark Development Research Group focuses on producing data that can be used to "benchmark" and verify estimates produced by catastrophe risk models. The group currently focuses on developing estimates of the exceedance probability of tropical cyclone winds along the Gulf and East Coasts of the United States. The Forecasting Research Group studies a variety of climate and weather data to develop and improve long-term and seasonal forecasts, mainly for landfalling tropical cyclones. The Emerging Markets Research Group develops products that can be used for the development of new catastrophe risk models. To date, this group has supported research on using satellite data to develop estimates of tropical cyclone wind fields.

The focus of the research groups has evolved over time in response to sponsor interests and market conditions. For example, at one time the Emerging Markets Research Group was interested in weather derivatives. Also, the structure of RPI has evolved over time. There were no research groups during RPI's early years. The groups were formed in 1998 in response to changes in RPI sponsor interests and the soft market for catastrophe reinsurance.

16.4 Overview of RPI-funded research results

One focus of RPI's research program has been supporting the development of the field of paleotempestology – the study of prehistoric tropical cyclones through the use of geologic proxies. Useful overviews of the field are provided by Liu (2004) and Donnelly and Webb (2004). The RPI has supported studies at numerous locations along the Gulf and East Coasts of the United States (Donnelly, 2005; Donnelly et al., 2001a, 2001b, 2004; Liu and Fearn, 2000a, 2000b, 2002; Liu et al., 2003; Lu and Liu, 2005). It is interesting to note that RPI sponsor interest in site location differed from the scientists' assumptions.

Initially, the scientists assumed that RPI sponsors would be most interested in areas that might be expected to experience the greatest loss; for example, near Miami, New York City, and New Orleans. The New York City area was and is of great interest, but Miami and New Orleans were of less interest because the insurance industry was fairly certain that these areas were at high risk and coverage was priced accordingly. In contrast, there is a fair amount of uncertainty in the wind-speed exceedance probabilities for the New York City area because of the relatively small number of strong storms that pass through the region. Paleotempestological studies are of great value in this case because they can extend the historic record and collect a larger sample of landfalling events. Unfortunately, uncertainty increases with time and the influence of changes in climate and sea level can become important.

An additional approach for extending the historical record is through archival research. Numerous records of tropical cyclone impacts exist in the archives and museums of many countries. The RPI has supported selected projects aimed at creating records for China (Louie and Liu, 2003, 2004) and for landfalling typhoons in Japan.

Essentially all hurricane catastrophe risk models are based on best-track data (Jarvinen *et al.*, 1984). The best-track data have known deficiencies (Landsea *et al.*, 2004, 2006) and as a result the RPI has supported a number of projects aimed at improving best-track data (Dunion *et al.*, 2003; Landsea *et al.*, 2004). The RPI has also supported an effort to improve data on tropical cyclone winds via development of algorithms for estimating tropical cyclone wind radii from satellite observations (Kossin *et al.*, 2007; Mueller *et al.*, 2007).

Another long-term emphasis for RPI has been support of research aimed at developing seasonal hurricane forecasts and for studies of the relationships between modes of climate variability and tropical cyclone activity. Prior to RPI support, seasonal hurricane forecasts were directed at estimating basin-wide statistics. The RPI was the first to support research aimed at developing seasonal forecasts of hurricane landfall probability (Lehmiller *et al.*, 1997) and has supported efforts to develop seasonal forecasts in other basins (Chan *et al.*, 1998; Liu and Chan, 2003). A natural extension of this work is the study of the relationships between tropical cyclone activity and modes of climate variability such as the El Niño/Southern Oscillation (ENSO) and the North Atlantic Oscillation (NAO) (Bove *et al.*, 1998; Elsner and Bossak, 2001, 2004; Elsner and Jagger, 2004, 2006; Elsner and Kocher, 2000; Elsner and Liu, 2003; Elsner *et al.*, 1998, 1999, 2000a,b,c, 2001, 2006a,b,c; Jagger and Elsner, 2006; Jagger *et al.*, 2001, 2002; Kimberlain and Elsner, 1998; Landsea, 2000a,b; Pielke and Landsea, 1999).

The RPI has supported a variety of other research projects. One of the most important for modeling insured losses from European windstorms is a study by Mailier *et al.* (2006). Other studies have looked at the response of tropical cyclone intensity to climate change (Druyan *et al.*, 1999; Shen *et al.*, 2000). Finally, for several years RPI had a research group focused on the weather derivative market. This interest resulted in a series of papers aimed at better understanding the urban heat island effect and developing techniques for testing data homogeneity (Allen and DeGaetano, 2000, 2001; Cai and Kalnay, 2004; DeGaetano and Allen, 2002; Kalnay and Cai, 2003). In addition, RPI support resulted in a NOAA product that downscales regional climate forecasts to degree-day forecasts for specific urban weather stations (National Weather Service, 2006).

16.5 Melding science into the catastrophe risk business

Catastrophe risk models play a major role in the world of insurance for catastrophe risk. Risk models cover a variety of natural hazards as well as terrorism. The type and geographic range of commercial risk models are continually expanding. Hazards in countries with large amounts of insurance include earthquakes in Japan and the United States; tropical cyclones in the United States, Japan, and Australia; and winter storms and floods in Europe, as well as smaller-scale events such as tornadoes and winter storms in the United States. In addition, risk modeling companies produce catastrophe risk models for countries with smaller insurance markets, such as for earthquakes in Chile or typhoons in Guam.

Risk models are a complex assemblage of scientific understanding of a natural hazard, engineering knowledge regarding the response of structures to the forces produced by a hazard, and financial and economic awareness of the factors associated with repairing and replacing structures, details on insurance policies and reinsurance treaties, and estimates of the effects of phenomena such as demand surge produced by large events that significantly affect the economy. Given the complexity of the problem, it is not difficult to imagine that integrating science into the risk models can be a challenge. The RPI continues to make science understandable, available, and useable by its sponsors, but the best way for this to occur is to incorporate the science into a risk model. However, companies still want to verify risk model estimates. The proxy work that extends the historical record of hurricane landfalls provides one of the few methods of independently verifying wind-speed exceedance probabilities.

Another factor that acts to slow the integration of science into the catastrophe risk business is the difference in timescales associated with the different sectors displayed in Figure 16.1. The private sector must respond on very short timescales. In addition, the typical reinsurance contract is renewed on an annual basis. This cycle is much shorter than the typical academic timescale for scientific research. When a scientist first proposes to undertake a research project, it is generally based on a timescale appropriate for someone undertaking a research project for a master's or doctoral degree. In addition, collaborations with scientists in the federal government generally require working within the federal budget cycle; this involves a lead-time of several years. Melding the disparate timescales can present a challenge. A final temporal factor involves the time lag between when a scientist is ready to put the results from a discovery or scientific program in writing and when the knowledge is incorporated into a risk model. The lag is due to the time required for writing the manuscript; its submission, review, and acceptance by a journal; the publication of the manuscript; and the incorporation of the scientific finding into a risk model. One measure of the time involved in this process is the six years between the RPI workshop on the extratropical transition of tropical cyclones, an upsurge in scientific interest in extratropical transitions, and the release of a risk model that incorporated extratropical transitions into its loss calculations.

The hazard catalogs used in catastrophe risk models generally assume a stable climate. However, there are now wind catalogs that account for climate variability associated with the ENSO and the NAO and that consider the multidecadal variability in Atlantic hurricane frequency associated with the Atlantic Multidecadal Oscillation (AMO). Researchers are now starting to explore the potential impacts of climate change on hurricane losses. For example, two chapters in this volume elaborate on presentations given at the October workshop and describe the use of catastrophe risk models for estimating the costs of climate change and for normalizing past catastrophic losses. These presentations provided an important link between scientific studies of extreme events and how this information can be used by the property catastrophe insurance industry.

Several conclusions drawn by participants in the October workshop relate to efforts to model the impacts of extreme events. First, there was an awareness of the potential for the frequency and intensity of extremes to change as a result of changes in climate regardless of whether the changes were due to natural or anthropogenic causes. Participants recognized that the record of extreme events is too short for developing robust statistics on how extreme events respond to climate change. Second, problems associated with data limitations force the use of models to better understand and estimate the statistics on extreme events and how they might change in response to climate variability. Improvements in models and computer hardware offer the best hope for developing this information. Third, Hurricane Katrina was a significant event for the world of catastrophe risk modeling. The losses from an intense hurricane striking New Orleans were expected to be large, but actual losses were much greater than expected. The unexpectedly large magnitude was caused in part by (nonlinear) interactions that were not modeled. For example, the extensive flooding of New Orleans was partly caused by the interactions between the design of levees, the loss of power, and the abandonment of pump houses.

16.6 The RPI as an example for other efforts

The history of RPI over the past 12 years offers several insights that might be of value to similar efforts. A key insight is that there must be a mechanism for identifying scientific problems that can be addressed on a timescale that is relevant to the business sector. The identification mechanism requires one or more people with a good intuition for what projects will quickly produce results on a limited budget. The timescale requirement means that progress must be made relatively quickly because there is no guarantee that the business person or company that initiates a research project will be able to see it to completion. Contacts within a company can (and do) leave or receive promotions, and their replacements may not provide the same interest. Nobody's bonus depends on what they do with RPI. On a corporate level, mergers, changes in a company's business strategy, or cutbacks mean that progress must come fairly quickly so that a company will still be around to see the benefits.

In addition, the structure of an organization must be flexible and able to evolve in response to changes in the business environment. Business conditions are not static, and a company's interests will change over time. An organization must be willing to adapt to changing needs.

Other entities that intend to follow the RPI model should consider developing a clear metric for performance. Within the RPI, there is no real mechanism for determining a company's satisfaction other than their renewing their sponsorship annually. This is not an ideal situation, as the unanticipated loss of a sponsor can seriously disrupt a budget. As a result, the RPI instituted an informal requirement that companies provide one year's notice of their intent to leave the program. This notice provides sufficient time to adjust the research budget and/or to find new sponsors.

The events that interest RPI sponsors in the property catastrophe reinsurance industry are mainly of large scale, high intensity, and short duration (see

Table 16.1). Landfalling tropical cyclones fit this description. The interest in tropical cyclone losses (Table 16.1) has been enhanced by recent studies suggesting an increase in hurricane intensity (Emanuel, 2005; Webster *et al.*, 2005) and a relationship with recent oceanic and atmospheric warming (Elsner, 2006; Hoyos *et al.*, 2006). However, other scientists dispute the findings for a variety of reasons, including data quality (Klotzbach, 2006; Landsea, 2005; Landsea *et al.*, 2006). Nevertheless, other meteorological hazards that cause large insured losses could form the basis for science–business partnerships. The best candidates are perhaps European windstorms and flooding, and tornadoes and winter storms in the United States. Scientists interested in hazards that produce significant losses of life and/or uninsured losses will likely need to look outside the insurance industry for support.

16.7 Conclusions

The RPI has been working with companies active in the catastrophe reinsurance market since 1994. The goal of the RPI is to support scientific research on topics of interest to its sponsors and to provide connections between the scientific and business communities. An indication of the importance of science and engineering for the business of catastrophe reinsurance is the growing number of scientists and engineers with expertise related to natural hazards that have been hired by past and current sponsors of the RPI and other companies. The services RPI provides can become more valuable to its sponsors when a company has a pool of technically sophisticated talent who understand and use the results of RPI-supported scientific research and who exploit the opportunities to interact with leading experts in a field.

The RPI has sponsored a significant number of workshops that bring together people who do not generally work together but who are involved with related aspects of natural hazards. These interactions offer opportunities for a cross-fertilization of ideas that can lead to new collaborations and allow scientists to learn directly from insurers what aspects of their research might be of interest and use to the private sector. The topics of many of the chapters in this volume are derived from material presented at the RPI-sponsored workshop held in Bermuda in October of 2005.

The RPI has offered important long-term support for the nascent field of paleotempestology. A significant amount of support is directed towards studies that will enhance our understanding of the relationships between tropical cyclone activity and climate variation. Work on improving best-track data, extending the historical record through archival research, and defining the distribution of winds around a tropical cyclone continues to receive funding as well. Past topics that have received support include studies on European storms, temperature records important for the weather-derivative industry, and the effects of future climate change on tropical cyclone intensity. Scientists supported by RPI have produced more than 50 publications in leading scientific journals.

Chapters in this volume document the important impacts of climate and weather extremes on society, how these extremes have varied in the past, and how they might change in the future. Our ability to "weather" these extremes could be improved, but challenges would be involved with any improvement. A thorough understanding of the natural variability in extremes, and how the extremes might change in the future, will be needed to help identify cost-effective strategies for reducing our vulnerability to weather and climate extremes. Partnerships between scientists and the private sector can help us develop this understanding.

Acknowledgments

The authors acknowledge the support of the current sponsors of RPI (Arch Reinsurance Ltd., Aspen Insurance, AXIS Capital Holdings Specialty Ltd, Flagstone Re, PartnerRe, Renaissance Reinsurance Ltd., Risk Management Solutions, State Farm Fire and Casualty Company, and XL Capital Re Ltd.), as well as that of past RPI sponsors. This is BIOS contribution number 1685.

References

Allen, R. J., and DeGaetano, A. T. (2000). A method to adjust long-term temperature extreme series for non-climatic inhomogeneities. *Journal of Climate*, **13**, 3680–95.
Allen, R. J., and DeGaetano, A. T. (2001). Estimating missing daily temperature extremes using an optimized regression approach. *International Journal of Climatology*, **21**, 1305–19.
Bove, M. C., Elsner, J. B., Landsea, C. W., and O'Brien, J. J. (1998). Effect of El Niño on U.S. landfalling hurricanes, revisited. *Bulletin of the American Meteorological Society*, **79**, 2477–82.
Cai, M., and Kalnay, E. (2004). Response to the comments by Rose *et al.* and Trenberth. *Nature*, **427**, 214, doi:210.1038/427214a.
Chan, J. C. L., Shi, J. E., and Lam, C. M. (1998). Seasonal forecasting of tropical cyclone activity over the western North Pacific and the South China Sea. *Weather Forecasting*, **13**, 997–1004.
DeGaetano, A. T., and Allen, R. J. (2002). A homogenized historical temperature extreme dataset for the United States. *Journal of Technology*, **19**, 1267–84.
Donnelly, J. P. (2005). Evidence of past intense tropical cyclones from backbarrier salt pond sediments: a case study from Isla de Culebrita, Puerto Rico, USA. *Journal of Coastal Research*, **SI**(42), 201–10.

Donnelly, J. P., and Webb, T., III (2004). Backbarrier sedimentary records of intense hurricane landfalls in the northeastern United States. In *Hurricanes and Typhoons: Past, Present and Future*, ed. R. J. Murnane and K.-b. Liu. New York: Columbia University Press, pp. 58–96.

Donnelly, J. P., Bryant, S. S., Butler, J., *et al.* (2001a). 700 yr sedimentary record of intense hurricane landfalls in southern New England. *Bulletin of the Geological Society of America*, **113**, 714–27.

Donnelly, J. P., Butler, J., Roll, S., Wengren, M., and Webb, T., III (2004). A backbarrier overwash record of intense storms from Brigantine, New Jersey. *Marine Geology*, **210**, 107–21.

Donnelly, J. P., Roll, S., Wengren, M., *et al.* (2001b). Sedimentary evidence of intense hurricane strikes from New Jersey. *Geology*, **29**, 615–18.

Druyan, L. M., Lonergan, P., and Eichler, T. (1999). A GCM investigation of global warming impacts relevant to tropical cyclone genesis. *International Journal of Climatology*, **19**, 607–17.

Dunion, J. P., Landsea, C. W., Houston, S. H., and Powell, M. D. (2003). A reanalysis of the surface winds for Hurricane Donna of 1960. *Monthly Weather Review*, **131**, 1992–2011.

Elsner, J. B. (2006). Evidence in support of the climate change: Atlantic hurricane hypothesis. *Geophysical Research Letters*, **33**, doi:10.1029/2006GL026869.

Elsner, J. B., and Bossak, B. H. (2001). Bayesian analysis of U.S. hurricane climate. *Journal of Climate*, **14**, 4341–50.

Elsner, J. B., and Bossak, B. H. (2004). Hurricane landfall probability and climate. In *Hurricanes and Typhoons: Past, Present, and Future*, ed. R. J. Murnane and K.-b. Liu. New York: Columbia University Press, pp. 333–53.

Elsner, J. B., and Jagger, T. H. (2004). A hierarchical Bayesian approach to seasonal hurricane modeling. *Journal of Climate*, **17**, 2813–27.

Elsner, J., and Jagger, T. H. (2006). Prediction models for annual U.S. hurricane counts. *Journal of Climate*, **19**, 2935–52.

Elsner, J. B., and Kocher, B. (2000). Global tropical cyclone activity: a link to the North Atlantic Oscillation. *Geophysical Research Letters*, **27**, 129–32.

Elsner, J. B., and Liu, K.-b. (2003). Examining the ENSO-typhoon hypothesis. *Climate Research*, **25**, 43–54.

Elsner, J. B., Bossak, B. H., and Niu, X.-F. (2001). Secular changes to the ENSO-U.S. hurricane relationship. *Journal of Geophysical Research*, **28**, 4123–6.

Elsner, J. B., Jagger, T., and Niu, X. (2000a). Shifts in the rates of major hurricane activity over the North Atlantic during the twentieth century. *Geophysical Research Letters*, **27**, 1743–6.

Elsner, J., Jagger, T. H., and Tsonis, A. A. (2006a). Estimated return periods for Hurricane Katrina. *Geophysical Research Letters*, **33**, L08704, doi:08710.01029/02005GL025452.

Elsner, J. B., Jagger, T., and Xiu, X.-F. (2000b). Changes in the rates of North Atlantic major hurricane activity during the twentieth century. *Geophysical Research Letters*, **27**, 1743–6.

Elsner, J. B., Kara, A. B., and Owens, M. A. (1999). Fluctuations in North Atlantic hurricane frequency. *Journal of Climate*, **12**, 427–37.

Elsner, J. B., Liu, K.-b., and Kocher, B. (2000c). Spatial variations in major U.S. hurricane activity: statistics and a physical mechanism. *Journal of Climate*, **13**, 2293–305.

Elsner, J., Murnane, R. J., and Jagger, T. H. (2006b). Forecasting U.S. hurricanes 6 months in advance. *Geophysical Research Letters*, **33**, L10704, doi:10710.11029/12006GL025693.

Elsner, J. B., Niu, X. F., and Tsonis, A. A. (1998). Multi-year prediction of North Atlantic hurricane activity. *Meteorology and Atmospheric Physics*, **68**, 43–51.

Elsner, J., Tsonis, A., and Jagger, T. H. (2006c). High-frequency variability in hurricane power dissipation and its relationship to global temperature. *Bulletin of the American Meteorological Society*, **87**, 763–8.

Emanuel, K. (2005). Increasing destructiveness of tropical cyclones over the past 30 years. *Nature*, **436**, 686–8, doi:610.1038/nature03906.

Hoyos, C. D., Agudelo, P. A., Webster, P. J., and Curry, J. A. (2006). Deconvolution of the factors contributing to the increase in global hurricane intensity. *Science*, **312**, 94–7, doi:10.1126/science.1123560.

Jagger, T. H., and Elsner, J. (2006). Climatology models for extreme hurricane winds near the United States. *Journal of Climate*, **19**, 3220–36.

Jagger, T., Elsner, J. B., and Niu, X. (2001). A dynamic probability model of hurricane winds in coastal counties of the United States. *Journal of Applied Meteorology*, **40**, 853–63.

Jagger, T. H., Niu, X., and Elsner, J. B. (2002). A space-time model for seasonal hurricane prediction. *International Journal of Climatology*, **22**, 451–65.

Jarvinen, B. R., Neumann, C. J., and Davis, M. A. S. (1984). A tropical cyclone data tape for the North Atlantic Basin, 1886–1983: contents, limitations, and uses. NOAA Technical Memorandum, National Weather Service, National Hurricane Center, Coral Gables, Florida.

Kalnay, E., and Cai, M. (2003). Impact of urbanization and land use on climate change. *Nature*, **423**, 528–31.

Kimberlain, T. B., and Elsner, J. B. (1998). The 1995–1996 North Atlantic hurricane seasons: a return of the tropical-only hurricane. *Journal of Climate*, **11**, 2062–9.

Klotzbach, P. J. (2006). Trends in global tropical cyclone activity over the past 20 years (1986–2005). *Geophysical Research Letters*, **33**, doi:10.1029/2006GL025881.

Kossin, J. P., Knaff, J. A., Berger, I., *et al.* (2007). Estimating hurricane wind structure in the absence of aircraft reconnaissance. *Weather and Forecasting* (in press).

Landsea, C. W. (2000a). Climate variability of tropical cyclones: past, present and future. In *Storms*, ed. R. A. Pielke, Jr., and R. Pielke, Sr. New York: Routledge, pp. 220–41.

Landsea, C. W. (2000b). El Niño–Southern Oscillation and the seasonal predictability of tropical cyclones. In *El Niño: Impacts of Multiscale Variability on Natural Ecosystems and Society*, ed. H. F. Diaz and V. Markgraf. New York: Springer, pp. 149–81.

Landsea, C. W. (2005). Hurricanes and global warming. *Nature*, **438**, E11–E13, doi:19,1938.nature04477.

Landsea, C. W., Anderson, C., Charles, N., *et al.* (2004). The Atlantic hurricane database re-analysis project: documentation for 1851–1910 alterations and additions to the HURDAT database. In *Hurricanes and Typhoons: Past, Present, and Future*, ed. R. J. Murnane and K.-b. Liu. New York: Columbia University Press, pp. 177–221.

Landsea, C. W., Harper, B. A., Hoarau, K., and Knaff, J. A. (2006). Climate change: can we detect trends in extreme tropical cyclones? *Science*, **313**, 452–4, doi:410.1126/science.1128448.

Lehmiller, G. S., Kimberlain, T. B., and Elsner, J. B. (1997). Seasonal prediction models for North Atlantic basin hurricane location. *Monthly Weather Review*, **125**, 1780–91.

Liu, K.-b. (2004). Paleotempestology: principles, methods, and examples from Gulf Coast lake sediments. In *Hurricanes and Typhoons: Past, Present, and Future*, ed. R. J. Murnane and K.-b. Liu. New York: Columbia University Press, pp. 13–57.

Liu, K.-b., and Fearn, M. L. (2000a). Holocene history of catastrophic hurricane landfalls along the Gulf of Mexico coast reconstructed from coastal lake and marsh sediments. In *Current Stresses and Potential Vulnerabilities: Implications of Global Change for the Gulf Coast Region of the United States*, ed. Z. H. Ning and K. K. Abdollahi. Baton Rouge, Louisiana: Gulf Coast Regional Climate Change Council, Franklin Press, pp. 38–47.

Liu, K.-b., and Fearn, M. L. (2000b). Reconstruction of prehistoric landfall frequencies of catastrophic hurricanes in northwestern Florida from lake sediment records. *Quaternary Research*, **54**, 238–45.

Liu, K.-b., and Fearn, M. L. (2002). Lake sediment evidence of coastal geologic evolution and hurricane history from Western Lake, Florida. Reply to Otvos. *Quaternary Research*, **57**, 429–31.

Liu, K.-b., Lu, H. Y., and Shen, C. M. (2003). Assessing the vulnerability of the Alabama Gulf Coast to intense hurricane strikes and forest fires in the light of long-term climatic changes. In *Integrated Assessment of the Climate Change Impacts on the Gulf Coast Region*, ed. Z. H. Ning, R. E. Turner, T. Doyle, and K. Abdollahi. Baton Rouge, Louisiana: Gulf Coast Regional Climate Change Council, pp. 223–30.

Liu, K. S., and Chan, J. C. L. (2003). Climatological characteristics and seasonal forecasting of tropical cyclones making landfall along the South China coast. *Monthly Weather Review*, **131**, 1650–62.

Louie, K. S., and Liu, K.-b. (2003). Earliest historical records of typhoons in China. *Journal of Historical Geography*, **29**, 299–316.

Louie, K. S., and Liu, K.-b. (2004). Ancient records of typhoons in Chinese historical documents. In *Hurricanes and Typhoons: Past, Present, and Future*, ed. R. J. Murnane and K.-b. Liu. New York: Columbia University Press.

Lu, H. Y., and Liu, K.-b. (2005). Phytolith indicators of coastal environmental changes and hurricane overwash deposition. *The Holocene*, **15**, 965–72.

Mailier, P. J., Stephenson, D. B., Ferro, C. A. T., and Hodges, K. I. (2006). Serial clustering of extratropical cyclones. *Monthly Weather Review*, **134**, 2224–40.

Malmquist, D. (1997). *Tropical Cyclones and Climate Variability: a Research Agenda for the Next Century*. St. George's: Risk Prediction Initiative, Bermuda Biological Station for Research, Inc.

Malmquist, D. L., ed. (1998). *Hurricane Winds over Land: Recommendations for Research*. Hamilton, Bermuda: Risk Prediction Initiative.

Malmquist, D. L., ed. (1999). *European Windstorms and the North Atlantic Oscillation: Impacts, Characteristics, and Predictability*. Hamilton, Bermuda: Risk Prediction Initiative.

Malmquist, D. L., ed. (2000). *Tornadoes, Hail, and Insurance*. Hamilton, Bermuda: Risk Prediction Initiative.

Mueller, K. J., DeMaria, M., Knaff, J. A., Kossin, J. P., and Haar, T. H. V. (2007). Objective estimation of tropical cyclone wind structure from infrared satellite data. *Weather and Forecasting* (in press).

Murnane, R. J., ed. (2003). *Weather Extremes and Atmospheric Oscillations: A Science-Business Research Agenda*. St. George's, Bermuda: Risk Prediction Initiative.

Murnane, R. J. (2004). Climate research and reinsurance. *Bulletin of the American Meteorological Society*, **85**, doi:10.1175/BAMS-1185-1175-1697, 1697–707.

Murnane, R. J., and Diaz, H. F. (2006). Assessing, modeling, and monitoring the impacts of extreme climate events. *Eos, Transactions of the American Geophysical Union*, **87**, 25.

Murnane, R. J., Barton, C., Collins, E., *et al.* (2000). Model estimates hurricane wind speed probabilities. *Eos, Transactions of the American Geophysical Union*, **81**, 433, 438.

Murnane, R. J., Crowe, M., Eustis, A., *et al.* (2002). The weather risk management industry's climate forecast and data needs: a workshop report. *Bulletin of the American Meteorological Society*, **83**, 1193–8.

National Weather Service, Climate Prediction Center (2006). *CPC Outlooks for Major U.S. Cities*. Washington, D.C.: National Weather Service.

Pielke, R. A., Jr., and Landsea, C. W. (1999). La Niña, El Niño, and Atlantic hurricane damages in the United States. *Bulletin of the American Meteorological Society*, **80**, 2027–33.

Shen, W., Ginis, I., and Tuleya, R. E. (2000). A sensitivity study of the thermodynamic environment on GFDL model hurricane intensity: implication for global warming. *Journal of Climate*, **13**, 109–21.

Webster, P. J., Holland, G. J., Curry, J. A., and Chang, H.-R. (2005). Changes in tropical cyclone number, duration, and intensity in a warming environment. *Science*, **309**, 1844–6, doi:1810.1126/science.1116448.

Zanetti, A., and Schwarz, S. (2006). Sigma 2/2006. *Natural Catastrophes and Man-made Disasters 2005: High Earthquake Casualties, New Dimension in Windstorm Losses*. Zürich: Swiss Reinsurance Company.

Index